Proteomics

From protein sequence to function

Proteomics

From protein sequence to function

S.R. Pennington
Department of Human Anatomy and Cell Biology, University of Liverpool, UK

and

M.J. Dunn
Heart Science Centre, Harefield Hospital, Harefield, UK

S.R. Pennington and M.J. Dunn
Department of Human Anatomy and Cell Biology, University of Liverpool, UK and Heart Science Center, Harefield Hospital, Harefield, UK

Published in the United States of America, its dependent territories and Canada by arrangement with BIOS Scientific Publishers Ltd, 9 Newtec Place, Magdalen Road, Oxford OX4 1RE, UK

A CIP catalogue record for this book is available from the British Library.

ISBN 0-387-91589-3 Springer-Verlag New York Berlin Heidelberg SPIN 10755194

Springer-Verlag New York Inc.
175 Fifth Avenue, New York
NY 10010–7858, USA

Production Editor: Andrea Bosher
Typeset by Paston PrePress Ltd, Beccles, UK
Printed by The Cromwell Press, Trowbridge, UK

Contents

Contents vii

Contributors

Aebersold, Ruedi. Department of Molecular Biotechnology, University of Washington, Box 357730, Seattle, WA 98195, USA

Blakemore, Steve. Genomics Unit, Glaxowellcome Medicine Research Centre, Gunnels Wood Road, Stevenage, SG1 2NY, UK

Cahill, Dolores J. Max-Planck-Institute of Molecular Genetics, Ihnestrasse 73, D-14195 Berlin-Dahlem, Germany

Cahill, Michael A. Department of Molecular Biology, Institute for Cell Biology, University of Tuebingen, Auf der Morgenstelle 15, Tuebingen, D-72076, Germany

Davison, Matthew. 13S28 Mereside, Alderley Park, Macclesfield, SK10 4TG, UK

Dunn, Michael J. National Heart and Lung Institute, Imperial College of Science, Technology and Medicine, Heart Science Centre, Harefield Hospital, Harefield UB9 6JH, UK

Eickhoff, Holger. Max-Planck-Institute of Molecular Genetics, Ihnestrasse 73, Berlin, D-14195, Germany

Evers, Stefan. F. Hoffmann-La Roche Ltd, PRPI-D, Building 69/16, Basel, CH-4070, Switzerland

Goodlett, David R. Department of Molecular Biotechnology, University of Washington, Box 357730, Seattle, WA 98195, USA

Görg, Angelika. Technische Universität München, Institut für Lebensmitteltechnologie und Analytische Chemie, FG Protemik, D-85350 Freising, Weihenstephan, Germany

Gray, Christopher. F. Hoffmann-La Roche Ltd, PRPI-D, Building 69/16, Basel, CH-4070, Switzerland

Humphrey-Smith, Ian. Department of Microbiology, Centre for Proteome Research and Gene-Product Mapping, Suite G12, National Innovation Centre, Australian Technology Park, Eveleigh, NSW, Australia

James, Peter. Wallenberg Laboratory, Lund University, Lund, Sweden

Jenkins, Rosalind. Department of Human Anatomy and Cell Biology, New Medical School, Ashton Street, Liverpool L69 3GE, UK

Johnson, Kevin. The Science Park, Melbourne, Cambridge SG8 6JJ, UK

Keatinge, Lucy. Department of Biological Sciences, Division of Environmental and Life Sciences, Maquarie University, NSW 2109, Australia

Klose, Joachim. Humboldt-Universität, Charité, Virchow-Klinikum, Institüt für Humangenetik, Augustenburger Platz, 13353 Berlin, Germany

Lehrach, Hans. Max-Planck-Institute of Molecular Genetics, Ihnestrasse 73, Berlin, D-14195, Germany

Lopez, Mary. 22 Alpha Road, Chelmsford, MA 01824-4, USA

Nordheim, Alfred. Department of Molecular Biology, Institute for Cell Biology, University of Tuebingen, Auf der Morgenstelle 15, Tuebingen, D-72076, Germany

Nordhoff, Eckhard. Max-Planck-Institute of Molecular Genetics, Ihnestrasse 73, Berlin, D-14195, Germany

O'Brien, John. Max-Planck-Institute of Molecular Genetics, Ihnestrasse 73, Berlin, D-14195, Germany

Oswald, Helmut. German Heart Institute, Augustenburger Platz, Berlin, D-13353, Germany

Packer, Nicolle H. Proteome Systems Ltd, 1/35–41 Waterloo Road, North Ryde, Sydney, NSW 2113, Australia.

Patterson, Scott D. Amgen Inc., Amgen Centre, MS 14-2-E, One Amgen Centre Drive, Thousand Oaks, California, CA 91320-1789, USA

Patton, Wayne F. Molecular Probes Inc., 4849 Pitchford Avenue, Eugene, OR 97402, USA

Pennington, Stephen R. Department of Human Anatomy & Cell Biology, New Medical School, University of Liverpool, Liverpool L69 3GE, UK

Pleissner, Klaus-Peter. Institute for Plant Genetics and Crop Plant Research, Corrensstr 3, Gatersleben, D-06466, Germany

Plomion, Christophe. Laboratoire de Génétique et Amélioration des Arbres Forestiers, INRA, France

Pritchard, Kevin. Cambridge Antibody Technology Ltd, The Science Park, Melbourne, Cambridge SG8 6JJ, UK

Quadroni, Manfredo. Institute of Biochemistry, University of Lausanne, Epalinges, Switzerland

Thiellement, Herve. Department de Biologie Vegetale, Universite de Geneve, 3 Place de l'Universite, Geneve 4, CH-1211, Switzerland

Trower, Michael K. Genomics Unit, Medicines Research Centre, Gunnels Wood Road, Stevenage SG1 2NY, UK

Valge-Archer, Viia. Cambridge Antibody Technology Ltd, The Science Park, Melbourne, Cambridge SG8 6JJ, UK

Wallace, Don M. Medicines Research Centre, Glaxo Wellcome Research and Development, Stevenage SG1 2NY, UK

Wegner, Susan. German Heart Institute, Augustenberger Platz 1, Berlin, D-13353, Germany

Wilkins, Marc. Proteome Systems Ltd, Locked Bag 2073, North Ryde, Sydney, NSW 1670, Australia

Zivy, Michel. Station de Génétique Végétale, Gif-sur-Yvette, France

Abbreviations

ABA	abscissic acid
AFLP	amplified fragment length polymorphism
ANS	1-anilino-8-napthalene sulphonate
APAF	Australian Proteome Analysis Facility
ARRM	Advanced Rapid Robotics Manufacturing
ASRI	ABA/water stress/ripening related protein
bis-ANS	bis(8-toluidino-1-naphthalene sulfonate)
BLAST	basic local alignment search tool
BSA	bovine serum albumin
CAD	collision-activated dissociation
Cam	Chloramphenical
CCD camera	charge-coupled device camera
cDNA	complementary DNA
CE	capillary electrophoresis
CG	candidate gene
CHAPS	3[(cholamidopropyl)dimethylammonio]-1-propane sulphonate
CID	collision-induced dissociation
CP	candidate protein
CS	Chinese spring
CSF	cerebrospinal fluid
CyA	cyclosporine A
CZE	capillary zone electrophoresis
dbEST	expressed sequence tag database
DD-PCR	differential display PCR
1-DE	one-dimensional polyacrylamide gel electrophoresis
2-DE	two-dimensional polyacrylamide gel electrophoresis
DGE	differential gene expression
DHFR	dihydrofolate reductase
2D-PP	two-dimensional phosphopeptide
DMAA	N,N-dimethyl acrylamide
DNA	deoxyribonucleic acid
DSN	database spot number
DTT	dithiothreitol
EDTA	ethylenediaminetetraacetic acid
ELISA	enzyme-linked immunosorbant assay
ESI	electrspray ionization
EST	expressed sequence tags

FACS	fluorescence-activated cell sorter
FITC	fluorescein isothiocyante
FPR	formyl peptide receptor
FSH	follicle stimulating hormone
FT-ICR-MS	fourier transform ion cyclotron resonance mass spectrometry
Fus	fusidic acid
FWHM	full-width half maximum
GAPDH	glyceraldehyde-3-phosphate dehytrogenase
GPI	glycosyl phosphatidylinositol
hCG	human chorionic gonadotrophin
HCl	hydrochloric acid
HIS-tagged	histidine-tagged
HIV	human immunodeficiency virus
HPLC	high-performance liquid chromatography
HRP	horseradish peroxidase
HSP	heat shock protein
HUVEC	human umbilical vein endotheilial cells
ICAT	isotope-coded affinity tagging
ICC	immunocytochemistry
ICE	interleukin converting enzyme
ICL	instrument control language
IEF	isoelectric focusing
IgG	immunoglobulin G
IMAC	immobilized metal affinity chromatography
IMAGE	integrated molecular analysis of genomes and their expression
IMS	ion mobility spectrometry
IP	isoelectric point
IPG	immobilized pH gradient
IPG-DALT	2-DE with immobilized pH gradients in the first dimension
IPTG	isopropyl-β-D-thiogalactopyranoside
IR	infrared
IRMPD	infrared multiphoton dissociation
IT	ion trap
Kan	kanamycin
KCl	potassium chloride
LC	liquid chromatography
LCM	laser capture microdissection
LC-MS/MS	liquid chromatography tandem mass spectrometry
LEA	late embryogenesis-abundant
LH	luteinising hormone
LIMS	laboratory information management
LMW	low molecular weight
LOG	Laplacian of Gaussian
m/z	mass-to-charge ratio
MALDI-MS	matrix-assisted laser desorption-ionization mass spectrometry

MALDI-TOF	matrix-assisted laser desorption-ionization time-of-flight
MAP	mitogen-activated protein
MDPF	2-methoxy-2,4-diphenyl-3(2H)furanone
MeCN	acetonitrile
MIP	molecularly imprinted polymer
MPI	minimal protein identifier
Mr	relative molecular mass
mRNA	messenger RNA
MS	mass spectrometry
MS/MS	tandem mass spectrometry
NCBI	National Centre for Biotechnology Information
NEPHGE	non-equilibrium pH gradient electrophoresis
NP-40	Nonidet P40
OD	optical density
PBS	phosphate buffered saline
P/A	presence/absence
PAGE	polyacrylamide gel electrophoresis
PCR	polymerase chain reaction
PDA	piperazine diacrylamide
ppm	parts per million
PQL	protein quantity loci
PS	position shift
PSD	postsource decay
PVDF	polyvinylidene difluoride
QCM	quartx crystal microbalance
QTL	quantitative trait loci
RAPD	random amplified polymorphic DNA
RCE	relative collision energy
RF	radio frequency
RFLP	restriction fragment length polymorphism
rHu G-CSF	human granulocyte colony-stimulating factors
RNA	ribonucleic acid
RP-HPLC	reversed-phase high performance liquid chromatography
RT-PCR	reverse transcriptase polymerase chain reaction
SAGE	serial analysis of gene expression
SAM	self-assembled monolayer
SB	sulphobetaine
SCA	synthetic carrier ampholytes
scFv	single chain variable fragment
SCX	cation exchange
SDS	sodium dodecyl sulfate
SDS-PAGE	sodium dodecyl sulfate polyacrylamide gel electrophoresis
SELDI	surface-enhanced laser desorption/ionization
SIM	selected ion monitoring
Smz	sulfamethoxazole

SP	candidate protein
SPR	surface plasmon resonance
SSP	standard spot number
Str	streptomycin
TBP	tributyl phosphine
TCA	trichloroacteic acid
TEA	triethylamine
TEMED	N, N, N', N'-tetramethylethylenediamene
Tet	tetracycline
TFA	trifluroracetic acid
TGF	transforming growth factor
THF	tetrahydrofuran
TIC	total ion count
TIGR	The Institute for Genomics Research
TLC	thin-layer chromatography
TM	transmembrane
Tmp	trimethoprim
TNF	tumor necrosis factor
TQ	triple quadrupole
tRNA	transfer RNA
UV	ultraviolet
VCAM	vascular cell adhesion molecule
WST	watershed transformation
www	World Wide Web
ZP	zona pellucida

Preface

Recent developments in analytical methods for protein characterization, and the growing rate at which whole genome sequencing projects are being completed, have combined to support the emergence and development of the new field of 'proteomics'. Proteomics, the study of protein expression and function on a genome scale, is the amalgamation of very many different experimental approaches ranging from the analysis of gene function by mRNA expression profiling with cDNA arrays, analysis of protein:protein interactions by genome-wide two hybrid screening, to more direct analysis of protein expression, sequence and structure.

Proteomics: from protein sequence to function covers many aspects of this emerging field. In particular, it provides a detailed overview of the current methods used to undertake two-dimensional polyacrylamide gel electrophoresis based measurement of protein expression and mass spectrometry based protein identification, and characterization and their associated informatics. The contributors include many leaders in the field including those who have played pivotal roles in the development of two-dimensional polyacrylamide gel electrophoresis, detection of proteins, gel image analysis, analysis of proteins by mass spectrometry, the generation and application of phage-displayed antibodies and the integration of methods to support proteomics. This book is intended as a guide to these and other methods for all who have an interest in proteomics: newcomers and experienced practitioners alike.

The book begins with two important chapters: one covers the integration of genomics and proteomics and the second describes current approaches to the measurement of mRNA expression. Increasingly, data from proteome analyses are being integrated with those from other DNA based approaches and so these chapters form a vital platform for the subsequent chapters that describe various aspects of the proteomics workflow. Chapters on two-dimensional polyacrylamide gel electrophoresis, protein detection, mass spectrometry based protein characterization including a chapter on recent developments in mass spectrometry that support quantitative analysis of protein expression and image analysis follow. Much of the proteomics workflow is at present laborious, time consuming and generates data on a scale that requires the application of software for data management. These are the subjects of two chapters on approaches to automation of the proteomics workflow. Although no one can deny that two-dimensional gel electrophoresis currently provides the most powerful platform for the analysis of protein expression in both simple and complex organisms, it does have significant limitations and so a chapter on potential alternatives to the technique follows. Further chapters describe the application of proteome analysis to drug development; the use of phage antibodies as tools for proteomics; glycobiology and proteomics; the establishment of proteomics facilities in academic laboratories and the use of proteomics in plant genetics and breeding.

Clearly, there are many other aspects of proteomics that could have been included; we hope that the selection we have chosen (which inevitably reflects our own interests) will provide a strong foundation for those wishing to learn more about proteomics. We are very aware that we have not included the exciting developments, both practical and bioinformatic, in the elucidation of protein tertiary structure; these would arguably require a book in their own right if they were to be covered in sufficient detail.

There have been many people who have helped to bring this book to completion - a feat that at times it seemed might not be achieved – and our sincere thanks go to all of them. We are very grateful indeed to the contributors, their interest and enthusiasm in the project has been much appreciated. We are especially grateful to Scott Patterson, Marc Wilkins and Peter James who provided invaluable advice and help. We also thank all the team at BIOS and those who supported our 'logistics' activities including Jenni Brown, Lisa Crimmins and Jane Hamlett.

The role of proteomics in meeting the post-genome challenge

The rapidly emerging field of proteomics has now established itself as a credible approach for furthering our understanding of the biology of whole organisms – from simple unicellular organisms to those as complex as man. The readily available experimental tools for measurement of protein expression by two-dimensional gel electrophoresis (2-DE), and for protein identification and characterization by mass spectrometry-based methods, have already made a significant impact on proteomics. The growing interest in the field looks set to accelerate the development and implementation of improved strategies for the analysis of protein expression and function on a genome-wide scale.

The advent of proteomics has, in substantial part, been dependent on the success of whole genome sequencing projects, not least because the completion of these projects has resulted in the more widespread appreciation that in themselves they reveal little about the biology of the organism. Instead, they provide an essential platform for a wide range of complementary experimental approaches that will support the characterization of the genes encoded within the genomes, and ultimately the understanding of how the products of these genes act together to regulate the activity of the organism. Thus, it seems evident that genomics and proteomics will best serve the community of biologists if these two fields synergistically develop their co-dependence.

The success of whole genome sequencing projects has been both remarkable and exciting. The first complete genome for a free-living organism was published in 1995 (Fleischmann *et al.*, 1995) and as of July 2000, there were 40 complete genome sequences in the public domain, with a further 127 prokaryotic and 95 eukaryotic genomes under analysis (wit.integratedgenomics.com/GOLD/). So far, chromosomes 5, 16, 19, 21 and 22 of the human genome have been completed and the release of the complete, corrected and mapped human genome has been predicted for 2003. This is likely to have its initial impact on medical diagnostics and indeed a Consortium to map single nucleotide polymorphisms (SNPs) has been initiated (see *Table 1* for websites). The list of complete genome sequences continues to grow at an accelerating pace and is being matched by the unprecedented public access to databases, search tools and associated electronic information sources that are available (see *Table 1*). The sequence information has already been exploited in ways that would have been almost inconceivable to most just a decade ago. For example, comparative genomics compares all the gene sequences of a particular organism with all other genomes in order to identify differences that may account for defined and important properties, such as pathogenicity. The new field of structural genomics aims to expedite the determination of protein structure via

Table 1. Web addresses for a selection of sequence databases and analysis tools

Website	Organization	Information available
ensembl.ebi.ac.uk	European Molecular Biology Laboratory, Heidelberg, Germany/ Sanger Centre, Cambridge, UK	Annotation of human, mouse and worm genomes
genome.wustl.edu/gsc/	Genome Sequencing Center, Washington University School of Medicine, St. Louis, USA	Human and model organism sequencing projects, EST projects, protocols and technical help
www.hgmp.mrc.ac.uk/	Human Genome Mapping Project Resource Centre, Wellcome Trust Genome Campus, Cambridge, UK	Sequence databases and search engines, phylogenetic linkage analysis, links to useful websites
www.hgsc.bcm.tmc.edu/	Human Genome Sequencing Center, Baylor College of Medicine, Houston, Texas, USA	Human, mouse and Drosophila sequencing projects, human transcript database
wit.integratedgenomics.com/ GOLD	Integrated Genomics Inc., Chicago, Illinois, USA	Monitors complete and ongoing genome sequencing projects and links to relevant sites and publications
www.jgi.doe.gov/	Joint Genome Institute, Walnut Creek, California, USA	Human and microbial sequencing and mapping, functional genomics programme
star.scl.genome.ad.jp/kegg	Kyoto Encyclopaedia of Genes and Genomes, Institute for Chemical Research, Kyoto University, Japan	Sequence databases and current knowledge on molecular interactions
www.ornl.gov/hgmis/	Life Sciences Division, Department of Energy, Oak Ridge National Laboratory, Tennessee, USA	Links to progress reports, publications, meetings etc, particularly with regards to the Human Genome Project
www.ncbi.nlm.nih.gov/genome/ seq/	National Center for Biotechnology Information, National Institutes of Health, Bethesda, Maryland, USA	Sequence, SNP and literature databases, tools for data mining, human and mouse genetic and physical maps
www.sanger.ac.uk/	Sanger Centre, Wellcome Trust Genome Campus, Cambridge, UK	Progress of the human sequencing project plus many of the other prokaryotic and eukaryotic projects
www.tigr.org/tdb/	The Institute for Genome Research, Rockville, MD, USA	Sequence, function and taxonomy databases for microbes, plants and humans
snp.cshl.org	The SNP Consortium Ltd., Cold Spring Harbour Laboratory, Cold Spring Harbour, New York, USA	SNP map of the human genome
www-genome.wi.mit.edu/	Whitehead Institute Center for Genome Research, Cambridge, Massachusetts, USA	Genetic and physical maps for human, mouse and rat plus human SNP database

X-ray crystallography, nuclear magnetic resonance and other methods to relate structure to gene sequence and function. Genome sequencing and allied projects have also spawned new approaches to mRNA expression analysis or transcriptomics (see Chapter 2; Chee *et al.*, 1996; Eisen *et al.*, 1998; Gerhold *et al.*, 1999) and techniques to undertake comprehensive approaches to the analysis of protein:protein interactions using the two-hybrid assay (Fields and Song, 1989; Fromont-Racine *et al.*, 1997; Lecrenier *et al.*, 1998). Notably, the use of DNA chips and microarrays for high throughput analysis of mRNA expression, sometimes on a genome-wide scale, is having a dramatic impact on the investigation of gene function (Hughes *et al.*, 2000; Young, 2000). More recent advances in microarray technology have increased still further the speed at which sequence information and differential gene expression data may be gathered (Bowtell, 1999; Young, 2000). Together, these approaches are likely to transform our ability to identify and hence target both the desirable and undesirable attributes of organisms.

The application of cDNA arrays to transcriptomics exploits several important features: the relative ease with which the analyses may be undertaken on a comprehensive scale; the ability to automate both the production, and the hybridization and scanning, of the arrays; and the availability of effective software for analysis of the results. As importantly, once mRNAs that alter in expression in response to the conditions under investigation have been identified, it is very straightforward to use the extensive and readily accessible tools of molecular biology to begin to elucidate the expression and function of the genes identified. However, the approach also suffers from several serious limitations, not least of which are the observations that (i) mRNA abundance does not always correlate well with protein abundance, (ii) the sensitivity and dynamic range of existing approaches are such that the lowest abundance mRNAs (potentially encoding the most important regulatory proteins) are not readily measured alongside the more abundant mRNAs, and (iii) the activity of the proteins encoded by mRNAs is regulated at several levels beyond their mRNA or protein expression by, for example, their subcellular localization and/or the extent to which they are post-translationally modified, neither of which are revealed by measurement of mRNA abundance. In addition, there are a number of important biological samples, particularly those that might be used for human diagnostics, such as urine, cerebrospinal fluid and blood plasma, that do not contain mRNA. Moreover, the analysis of mRNA expression in human biopsy and post-mortem samples is still a significant challenge given the potentially protracted time between the collection of the sample and the vulnerability of mRNA to degradation.

The current applications of cDNA arrays have other limitations. Whilst cDNA arrays may be readily available for those model organisms whose genomes have been sequenced, concerns still remain about the availability of tools (cDNAs) for the construction of arrays for non-model organisms for which a significant amount of biology is known. There are also organisms, such as *Plasmodium falciparum,* whose GC/AT ratio is such that the applicability/usefulness of cDNA arrays has yet to be established. Despite these limitations, the application of cDNA arrays and the use of other 'DNA sequence'-based approaches to the elucidation of gene expression and function, broadly classified as functional genomics, seems set to transform our understanding of biological organisms (Alizadeh *et al.*, 2000; DeRisi *et al.*, 1996; Perou *et al.*, 1999).

Hence it is very apparent that there is much excitement about the application of genome sequence data. However, amidst this excitement there is growing recognition and appreciation that such DNA sequence based approaches will have to be complemented by the direct analysis of the products encoded by the genes – the proteins. For example, knowing the sequence of a genome does not at present imply that the proteins encoded by the genome are immediately recognizable. Thus, although genome sequences can be used to predict open reading frames, such predictions are not infallible. It is also apparent that the application of DNA sequence data and existing knowledge of the relationships between DNA sequence and protein function do not currently support the assignment of function to the proteins encoded by the genes based on knowledge of their DNA sequence alone. It is difficult, if not impossible, to establish the number of protein products to which each gene will give rise, and DNA sequence information does not as yet give an unambiguous insight into the post-translational modifications to which each protein product is subjected. There are more than 400 possible chemical modifications a protein may undergo with consequences for its function, localization and activity. So, DNA sequence information has significant limitations in supporting our understanding of the protein constituents encoded by an individual genome – the proteome. For the human genome, the predicted 50 000–100 000 genes could encode as many as 250 000–500 000 individual protein moieties, so the scale of the protein discovery task ahead is very large indeed. Ultimately, the assignment of protein function will require detailed and direct analysis of the expression, localisation (tissue, cell and sub-cellular) and structure of the proteins encoded by genomes (Dove, 1999; Pennington *et al.*, 1997; Wilkins *et al.*, 1995).

The proteome, because of its highly dynamic nature, is difficult to define. On one level, proteomics can be regarded as the identification of all the proteins encoded by the genome, cf. genome sequencing, whilst on another it is the comprehensive analysis of changes in protein expression between different conditions cf. transcriptomics. So in many ways there is parity between the general approaches suggested for genomics and proteomics, but as yet there is a large difference in the effectiveness of the two fields. Thus at present, there is no obvious way to unravel the proteomes of more than the simplest organisms, and methods to undertake the comparative analysis of protein expression lag considerably behind those available for the analysis of mRNA expression.

Currently, the starting point for many attempts to investigate changes in protein expression involves the resolution of proteins in complex mixtures by 2-DE (see Chapter 3; Anderson and Anderson, 1996) and their subsequent identification using a growing array of powerful analytical methods (see Chapter 5; Andersen *et al.*, 1996; Eckerskorn *et al.*, 1997; Figeys *et al.*, 1998; Pappin *et al.*, 1993; Roepstorff, 1997; Scheler *et al.*, 1998; Yates, 1998). Relatively recent advances in the characterization of proteins have come from the application of matrix-assisted laser desorption-ionization (MALDI) and electrospray ionization (ESI) mass spectrometry (MS) to peptide mass fingerprinting and sequencing, respectively (see Chapters 5 and 7; Andersen *et al.*, 1996; Figeys *et al.*, 1998; Pappin *et al.*, 1993; Roepstorff, 1997; Yates, 1998). 2-DE can separate several thousand proteins simultaneously (Klose, 1999) and so, whilst the method has significant limitations (see Chapter 10), it seems likely to remain unrivalled as a method to resolve large numbers of proteins for expression profiling and subsequent identification for quite some time to come.

The case for applying a proteomics approach to post-genomic discovery has been proven and the only question now is how this might be achieved. A platform technology is in place, namely 2-DE, but it requires continued refinement. Here we have attempted to bring together a series of chapters that together should enable the reader to appreciate the scope of current methods, and their applications and limitations. Existing methods and new developments for protein visualization (Chapter 4), image analysis (Chapter 6), quantitative mass spectrometry (Chapter 7), the automation of the 2-DE and protein identification workflow (Chapters 8 and 9) and integration with informatics (Chapter 9) are all described in an accessible and informative manner. The application of proteomics to drug discovery and toxicology (Chapter 11) and plant genetics (Chapter 15) are included, as are chapters on proteomics and glycobiology (Chapter 13) possible future alternatives to 2-DE (Chapter 10) and development of complementary proteomics platforms, including the generation and application of phage-displayed antibodies (Chapter 12).

We trust you will agree that the authors have provided accessible information to those new to proteomics whilst at the same time providing detailed and valuable insights for those more experienced in this fast-moving and rapidly expanding field. For this we offer them our sincere gratitude.

Stephen R. Pennington and Michael J. Dunn

References

Alizadeh, A.A., Eisen, M.B., Davis, R.E., *et al.* (2000) Distinct types of diffuse large B-cell lymphoma identified by gene expression profiling. *Nature* **403:** 503–511.

Andersen, J.S., Svensson, B. and Roepstorff, P. (1996) Electrospray ionization and matrix assisted laser desorption/ionization mass spectrometry: Powerful analytical tools in recombinant protein chemistry. *Nature Biotechnol.* **14:** 449–457.

Anderson, N.G. and Anderson, N.L. (1996) Twenty years of two-dimensional electrophoresis: Past, present and future. *Electrophoresis* **17:** 443–453.

Bowtell, D.D.L. (1999) Options available – from start to finish – for obtaining expression data by microarray. *Nature Genet.* **21:** 25–32.

Chee, M., Yang, R., Hubbell, E., *et al.* (1996) Accessing genetic information with high-density DNA arrays. *Science* **274:** 610–614.

DeRisi, J., Penland, L., Brown, P.O., Bittner, M.L., Meltzer, P.S., Ray, M., Chen, Y., Su, Y.A. and Trent, J.M. (1996) Use of a cDNA microarray to analyse gene expression patterns in human cancer. *Nature Genet.* **14:** 457–460.

Dove, A. (1999) Proteomics: translating genomics into products? *Nature Biotechnol.* **17:** 233–236.

Eckerskorn, C., Strupat, K., Schleuder, D., Hochstrasser, D., Sanchez, J.C., Lottspeich, F. and

Hillenkamp, F. (1997) Analysis of proteins by direct scanning infrared-MALDI mass spectrometry after 2D PAGE separation and electroblotting. *Anal. Chem.* **69**: 2888–2892.

Eisen, M.B., Spellman, P.T., Brown, P.O. and Botstein, D. (1998) Cluster analysis and display of genome-wide expression patterns. *Proc. Natl. Acad. Sci. USA* **95**: 14 863–14 868.

Fields, S. and Song, O. (1989) A novel genetic system to detect protein-protein interactions. *Nature* **340**: 245–246.

Figeys, D., Gygi, S.P., Zhang, Y., Watts, J., Gu, M. and Aebersold, R. (1998) Electrophoresis combined with novel mass spectrometry techniques: powerful tools for the analysis of proteins and proteomes. *Electrophoresis* **19**: 1811–1818.

Fleischmann, R.D., Adams, M.D., White, O., *et al.* (1995) Whole-genome random sequencing and assembly of haemophilus-influenzae Rd. *Science* **269**: 496–512.

Fromont-Racine, M., Rain, J.C. and Legrain, P. (1997) Toward a functional analysis of the yeast genome through exhaustive two-hybrid screens. *Nature Genet.* **16**: 277–282.

Gerhold, D., Rushmore, T. and Caskey, C.T. (1999) DNA chips: promising toys have become powerful tools. *TIBS* **24**: 168–173.

Hughes, T.R., Marton, M.J., Jones, A.R., *et al.* (2000) Functional discovery via a compendium of expression profiles. *Cell* **102**: 109–126.

Klose, J. (1999) Large-gel 2D electrophoresis. *Meth. Mol. Biol.* **112**: 147–172.

Lecrenier, N., Foury, F. and Goffeau, A. (1998) Two-hybrid systematic screening of the yeast proteome. *Bioessays* **20**: 1–5.

Pappin, D.J.C., Hojrup, P. and Bleasby, A.J. (1993) Rapid identification of proteins by peptide-mass fingerprinting. *Current Biology* **3**: 327–332.

Pennington, S.R., Wilkins, M.R., Hochstrasser, D.F. and Dunn, M.J. (1997) Proteome analysis: from protein characterisation to biological function. *Trends Cell Biol.* **7**: 168–173.

Perou, C.M., Jeffrey, S.S., van de Rijn, M., *et al.* (1999) Distinctive gene expression patterns in human mammary epithelial cells and breast cancers. *Proc. Natl. Acad. Sci. USA* **96**: 9212–9217.

Roepstorff, P. (1997) Mass spectrometry in protein studies from genome to function. *Curr. Opin. Biotechnol.* **8**: 6–13.

Scheler, C., Lamer, S., Pan, Z., Li, X.P., Salnikow, J. and Jungblut, P. (1998) Peptide mass fingerprint sequence coverage from differently stained proteins on two-dimensional electrophoresis patterns by matrix assisted laser desorption/ionization-mass spectrometry (MALDI-MS). *Electrophoresis* **19**: 918–927.

Wilkins, M.R., Sanchez, J.-C., Gooley, A.A., Appel, R.D., Humphery-Smith, I., Hochstrasser, D.F. and Williams, K.L. (1995) Progress with proteome projects: why all proteins expressed by a genome should be identified and how to do it. *Biotechnology* **13**: 19–50.

Yates, J.R. (1998) Mass spectrometry and the age of the proteome. *J. Mass Spectrometry* **33**: 1–19.

Young, R.A. (2000) Biomedical discovery with DNA arrays. *Cell* **102**: 9–15.

Chapter 1

Bridging genomics and proteomics

Dolores J. Cahill, Eckhard Nordhoff, John O'Brien, Joachim Klose, Holger Eickhoff and Hans Lehrach

1. Introduction

Proteomics can be most broadly defined as the systematic analysis and documentation of the proteins in biological samples. It is a field that has leapt to prominence within the last 2 years and is widely expected to have a major impact on biotechnology; it has recently been reviewed by Blackstock and Weir (1999).

Proteomics can be seen as a mass-screening approach to molecular biology, which aims to document the overall distribution of proteins in cells, identify and characterize individual proteins of interest, and ultimately to elucidate their relationships and functional roles. Such direct, protein-level analysis has become necessary because the study of genes, by genomics, cannot adequately predict the structure or dynamics of proteins, since it is at the protein level that most regulatory processes take place, where disease processes primarily occur and where most drug targets are to be found. There is, however, a strong and synergistic relationship between proteomics and genomics as the two disciplines investigate the molecular organization of the cell at complementary levels (proteins and genes) and each discipline provides information that increases the effectiveness of the other. This chapter will review the current techniques used to analyze gene and protein expression on high density arrays, and how we propose to combine these approaches to bridge genomics and proteomics. This chapter is divided into four main sections, as follows:

(i) The generation of cDNA expression libraries, their robotic arraying, construction of polymerase chain reaction (PCR) filters, generation of a nonredundant set by oligonucleotide fingerprinting and complex hybridizations on DNA chips.

(ii) The generation of an expression subset of a cDNA library, and a nonredundant Unigene–Uniprotein set, expression of these proteins in *E. coli* in high-throughput and production of high-density protein arrays and micro-arrays.

(iii) Description of the characterization of the protein complement of a specific cell type or tissue at a certain time (proteome) by high-resolution two-dimensional electrophoresis (2-DE).

(iv) Bridging the current proteomic and genomic approaches by using mass spectrometry to identify native protein populations separated by high-resolution two-

Proteomics, edited by S.R. Pennington and M.J. Dunn

dimensional electrophoresis and to correlate these to proteins in the Uniprotein set and vice versa.

Currently, individual proteins are identified in sequence databases using mass spectrometrically determined peptide maps, sequence tags, or fragment-ion fingerprints of individual proteolytic cleavage products. Obviously, this approach is most reliable for relatively few known proteins. To overcome this limitation, we have developed a novel concept: each protein is specified by a minimal set of structural information readily accessible by mass spectrometry, which we have hereby designated the 'minimal protein identifier' (MPI). MPIs contain accurate molecular masses of enzymatic cleavage products in conjunction with fragment-ion data, and are recorded by the use of high-throughput matrix-assisted laser desorption ionization mass spectrometry (MALDI-MS). MPIs can be generated from excised 2-DE gel spots and from recombinant proteins, such as from a Uniprotein set, and can be used to identify homologous proteins on 2-DE gels and vice-versa (*Figure 1*).

Once recorded, MPIs allow rapid recognition of known, as well as unknown, gene products. At the same time, MPIs allow identification of proteins in sequence databases. Equipped with these features, MPIs enable safe comparison of 2-DE gels run with different biological samples independent of their format, resolution and the separation technology used. This approach results in more reliable protein identification as measured MPIs from 2-DE gel spots are compared with measured MPIs from the expressed proteins, instead of DNA sequence-predicted MPIs (*Figure 2*). In addition, the recombinant protein libraries provide a perfect source for mass spectrometric quantification of proteins from 2-DE gels by stable-isotope labeling (e.g. *E. coli* cells grown in ^{15}N medium), a key technique which has been demonstrated (see Chapter 7).

This method requires the generation and availability of a Unigene–Uniprotein set (Section 3; *Figure 3*), which not only provides an immortal source for generating cDNA micro-arrays, which can be used to profile mRNA levels by complex hybridization (Section 2). In addition, each protein of the Uniprotein set will have its MPI determined and stored in a database.

The success of this procedure requires not only the application of established technologies, such as high-performance microarraying, large-scale protein expression and high-resolution 2-DE, but also the development of new technologies. These include automated 2-DE, gel imaging, spot recognition, spot excision followed by protein proteolysis, sample purification and mass spectrometric analysis and software for closed-loop data acquisition and processing.

2. The generation of cDNA expression libraries, their robotic arraying, construction of PCR filters, generation of a nonredundant set by oligonucleotide fingerprinting, complex hybridizations on DNA chips

With the human genome project well underway and the deadline for completion approaching, the challenges of understanding the function of newly discovered genes have to be addressed. Initial attempts at sequencing the large and complex human

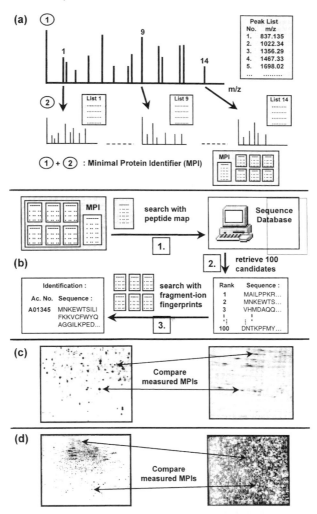

Figure 1. (a) Acquisition of minimal protein identifiers (MPIs) by MALDI-MS. The proteins are digested with a specific protease (e.g. trypsin), and the cleavage products' molecular masses are determined. Subsequently, for each protein fragment-ion spectra of a selection of prominent cleavage peptides are recorded. The peptide mass map extracted from the first spectrum provides a fingerprint of the protein's primary structure, whereas the fragment-ion peak lists yield fingerprints of the cleavage peptides' amino-acid sequences. These data are combined and stored as MPIs, one for each protein. (b) Suggested strategy for identifying proteins in sequence databases. Searching the database for a specific peptide mass map retrieves a list of candidate protein sequences (e.g. 100 sequences). This list is searched for cleavage peptides that match the recorded fragment-ion fingerprints and ranked accordingly. The advantage of the proposed sequential strategy is high search specificity and short search times, since the second selection round is applied only to small subset of the whole database. (c) Suggested strategy for comparing 2-DE protein gels. For assigning protein spots, instead of their patterns their recorded MPIs are compared *in silico*. This assignment is independent of the gel formats used, the separation technique applied and the 2-DE protocol followed. (d) Correlation of 2-DE protein spot patterns and ordered protein microarrays. For all recombinant proteins spotted onto the array, MPIs were recorded previously and stored in a database. Native proteins separated by 2-DE can now be assigned to their recombinant derivatives by comparing their determined MPIs with the above database entries.

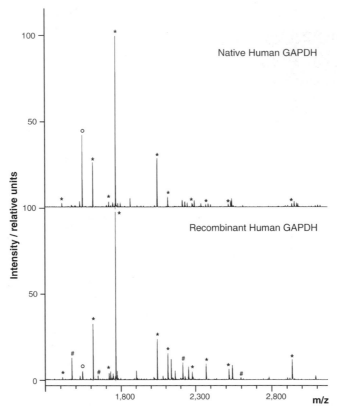

Figure 2. MALDI-TOF-MS tryptic peptide maps of native and recombinant human GAPDH. Native GAPDH was isolated from total human fetal brain protein extract by large-format 2-DE and digested *in situ*. The spectrum (top panel) was obtained from a 5 μl aliquot of purified overnight digestion supernatant. Recombinant human GAPDH equipped with an RGSHis$_6$-tag at the N-terminus was expressed in *E. coli*. Tagged proteins were metal-chelate affinity purified from crude cell extract using NTA-ligands immobilized on agarose (Qiagen, Germany) under denaturing conditions. The purified proteins were digested *in situ*. The spectrum (bottom panel) was obtained from 0.5 μl of a total of 150 μl digestion supernatant. Marked signals: (*) tryptic cleavage peptides detected in the digestion supernatant of native GAPDH according to the NCBI database (accession number: 120 649, release 5 May 1999). All these peptides were also detected in the digestion supernatant of recombinant GAPDH. (#) Additional tryptic cleavage peptides detected in the digestion supernatant of recombinant GAPDH. ° Peptide detected in both digestion supernatants that could not be assigned to GAPDH and not to any trypsin autolysis product.

genome were intentionally focused on expressed regions, as represented by cDNA repertoires. Estimates of the total gene number vary from 60 000 to over 140 000 (Aparicio, 2000) in the human genome. While the majority of the total number of human genes are now represented as expressed sequence tags (ESTs) in the dbEST database, only a tiny minority have yet been assigned a function. For example, in the 7 July 2000 release, the number of entries for human was 2 121 173 (http://www.ncbi.nlm.nih.gov/dbEST/index.html) (Wolfsberg and Landsman, 1997), corresponding to 81 967 clusters in the UniGene set (www.ncbi.nlm.gov/UniGene/

Gene Expression Analysis   **Protein Analysis**

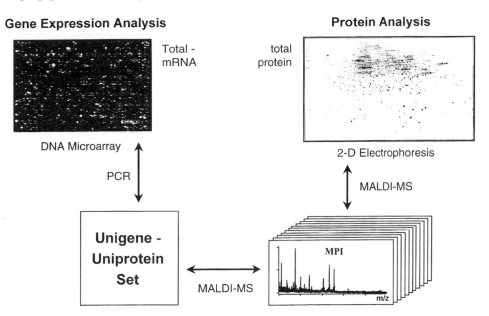

Total -
mRNA

total
protein

DNA Microarray

2-D Electrophoresis

PCR

MALDI-MS

**Unigene -
Uniprotein
Set**

MALDI-MS

MPI

m/z

Figure 3. The proposed concept, 'The Bridge'. Native proteins are correlated to their genes and RNA expression levels by the use of minimal protein identifiers (MPIs, see *Figure 1*) determined by mass spectrometry. A Unigene–Uniprotein set extracted from cDNA libraries provides both unique gene representatives via PCR readily accessible for gene expression analysis on cDNA microarrays, and the corresponding expression products as $(His)_6$-fusion proteins ready for affinity purification. The purified proteins are proteolyzed and analyzed by MALDI. Native protein populations extracted from cell cultures or tissue are separated and characterized by 2-DE followed by *in situ* proteolysis and MALDI-MS. The collected MPIs are compared with the MPIs obtained from the recombinant protein library, and vice versa. Thousands of biologically active gene products are thereby linked to their genes. This linkage is independent of any sequence information.

Hs.stats.shtml), of which only 13 407 contained at least one known gene. The most straightforward solution to this structure–function discrepancy seems to be the direct correlation between the functional status of a tissue and the expression of certain sets of genes. On the transcriptional level, the mRNA repertoire of a cell, as the first level of gene expression, most directly reflects gene activity and therefore mRNA profiling complements studying gene function. Technologies used to analyze gene expression patterns include serial analysis of gene expression (SAGE) (Velculescu *et al.*, 1995), oligonucleotide fingerprinting (Herwig *et al.*, 1999; Meier-Ewert *et al.*, 1993, 1998; Poustka, 1999; Radelof *et al.*, 1998) as well as sequencing and hybridization approaches with complex probes (Schena *et al.*, 1998). The latter technique allows the high-throughput analysis of genetic material on DNA chips to study gene expression and discover novel disease-related genes. High-density DNA and protein arrays are miniaturized-automated devices, which comprise small flat surfaces which are used for biological experimentation. Currently, DNA chips are used to study the relative expression levels of all genes in a given tissue, comparing these profiles to the expression patterns observed for example in healthy and diseased tissues.

A full understanding of the expression profile of a tissue or organism on the genomic and proteomic levels requires the screening of many samples in parallel, as rapidly as possible. In this section, those steps which have been automated and miniaturized in our laboratory to enable high-throughput and highly parallel hybridization profiling have recently been reviewed (Clark *et al.*, 1999; Eickhoff, 1998) and will be summarized below.

2.1 Generation of cDNA libraries

The analysis of large cDNA libraries from many different tissues and developmental stages is likely to give us access to most genes, before the genomic sequencing of any vertebrate genome is completed. Since mRNAs are often processed *in vivo* after transcription (splicing, editing), cDNAs often represent the only possibility for determining the sequence of correctly spliced transcripts and, therefore, the correct protein products. In addition to the identification of new genes, and of their correct splice products, cDNA libraries can also play an essential role in the analysis of the abundance level of different mRNAs in the starting material, based on the fraction of cDNA clones derived from each gene. The generation and analysis of cDNA libraries have recently been reviewed (Clark *et al.*, 1999). Although automation makes the handling and analysis of hundreds of thousands of clones feasible, even such large libraries will often not be sufficient to identify all clones representing very rare transcripts. The use of arrayed cDNA libraries should, therefore, be comple- mented by other techniques to identify genes, which are represented by very rare transcripts. Such techniques include screening of very large and randomly plated libraries, construction and use of cell/tissue-specific libraries by cDNA selection or exon trapping, and the cloning of RT-PCR products prepared with primers from within predicted exons. Despite their shortcomings, arrayed cDNA libraries of sufficient size from a variety of tissues/developmental stages will, however, often be enough to identify and analyze the majority of the genes of an organism.

2.2 Large-scale thermocycling and generation of high-density arrays

For DNA arrays, PCR techniques are well established and play a central role in large-scale genome analysis. Commercially available cycling devices can handle up to 1536 (4 × 384-well microtiter plate (MTP)) samples, in parallel. Our department in the Max-Planck-Institute of Molecular Genetics, in Berlin, has developed a laboratory thermocycling prototype for high-throughput DNA amplification based on large water baths. A basket filled with 135 × 384-well MTP (51 840 reactions) is moved with a pneumatically driven x/z-sliding stage between three water baths at three designated temperatures.

For the generation of protein arrays, we have cloned cDNA libraries into histidine-tagged expression systems, enabling the generation of large protein arrays on polyvinylidene difluoride (PVDF) membranes (Büssow *et al.*, 1998; Section 2). Several gridding robots have been built which are able to transfer clones, DNA and proteins from microtiter plates onto nylon or PVDF membranes and onto glass slides, in high-density grids. The spotting gadget carries a 384-pin head on a servo-controlled three-axis linear drive system which can be positioned with an accuracy of 5 μm. This allows densities of approximately 300 spots/cm^2. The tip diameter depends on the spotting application and varies between 150 and 450 μm (Lueking et al., 1999).

For routine applications, 27 648 clones, PCR products or proteins are spotted in duplicate onto a single 22.2 × 22.2 cm filter. Up to 15 filter copies can be produced in parallel. The spots are arranged in 5 × 5 blocks, each of which consists of a guide dot (usually central) and 12 pairs of duplicate spots (*Figure 3*). For DNA arrays, salmon sperm DNA is often used for the guide spot, while for protein arrays ink is used. The guide dots and the duplicate spots eliminate errors in spot identification as well as contributing to the ease of image analysis.

DNA filters, where clones or PCR products are gridded onto nylon membranes, can be re-used at least 20 times without significant loss of signal intensity. The protein filters arrayed on PVDF membranes are routinely used only once.

Wet membranes in DNA hybridization procedures can be difficult to handle, thus adding considerably to the length of experiments, and they can be a serious obstacle to the performance of large-scale investigations. To address this we have developed a method which simplifies handling of high-density filters using rigid plastic laminates to give rigidity to membranes. This makes these filters easier to handle and in turn improves both the speed and efficiency of experimental work (Bancroft et al., 1997).

We routinely spot DNA or proteins to make high-density arrays in two ways (Clarke et al., 1999). One method involves needle or pin spotting, where the liquid containing the DNA fragment is delivered through adhesion to stainless steel pins. The second method utilizes piezo-ink-jet technology, where cDNAs, for example, are transferred without touching the surface. We have implemented a multihead piezo-jet microarraying system, which permits the construction of large microarrays on a variety of surfaces. Since the spot density obtained by this system is greater than 2000 clones/cm^2, higher resolution detection systems, based on a laser scanning principle, were developed in cooperation with Perkin Elmer. As a further alternative to conventional needle spotting, a drop-on-demand technology has been developed. The genetic samples are pipetted with a multichannel microdispensing robot, which works on a similar principle to an ink jet printer to reduce the dimensions of the hybridization arrays by one or two orders of magnitude. Integrated image analysis routines decide whether a suitable drop is generated. If the drop is poorly formed, the nozzle tip is cleaned automatically. A second integrated camera defines positions for automated dispensing, for example filling of cavities in silicon wafers. Each head is capable of dispensing single or multiple drops with a volume of 100 pl. A magnetic bead-based purification system has recently been developed inside the dispensers. This allows concentration and purification of spotting probes prior to dispensing. The resulting spot size depends on the surface onto which the liquid is dispensed and varies between 100 and 120 μm in diameter. The density of the arrays can be increased to 3000 spots/cm^2. The micro-dispensing system has the ability to dispense 'on-the-fly' and takes less than 3 min to dispense 100 × 100 spots, in a square, with

100 μm diameter and with 230 μm distance between the center of each spot. At this density, it is possible to immobilize a small cDNA library consisting of 14 000 clones, on a microscope-slide surface. This offers a higher degree of automation since glass slides are more rigid and easier to handle than membranes.

Another application of the microdispensing technology is the preparation of high-density arrayed probes on target plates for MALDI-MS. Mass spectrometry allows an enormous speed for high-throughput applications in DNA and protein analysis. In a commercially available MS instrument, it is possible to analyze one sample in seconds and, theoretically, several thousand samples can be studied in detail per day. Until now, only prototypes have been developed that use micro-dispensed arrays of several thousand DNA fragments or proteins on a 1 cm^2 MALDI-MS target (Section 4).

2.3 *Laboratory automation and recent developments in high-density arraying*

The first step towards the analysis of large genetic libraries is the picking of randomly distributed clones from agar plates and arraying these clones into microtiter plates. Picking robots have been developed in our laboratory, which have a picking rate of 3500 clones per hour. The colonies are checked by an image analysis system to address the position for picking. The software developed identifies clone positions and translates the position into robot movement. The latest version of the software is able to recognize clones as small as 0.5 mm in diameter and, independently of the selection algorithm, blue/white selections can be made with nearly 100% accuracy. Developments have been made in further miniaturizing arraying and microarraying technologies (Eickhoff, 1998). We have described the construction and performance of a fully automated multicapillary electrophoresis system for the analysis of fluorescently labeled biomolecules such as DNA (Behr et al., 1999). In this device, a special detection system allows the simultaneous spectral analysis of all 96 capillaries, which has no moving parts, is robust, and is fully compatible with existing systems. The device can process up to 40 microtiter plates (96- and 384-well) without human interference, which means up to 15 000 samples before it has to be reloaded.

To allow the systematic generation, collection and redistribution of such data, the department started, in 1989 (Lehrach et al., 1990), to distribute filter grids prepared from a number of different libraries. To collect the data generated on these materials a separate database was developed; the Reference Library Database (Zehetner and Lehrach, 1994). This system has been continued in the form of the Resource Centre (RZPD; http://www.rzpd.de/) of the German Human Genome Project and has been the model for other efforts to systematically generate data on arrayed libraries (e.g. the IMAGE library collection, Lennon et al., 1996). Logically, the next step will be the profiling of protein products encoded by differentially expressed cDNA clones. To this end, we have developed a highly parallel approach to protein expression analysis, including the simultaneous expression of large numbers of cDNA clones in an appropriate vector system and high-speed arraying of protein products (Büssow et al., 1998). These automated systems have also been used to generate high-density

DNA and protein microarrays (Lueking *et al.*, 1999) and will be further described in Section 3.

2.4 Library normalization and generation of a Unigene set by oligonucleotide fingerprinting

The determination of the genomic sequences of higher organisms, including humans, is now an attainable goal, but represents only one level in the analysis of genetic complexity. The determination of the expression profiles of biological information represents another level of complexity, which is as important as sequencing for understanding function in organisms. To be able to identify the genes, and to count their abundance, each cDNA clone has to be assigned to its specific gene, typically by some form of sequencing process. A technique involving a highly efficient partial sequencing approach based on hybridization of short oligonucleotides, called oligonucleotide fingerprinting, has been developed to characterize cDNA libraries (Herwig *et al.*, 1999; Meier-Ewert *et al.*, 1998; Poustka *et al.*, 1999; Radelof *et al.*, 1998). This involves sequentially hybridizing 200 oligonucleotides to amplified and arrayed cDNA inserts. Hybridization with different known probes can define an unknown target, as the hybridization profile of each clone produces a DNA sequence-based 'fingerprint' which allows the grouping together, or clustering, of related clones. After construction of theoretical fingerprints from DNA sequence databases and comparison with each cluster's consensus fingerprint, cDNAs from known, and from novel, genes can be identified. If the library is cloned in an expression vector prior to the oligonucleotide fingerprinting, not only a Unigene set, but also a Uniprotein set is generated (Cahill, 2000).

2.5 Differential expression profiling by complex hybridizations on DNA microarrays

On generation of a Unigene set, the PCR products are arrayed at high density on glass slides as described in Section 2.2. These microarrays will allow the simultaneous measurement of the expression level and, therefore, an indication of the level of activity at the time point sampled of all genes represented in the array in any tissue investigated. When complex mixtures of RNAs or cDNAs from different tissues or developmental stages are hybridized to these DNA chips, it is possible to determine differences in gene expression profiles (Schena *et al.*, 1998). Automated technology allows such high-throughput gene activity monitoring by analysis of complex expression patterns. Developing from high-density filter membranes (Lennon and Lehrach, 1991), DNA microarrays are now often small and dense enough to be referred to as gene chips (Jordan, 1998; Ramsay, 1998). Using microarrays, gene expression profiles have been attained for Arabidopsis (Schena *et al.*, 1995), *Saccharomyces cerevisiae* (Cho *et al.*, 1998; Chu *et al.*, 1998; DeRisi *et al.*, 1997; Shalon *et al.*, 1996), including large-scale yeast two-hybrid screens (Uetz *et al.*, 2000), human lymphoid cells and tissues like bone marrow, brain, prostate and heart (Schena *et al.*, 1996). cDNA microarrays have already been used to profile complex diseases such as rheumatoid arthritis (DeRisi *et al.*, 1996; Heller *et al.*, 1997; Trent *et al.*, 1997) and Ewing's sarcoma (Welford *et al.*, 1998).

Complex hybridizations are performed in our department as follows: instead of carrying out two differently labeled mRNA hybridizations on one glass slide, our approach is to label the mRNA with one label only and to perform the expression analysis for each tissue on up to four separate glass slides. For complex hybridizations, the glass format of choice is 2.5 × 7.5 cm for up to 10 000 genes. For higher numbers of genes that have to be analyzed, the format of choice will be 8 × 12 cm glass slides, where up to 50 000 genes can be covalently attached to the glass surface. This makes it possible to reuse the DNA microarrays at least five times. The signals representing hybridizations to each arrayed gene are analyzed to determine the relative abundance in the samples of mRNAs corresponding to each gene. Internal controls include back hybridizations with short oligomers, the appearance of housekeeping genes in the array, as well as the use of *A. thaliana* control clones in the array and spiking the tissue RNA probe with RNA from *A. thaliana*.

2.6 *Automated image analysis*

An important feature in high-throughput laboratories is the automated large-scale characterization of positive clones in the libraries. The main requirements of such an automated analysis system are, firstly, the automatic grid-finding of spotted patterns on a filter and, secondly, the determination of positive clones. Unfortunately, different hybridized filters show different hybridization qualities. The picture quality obtained from these filters can be affected by several factors. Existing problems include the uneven distribution of spots on a picture due to uneven nylon membranes or a high hybridization background. Although human judgment is by far the best method for the decision whether a clone is positive or not, the algorithms for automatic spot finding are well developed in our department. At the current stage, nearly 80% of all positive clones can be scored automatically. For the screening on our high-density protein arrays of expressed HIS-tagged proteins (27 648 in duplicate on a 22.2 × 22.2 cm PVDF membrane), 80% of the proteins in the rearrayed expression subset showed expression. Using the same robot technology and the same image analysis software, we can analyze DNA and protein arrays and microarrays, and we are currently developing these systems for the analysis of 2-DE gels. The highly parallel data acquisition and analysis systems available will increase the speed of experimental work, and allow meaningful comparisons between for example, gene expression on the RNA level and profiling of the antibody repertoire in serum on protein chips and the determination of relativative the protein abundance as revealed by 2-DE patterns.

3. Protein arrays. Generation of cDNA expression libraries, use of automated technologies to generate protein arrays and chips, and the applications of protein chips in proteomics

Proteins translate genomic sequence information into function, thereby enabling biological processes. Therefore, a full understanding of the expression profile of a tissue or organism on the genomic and proteomic levels requires the screening of many samples in parallel, as rapidly as possible. Because a large number of proteins are involved in the homeostasis of even the simplest organisms, an automated approach with very high throughput is required for protein analysis (Uetz *et al.*, 2000). We have shown that cDNA libraries can be screened for protein expression on high-density filter membranes (Büssow *et al.*, 1998) of high-density filters for parallel DNA hybridization, protein expression and antibody screening. Using robot technology (Section 2), a human fetal brain cDNA expression library (hEx1) was arrayed in microtiter plates, and bacterial colonies were gridded onto PVDF filters. *In situ* expression of recombinant fusion proteins was induced and detected using an antibody against a His_6 tag-containing epitope. This approach has been extended to the automated spotting of protein microarrays from liquid expression cultures using a transfer stamp mounted onto a flat-bed spotting robot (Lueking *et al.*, 1999), which showed that protein microarrays provide the means for very sensitive gene expression and antibody specificity screening at high throughput. Using this transfer stamp, which has a diameter of 150 μm instead of 450 μm, 4800 samples can be placed onto a microscopic slide and simultaneously screened, applying minimal amounts of reagents. Sharp and well-localized signals allowed the detection of 250 attomol or 10 pg of a spotted test protein (GAPDH). The protein expression clones of the cDNA library are reliably detected, with a low rate (11%) of false positive clones expressing proteins in incorrect reading frames. These improved protein microarrays allow very sensitive protein expression and antibody specificity screening. Since protein microarrays are a useful tool for connecting gene expression analysis and molecular binding studies on a whole-genome level, if differentially expressed genes are identified using cDNA microarrays, the same clones can be analyzed simultaneously for protein expression in different cellular systems or by *in vitro* transcription/translation. On identical protein microarrays, expression clones can be screened for binding to other proteins (e.g. antibodies) or to diverse molecules from nucleic acids to small-molecule ligands. This versatility should make protein microarrays a multipurpose tool for diagnostic use. We have recently described their use as powerful tools for high-throughput ligand–receptor interaction studies, diagnosis and antibody specificity characterization (Walter *et al.*, 2000).

   In this section, we will describe the production of nonredundant protein arrays, so called Uniprotein arrays, and their applications in the context of genomics and proteomics will be described in Section 5. The application of high-density protein arrays and other technologies in proteomics has been described as 'second generation proteomics' (Humphery-Smith, 1999). This has been defined as 'array technologies used to detect the total protein complement of a genome without calling upon the separation sciences (for example, two-dimensional gel electrophoresis, mass spectrometry, column chromatography or capillary electrophoresis), but rather employing more traditional molecular biological approaches to conduct holistic analysis of cellular proteins'. While the use of protein arrays in proteomics described above is valid, it is different from the use of protein arrays to bridge genomics and proteomics described here. Our application involves the generation of an MPI catalog for all gene products encoded by the Unigene–Uniprotein set, and its use to bridge experiments on high-density expression arrays and 2-DE (Section 5; see *Figure 1*).

3.1 Generation and applications of a Uniprotein set and protein arrays

Until recently, it was not possible to analyze proteins using the same high-density array, automated approach. We have applied automation technologies to the high-throughput, and large-scale analysis of proteins, by generating cDNA expression libraries, high-density protein arrays (Büssow *et al.*, 1998; Cahill *et al.*, 1998 (patent)) and microarrays (Lueking *et al.*, 1999). Our approach is to make protein products encoded by cDNA clones available for further analysis and use, such as the generation of protein arrays. This requires a highly parallel approach to protein expression analysis, including the simultaneous expression of large numbers of cDNA clones in an appropriate vector system and high-speed arraying of protein products, recently reviewed in Eickhoff *et al.* (2000). Using robotic technology (Lehrach *et al.*, 1997), a human fetal brain cDNA expression library (hEx1) was arrayed in microtiter plates, and bacterial colonies were gridded onto PVDF filters. *In situ* expression of recombinant fusion proteins was induced and detected using an antibody against a 6 × His-tag-containing epitope. Using our approach, the genes in these libraries can be studied on the DNA and protein levels simultaneously, and provide sources of recombinant genes and proteins to make DNA and protein chips. This approach also achieves the large-scale systematic provision of recombinant proteins for functional studies by making and arraying cDNA expression libraries and by allowing the direct connection from DNA sequence information on individual clones to protein products and back again on a whole genome level (Büssow *et al.*, 1998). This makes translated gene products directly amenable to high-throughput experimentation and generates a direct link between protein expression and DNA sequence data.

As described in Section 2, generating a nonredundant Uniclone or a Unigene–Uniprotein set involves oligonucleotide fingerprinting an expression subset of a cDNA library which has been cloned into a protein expression vector, such as the pQe vectors (Qiagen), rearraying of the library to a nonredundant Unigene–Uniprotein set (Cahill and Lehrach, 1998 (patent)), expression of these proteins in *E. coli* at high throughput, followed by the generation of high-density protein arrays and microarrays. The Unigene set can be used for the generation of high-density DNA arrays, followed by expression profiling on DNA microarrays using techniques such as complex hybridizations and subtractive hybridizations. Differential expression profiling will allow the comprehensive study of biological function by analyzing genes and gene networks, and their expression profiles. All the proteins of the Uniprotein array can be expressed and purified in high throughput. They can subsequently be arrayed on an MS target, followed by the generation of an MS profile for each protein and its storage in a database. 2-DE can also be performed to profile the proteome, where protein spots on the gel are excised, enzymatically digested and characterized by mass spectrometry. Their DNA sequence can then be routinely determined, even for large protein complexes (Neubauer *et al.*, 1998).

This approach has been extended to produce the second type of protein array which involves the automated spotting of purified protein microarrays from liquid expression cultures using a new transfer stamp mounted onto a flat-bed spotting robot (Lueking *et al.*, 1999). This showed that such protein microarrays provide the means for very sensitive gene expression and antibody specificity screening in high throughput; and should provide powerful tools for characterization of antibodies, profiling of serum (Cahill, 2000), high-throughput ligand–receptor interaction studies, as well as the proteomic applications described here.

Clonal cDNA expression in mammalian cells and matching protein products to 2-DE patterns of cellular proteins has been described. Pooled clones from an ordered cDNA library have been expressed by *in vitro* transcription/translation and analyzed by two-dimensional gel electrophoresis, which will be described in Section 5. However, we believe that combining protein array and mass spectrometry to generate MPIs is a novel approach to bridge genomics and proteomics.

4. The characterization of the protein complement of a specific cell type or tissue at a certain time by high-resolution 2-DE

Functional analysis of the proteins expressed in a specific tissue and/or stage of development is an essential step towards understanding biological processes. Once produced, the protein can then potentially undergo post-translational modification, before assuming its role in cell activity. Such modifications can be caused by the addition of other molecules, for example by phosphorylation, glycosylation, protein processing and proteolytic cleavage. Since a protein, its structural modifications and any quantitative changes are not all controlled by the same gene, the active form of the individual proteins cannot be determined by reference to any single gene. Therefore, even if all the genes of a genome have been sequenced and translated into proteins, it remains a major task to identify and characterize the various functional forms of each protein, in a particular cell type under certain conditions. Changes in the structure or abundance of proteins can lead to disease. In order to understand the mechanisms of disease, it is therefore essential to understand how different proteins are expressed and how they function in a particular biological context. By quantifying combinations of proteins present in diseased tissue and comparing them with those from healthy tissue, it is possible to identify specific proteins related to a particular disease. Such disease-specific proteins have commercial potential as disease markers or as drug targets and their corresponding gene sequences may have significant value in screening for susceptibility. A major frontier of research and drug discovery is the study and analysis of proteomes because, using this approach, it is possible to identify specific proteins related to a particular disease.

4.1 Proteomics versus genomics

Proteomics is the analysis of the protein complement expressed by a genome or a cell or tissue type (Wilkins *et al.*, 1996), and has recently been reviewed (Blackstock and

Weir, 1999). The proteome is, therefore, dynamic as not all proteins are expressed at the same time in a particular cell, tissue or organism. The word proteomics derives from the term proteome, a word that was coined by analogy with the term genome, by Marc Wilkins and Keith Williams of Macquarie University, Sydney, Australia in 1994. The term proteome was first discussed in print by Wasinger et al. (1995), who defined it as the 'total protein complement of a genome'. Similar definitions of the proteome, such as 'the set of proteins encoded by a genome', may frequently be found in the literature. Strictly speaking, however, all definitions which define the proteome as the protein readout of the genome are inadequate because the genome does not explicitly encode the full structure and diversity of the proteins present in an organism. This has been recognized in other definitions of the proteome, as has been discussed by the protein complement of an organism, in which there is no suggestion of a simplistic linear relationship between an organism's DNA and its protein profile.

Unlike the genome, which is complete in its entirety in nearly all cells, the proteome is highly cell specific: different cells express different subsets of the total protein complement of an organism. Equally, while the genome is a static repository of information whose contents do not change with time, the proteome is highly dynamic, with the subset of proteins expressed changing from moment to moment according to the developmental and physiological state of the cell. Here again, the analogy with functional genomics is pertinent, for functional genomics is focused on mRNA which, like proteins, is cell-type and cell-state specific, and dynamically expressed. The dynamic and cell-specific features of the proteome present a major challenge to systematic analysis, since to document the proteome of a multicellular organism in full, it would be necessary to document the protein profile of each cell type, in each of its possible physiological and developmental states.

4.2 Two-dimensional polyacrylamide gel electrophoresis (2-DE)

The characterization of complex proteomes requires separation and detection of many thousands of different protein species. The aim of this approach is the characterization of many proteins on the basis of a broad spectrum of functional parameters. This includes occurrence in different organs, in different embryonic, post-natal and adult stages (including the stages of ageing), in different cell fractions and cell organelles (nuclei, mitochondria), and in the two different sexes, where appropriate (Bowden et al., 1996; Dean et al., 1998; Römer et al., 1997). By analyzing co- and post-translational modifications (e.g. glycoproteins, phospho-proteins), a biochemical spectrum of characteristics of the individual proteins is also added (Janke et al., 1996). The analysis of the proteome is performed by running high-resolution 2-D gel electrophoresis (z2-DE), analyzing and comparing resulting spot patterns, and identifying candidate proteins by combining in situ proteolysis with mass spectrometric techniques (Gauss et al., 1999; Jungblut et al., 1998). The necessary high resolution is provided by the large-gel 2-DE technique that is able to reveal more than 10 000 protein spots in one gel (Klose and Kobalz, 1995; Klose, 1999). This technique was based on the original 2-DE method (Klose, 1975;

O'Farrell, 1975) and still shows the highest resolution among 2-DE methods reported to date in the literature. For sample preparation, a detailed and highly standardized protocol has been established (Klose, 1999). This protocol avoids any loss of particular groups or classes of protein species. According to this protocol, protein extraction of a specific tissue results in three fractions (cytoplasm, membranes, chromatin). The 2-DE patterns of these three fractions represent the vast majority of the proteins of this tissue (10 000–15 000 different protein spots). In order to detect proteins present in low concentrations, 2-DE patterns from purified cell organelles (e.g. nuclei) are produced in addition and may reveal one or two thousand protein spots additionally to the three basic groups.

The protein spots from the 2-DE patterns are cut out from the gel, enzymatically digested and digested peptides analyzed by mass spectrometry. Each spot will become specified by a unique MPI (*Figure 3*); the MPIs will be stored in a database. In order to characterize thousands of individual protein species according to the criteria mentioned, the different biological and biochemical classes of proteins are matched and subtracted with respect to the presence or absence of 2-DE spots, and with respect to the different quantitative levels a protein may reveal in different tissues. Evaluation of 2-DE patterns is performed by laser scanning followed by software-assisted spot recognition and characterization (see Chapter 6). For presence/absence analysis of protein patterns a highly sensitive silver staining procedure is used. For quantification, Coomassie blue or newly developed fluorescence stains are employed (see Chapter 4). To detect post-translational modifications, glycostaining and phosphostaining procedures are available. Matching is performed by computer analysis in two ways: (1) using 2-DE images from standard patterns of the protein classes (see Section 4.3); and (2) using the MPI database, which allows the proteins to be matched independently from the gel pattern. The MPI data also facilitate the detection, within each 2-DE pattern, of protein spots, which represent modifications of the same protein (spot families). From each 2-DE pattern, as many protein spots as possible, including all the unknown proteins, are specified by MPI using mass spectrometry, 2-DE computer patterns, and MPI data from the same patterns are independently matched and subtracted. In this way, thousands of protein species become characterized according to a broad spectrum of biological and biochemical parameters.

It is necessary to combine these data with gene expression profiles (Section 2) to fully elucidate what happens in a particular cell or tissue, under normal, disease and activated (e.g. drug-treated) conditions. This information can be correlated to mRNA expression profiles from the same tissues (Section 2) by comparing the measured MPIs with those obtained from the corresponding protein expression library (Section 3). On the basis of the MPI data, we will link the individual protein species of the 2-DE pattern with the homologous proteins of the protein expression arrays and in this way also with the corresponding cDNAs. Consequently, genes become linked to their proteins, which are functionally characterized. However, it has been shown that the correlation between results in genomics, functional genomics and proteomics have been low (Anderson and Seilhamer, 1997). We propose that the approach described here to use MPIs to bridge proteomics and genomics may also contribute to the interpretation and analysis of such large data sets and high-throughput approaches.

A basic problem, however, remains, which is the detection of regulatory and modifying genes acting on a protein. A genetic approach to this problem is to perform genetic linkage studies based on structural and quantitative polymorphisms occurring in protein phenotypes (Breen *et al.*, 1994; Klose, 1999; Klose *et al.* (manuscript in preparation); Nock *et al.*, 1999). Methods for studying protein–protein interactions, such as the yeast two-hybrid system (Wanker *et al.*, 1997), also lead to genes related to a protein as well as its structural gene.

4.3 Federated 2-DE databases

In the last few years, federated 2-DE databases have been established to allow laboratories worldwide to share 2-DE data. In practice, however, matching 2-DE patterns from different laboratories has turned out to be difficult or impossible, due to the different techniques used (carrier ampholytes, immobilized pH gradients (IPGs), gel format, staining procedure and sample preparation). In the future, however, this problem will be overcome by the more commonly used various mass spectrometric techniques, allowing the laboratories to compare results at the level of MS data rather than by matching 2-DE patterns. In particular, MPI databases may become the platform to exchange data derived from 2-DE patterns (*Figure 1*). Consequently, standardizing the 2-DE technique used in all laboratories worldwide may no longer be a prerequisite for sharing data.

4.4 Automation in 2-DE analysis

In the near future it will become possible to analyze several hundreds of protein spots by mass spectrometry per day per laboratory, or even to grid a complete 2-DE gel into small pieces with the aim of analyzing complete gels. This development makes automation of 2-DE an urgent requirement. Pharmacia has simplified the Immobiline technique to a certain extent, but it is far from being an automated technique. Furthermore, we found that Immobiline gels cannot be used to reach the high level of resolution necessary to resolve complex proteomes (Klose and Kobalz, 1995).

Another central interest is the improvement of high-throughput techniques in proteome analysis, such as the development of robotics automation for recognizing and excision of protein spots from 2-DE gels, enzymatic digestion and transferring the digested 2-DE gel spots to a mass spectrometry target for MS analysis and generation of MPIs (see Chapters 8 and 9).

5. Bridging the current proteomic and genomic approaches by mass spectrometry

Our strategy aims to bridge genomics and proteomics and can be summarized as follows: a Unigene set is generated, each being a unique gene representative, and expressed as recombinant proteins, whose MPIs will be determined by mass spectrometry. Alternatively, the proteins from 2-DE patterns corresponding to the Unigene set are specified by MPIs. This provides a bridge that connects the proteins characterized by 2-DE, with their corresponding mRNAs and genes (cDNAs). All MPIs collected from 2-DE gels will be compared *in silico* with the MPIs obtained from the recombinant protein library, and vice versa. Thereby, thousands of biologically active gene products are linked to their genes. This linkage is independent of any sequence information and therefore also attractive for functional proteome analysis of other organisms. Another advantage of our strategy, compared with current strategies, is that protein identification becomes more reliable because mass spectrometric data are compared with mass spectrometric data, and not with data predicted from DNA or protein sequences. Major shortcomings of the latter approach are that substrate-dependent protease performance, peptide solubility and signal suppression in the mass spectrometric analysis are not considered.

As already described, MPIs contain accurate molecular masses of enzymatic cleavage products combined with ion fragmentation data, and will be used to identify proteins in predicted or experimentally recorded MPI databases. The latter comprise two main categories: (i) *in vivo* expressed proteins isolated by 2-DE; and (ii) recombinant cDNA, genomic DNA expression products.

To achieve broad applicability, a Unigene set will be generated, initially for human genes, each member of which being a unique gene representative (Section 2). These genes will be expressed as recombinant proteins (Section 3), whose MPIs will be determined and stored in a central database. We predict that this will facilitate a novel bridge, which will allow the direct connection between native protein populations (proteome) with mRNA pools (transcriptome), and their underlying genes (genome).

6. Future perspectives and developments

6.1 Genomics

With the on-going miniaturization process in biotechnology, new hardware tools have to be developed. All of the necessary handling steps required for a miniaturized hybridization approach, especially the existing detection systems, have also to be improved. Smaller spot sizes result in smaller amounts of targets, probes and samples being required and also imply that more sensitive detection systems, with an increased spatial resolution, will have to be developed. While radio-labeled methods have dominated biotechnology in the past, the next generation of detection methods will be based on optical principles, since compounds labeled with ^{32}P, ^{33}P or ^{35}S show a diffuse signal on the autoradiogram and the emitted radiation is delivered over 360°. Because of this limitation, optical methods such as laser scanning devices for large areas (22.2 × 22.2 cm) or other microscopy methods, including confocal laser scanning microscopes for areas as small as a few mm^2, are the methods of choice for the near future.

6.2 Proteomics

In order to achieve the throughput necessary to record several thousand MPIs per day, a fully automated workstation for proteolysis of many proteins in parallel present in a gel matrix or bound to particles (affinity purification of recombinant fusion proteins), needs to be developed. This station should include sample clean-up, concentration and subsequent sample preparation for mass spectrometry. Furthermore, mass spectrometric equipment is required that can generate several thousand MPIs per day. In terms of throughput and overall costs, MALDI-TOF-MS is currently the mass spectrometric technology of choice. In addition, new mass spectrometric methods will need to be explored for high sample throughput, including improved instrumentation for PSD analysis and, alternatively, MALDI for ion traps. However, there are some drawbacks that may limit the use of MALDI-MS in proteome and genome projects. Most notably, salts and detergents, for example sodium dodecylsulfate from SDS–PAGE gels or staining reagents, drastically increase the background during the measurement. This therefore requires the additional necessity of sample clean-up prior to mass spectrometry (see Chapter 9).

6.3 Developments required

The technologies that need to be further developed for this approach to succeed include:

(i)   new methods to enable the large-scale sub-cloning of cDNAs, in parallel with modified tagged expression vectors, such as development and modification of vectors to improve and expression and secretion systems in *S. cerevisiae* and *S. pombe*, and to express proteins that are not easily expressed in *E. coli* (Lueking *et al.*, 2000);

(ii) an automated workstation to purify His-tagged proteins (such as the BioRobot 8000, Qiagen);

(iii) the development of more automation in the 2-DE process, academic and industrial research in 2-DE, as a central technique in proteomics calls for simplified large-scale 2-DE and technical improvements in 2-DE, automated apparatus, which should concentrate on three problems, casting thin gels in long capillary tubes and large gel apparatus.

Other critical issues include: casting the gels; cleaning tubes and glass plates; sample application to the isoelectric focusing (IEF) gel; and transfer of the IEF gel to the SDS gel. Including the development of a spot picking robot for 2-DE gels to automate the cutting of hundreds and thousands of protein spots from 2-DE gels. At the Max-Planck-Institute of Molecular Genetics, a prototype of gel spot picking device carrying an eight-tube punch has been developed, but requires further improvement (e.g. in releasing the gel pieces) and adaption to a robot system for 20×30 cm gels.

7. Summary

Comparison of MPIs from 2-DE with the MPIs recorded for the recombinant protein library directly links the Unigene–Uniprotein library to all observed gene products and vice versa. Furthermore, protein identification is more reliable because measured MPIs from 2-DE gel spots are compared with measured MPIs from the expressed proteins, instead of DNA sequence-predicted MPIs (enzymatic cleavage products combined with ion fragmentation results). The advantage is that this linkage is independent of any sequence information and, therefore, also attractive for functional proteome analysis of other organisms, especially those that have not been sequenced. Once recorded, MPIs allow rapid recognition of known, as well as unknown, gene products. At the same time, MPIs allow identification of proteins in sequence databases. Equipped with these features, MPIs enable save comparison of 2-DE gels run with different biological samples independent of their format, resolution and applied separation technology. Comparison of MPIs, derived from 2-DE gels, with the MPIs recorded for the recombinant protein library directly links the Unigene library to all observed gene products and vice versa, thereby linking genomics and proteomics. In addition, direct access to the corresponding recombinant protein expression library opens up a perfect source for mass spectrometric quantification of proteins taken from 2-DE gels. Labeling by stable isotope (e.g. *E. coli* cells grown in ^{15}N medium) is the key to this approach, as has been demonstrated (see Chapter 8).

References

Anderson, L. and Seilhamer, J. (1997) A comparison of selected mRNA and protein abundances in human liver. *Electrophoresis* **18:** 533–537.

Aparicio, S.A.J.R. (2000) How to count human genes. *Nature Genet.* **25:** 129–130.

Bancroft, D. R., O'Brien, J. K., Guerasimova, A. and Lehrach, H. (1997) Simplified handling of high-density genetic filters using rigid plastic laminates. *Nucleic Acids Res.* **25:** 4160–4161.

Behr, S., Matzig, M., Levin, A., Eickhoff, H. and Heller, C. (1999) A fully automated multicapillary electrophoresis device for DNA analysis. *Electrophoresis* **20:** 1492–507.

Blackstock, W.P. and Weir, M.P. (1999) Proteomics: quantitative and physical mapping of cellular proteins. *Trends Biotechnol.* **17:** 121–127.

Bowden, L., Klose, J. and Reik, W. (1996) Search for parent-specific gene expression in early embryos and embryonic stem cells in the mouse using high-resolution two-dimensional electrophoresis of proteins. *Int. J. Dev. Biol.* **40:** 499–506.

Breen, M., Deakin, L., MacDonald, B. *et al.* (1994) Towards high resolution maps of the mouse and human genomes – a facility for ordering markers to 0.1 cM resolution. *Human Mol. Genet.* **3:** 621–628.

Büssow, K., Cahill, D., Nietfeld, W., Bancroft, D., Scherzinger, E., Lehrach, H. and Walter, G. (1998) A method for global protein expression and antibody screening on high-density filters of an arrayed cDNA library. *Nucleic Acids. Res.* **26:** 5007–5008.

Büssow, K., Nordhoff, E., Lübbert, C., Lehrach, H. and Walter, G. (2000) Genomics A human cDNA library for high-throughput protein expression screening.

Cahill, D.J. (2000) In Blackstock, M.M. (Ed.). *Proteomics: A Trend Guide*, pp. 49–53. Elsevier Science, Oxford.

Cho, R.J., Campbell, M.J., Winzeler, E.A. *et al.* (1998) A genome-wide transcriptional analysis of the mitotic cell cycle. *Mol. Cell.* **2:** 65–73.

Chu, S., DeRisi, J., Eisen, M., Mulholland, J., Botstein, D., Brown, P.O. and Herskowitz, I. (1998) The transcriptional program of sporulation in budding yeast. *Science* **282:** 699–705.

Clark, M. D., Panopoulou, G. D., Cahill, D. J., Bussow, K. and Lehrach, H. (1999) Construction and analysis of arrayed cDNA libraries. *Methods Enzymol.* **303:** 205–233.

Dean, W., Bowden, L., Aitchison, A., Klose, J., Moore, T., Meneses, J.J., Reik, W. and Feil, R. (1998) Altered imprinted gene methylation and expression in completely ES cell-derived mouse fetuses: association with aberrant phenotypes. *Development* **125:** 2273–2282.

DeRisi, J., Penland, L., Brown, P.O., Bittner, M.L., Meltzer, P.S., Ray, M., Chen, Y., Su, Y.A. and Trent, J.M. (1996) Use of a cDNA microarray to analyse gene expression patterns in human cancer. *Nature Genet.* **14:** 457–460.

DeRisi, J.L., Iyer, V.R. and Brown, P.O. (1997) Exploring the metabolic and genetic control of gene expression on a genomic scale. *Science* **278:** 680–686.

Eickhoff, H. (1998) *Drug Discovery Today*, Vol. 3, pp. 148–149.

Eickhoff, H., Ivanov, I. and Lehrach, H. (2000) In Saluz, H.P. (Ed.). *Technical System Management in Microsystem Technology: a powerful tool for biomolecular studies*. Birkhäuser Verlag.

Gauss, C., Kalkum, M., Lowe, M., Lerhrach, H. and Klose, J. (1999) Analysis of the mouse proteome. (I) – Brain proteins: Separation by two-dimensional electrophoresis and identification by mass spectrometry and genetic variation. *Electrophoresis* **20:** 575–600.

Heller, R.A., Schena, M., Chai, A., Shalon, D., Bedilion, T., Gilmore, J., Woolley, D.E. and Davis, R.W. (1997) Discovery and analysis of inflammatory disease-related genes using cDNA microarrays. *Proc. Natl. Acad. Sci. USA* **94:** 2150–2155.

Herwig, R., Poustka, A., Müller, C., Bull, C., Lehrach, H. and O'Brien, J. (1999) Large-scale clusterig of cDNA-fingerprinting data. *Genome Res.* **9:** 1093–1105.

Humphery-Smith, I. (1999) Replication-induced protein synthesis and its importance to proteomics. *Electrophoresis* **20:** 653–659.

Janke, C., Holzer, M., Klose, J. and Arendt, T. (1996) Distribution of isoforms of the microtube-associated protein tau in gray and white matter areas of human brain – a two-dimensional gel-electrophoretic analysis. *FEBS Lett.* **379:** 222–226.

Jordan, B.R. (1998) Large-scale expression measurement by hybridization methods: From high-density membranes to 'DNA chips'. *J. Biochem.* **124:** 251–258.

Jungblutt, P.R., Otto, A., Favor, J., Lowe, M., Muller, E.C., Kastner, M., Sperling, K. and Klose, J. (1998) Identification of mouse crystallins in 2D protein patterns by sequencing and mass spectrometry. Application to cataract mutants. *FEBS Lett.* **435:** 131–137.

Klose, J. (1975) Protein mapping by combined isoelectric focusing and electrophoresis of mouse tissues. *Humangenetik* **26:** 231–243.

Klose, J. (1999a) Genotypes and phenotypes. *Electrophoresis* **20:** 643–652.

Klose, J. (1999b) Fractionated extraction of total tissue proteins from mouse and human for 2-D electrophoresis. *Methods Mol. Biol.* **112:** 67–85.

Klose, J. (1999c) Large-gel 2-D electrophoresis. *Methods Mol. Biol.* **112:** 147–172.

Klose, J. and Kobalz, U. (1995) Two-dimensional electrophoresis of proteins: An updated protocol and implications for a functional analysis of the genome. *Electrophoresis* **16:** 1034–1059.

Klose, J., Nock, C., Büssow, K., Löwe, M., Schalkwyk, L.C., Rastan, S., Brown, S., Himmbelbauer, H. and Lehrach, H. (2000) Gene mapping by protein mapping. Connecting proteome and genome of the mouse. In preparation.

Lehrach, H., Drmanac, R., Hoheisel, J., Larin, Z., Lennon, G., Monaco, A.P., Nizetic, D., Zehetner G. and Poustka, A. (1990) Hybridisation fingerprinting in genome mapping and sequencing. *Genome Anal. Genet. Phys. Map.* **1:** 39–81.

Lehrach, H., Bancroft, D. and Maier, E. (1997) Robotics, computing, and biology. An interdisciplinary approach to the analysis of complex genomes. *Interdisciplinary Science Reviews* **22**(1): 37–44.

Lennon, G.G. and Lehrach, H. (1991) Hybridization analyses of arrayed cDNA libraries. *Trends Genet.* **7:** 314–317.

Lennon, G., Auffray, C., Polymeropoulus, M. and Soares, M.B. (1996) The IMAGE consortium: An Integrated Molecular Analysis of Genomes and their Expression. *Genomics* **33:** 151–152.

Lueking, A., Horn, M., Eickhoff, H., Bussow, K., Lehrach, H. and Walter, G. (1999) Protein microarrays for gene expression and antibody screening. *Anal. Biochem.* **270:** 103–111.

Meier-Ewert, S., Maier, E., Ahmadi, A., Curtis, J. and Lehrach, H. (1993) An automated approach to generating expressed sequence catalogues *Nature* **361**: 375–376.

Meier-Ewert, S., Lange, J., Gerst, H. *et al.* (1998) Comparative gene expression profiling by oligonucleotide fingerprinting. *Nucleic Acids Res.* **26**: 2216–2223.

Neubauer, G., King, A., Rappsilber, J., Calvio, C., Watson, M., Ajuh, P., Sleeman, J., Lamond, A. and Mann, M. (1998) Mass spectrometry and EST database searching allows characterisation of the multi-protein spliceosome complex. *Nature Genet.* **20**: 46–50.

Nock, C., Gauss, C., Schalkwyk, L.C., Klose, J., Lehrach, H. and Himmelbauer, H. (1999) Technology development at the interface of proteome research and genomics: Mapping nonpolymorphic proteins on the physical map of mouse chromosomes. *Electrophoresis* **20**: 1027–1032.

O'Farrell, P.H. (1975) High resolution two-dimensional gel electrophoresis of proteins. *J. Biol. Chem.* **250**: 4007–4021.

Poustka, A. J., Herwig, R., Krause, A., Hennig, S., Meier-Ewert, S. and Lehrach, H. (1999) Toward the gene catalogue of sea urchin development: The construction and analysis of an unfertilized egg cDNA library highly normalized by oligonucleotide fingerprinting. *Genomics* **59**: 122–133.

Radelof, U., Hennig, S., Seranski, P., Steinfath, M., Ramser, J., Reinhardt, R., Poustka, A., Francis, F. and Lehrach, H. (1998) Preselection of shotgun clones by oligonucleotide fingerprinting: an efficient and high throughput strategy to reduce redundancy in large-scale sequencing projects. *Nucleic Acids Res.* **26**: 5358–5364.

Ramsay, G. (1998) DNA chips: state-of-the-art. *Nature Biotechnol.* **16**: 40–44.

Römer, I., Reik, W., Dean, W. and Klose, J. (1997) Epigenetic inheritance in the mouse. *Curr. Biol.* **7**: 277–280.

Schena, M., Shalon, D., Davis, R.W. and Brown, P.O. (1995) Quantitative monitoring of gene expression patterns with a complementary DNA microarray. *Science* **270**: 467–470.

Schena, M., Shalon, D., Heller, R., Chai, A., Brown, P.O. and Davis, R.W. (1996) Parallel humqn genome analysis: Microarray-based expression monitoring of 1000 genes. *Proc. Natl. Acad. Sci.* **93**: 10 614–10 619.

Schena, M., Heller, R. A., Theriault, T. P., Konrad, K., Lachenmeier, E. and Davis, R. W. (1998) Microarrays: biotechnology's discovery platform for functional genomics. *Trends Biotechnol.* **16**: 301–306.

Shalon, D., Smith, S.J. and Brown, P.O. (1996) A DNA microarray system for analyzing complex DNA samples using two-color fluorescent probe hybridization. *Genome Res.* **6**: 639–645.

Trent, J.M., Bittner, M., Zhang, J. *et al.* (1997) Use of microgenomic technology for analysis of alterations in DNA copy number and gene expression in malignant melanoma. *Clin. Exp. Immunol.* **1**: 33–40.

Uetz, P., Giot, L., Cagney, G. *et al.* (2000) A comprehensive analysis of protein-protein interactions in *Saccharomyces cerevisiae*. *Nature* **403**: 623–627.

Velculescu, V.E., Zhang, L., Vogelstein, B. and Kinzler, K.W. (1995) Serial analysis of gene expression. *Science* **270**: 484–487.

Walter, G., Büssow, K., Cahill, D., Lueking, A. and Lehrach, H. (2000) Protein arrays for gene expression and molecular interaction screening. *Curr. Opin. Microbiol.* **3**: 298–302.

Wanker, E.E., Rovira, C., Scherzinger, E., Hasenbank, R., Walter, S., Tait, D., Colicelli, J. and Lehrach, H. (1997) Hip-I – a Huntingtin interacting protein isolated by the yeast two-hybrid system. *Human Mol. Genet.* **6**: 487–495.

Wasinger, V.C., Cordwell, S.J., Cerpa-Poljak, A., Yan, J.X., Gooley, A.A., Wilkins, K.L. and Humphery-Smith, I. (1995) Progress with gene-product mapping of the Mollicutes: Mycoplasma genitalium. *Electrophoresis* **16**: 1090–1094.

Welford, S.M., Gregg, J., Chen, E., Garrison, D., Sorensen, P.H., Denny, C.T. and Nelson, S.F. (1998) Detection of differentially expressed genes in primary tumor tissues using representational differences analysis coupled to microarray hybridization. *Nucleic Acids Res.* **26**: 3059–3065.

Wilkins, M.R., Sanchez, J.C., Gooley, A.A., Appel, R.D., Humphery-Smith, I., Hochstrasser, D.F. and Williams, K.L. (1996) Progress with proteome projects: why all proteins expressed by a gneome should be identified and how to do it. *Biotechnol. Genet. Engin. Rev.* **13**: 19–50.

Wolfsberg, T.G. and Landsman, D. (1997) A comparison of expressed sequence tags (ESTs) to human genomic sequences. *Nucleic Acids Res.* **25:** 1626–1632.

Zehetner, G. and Lehrach, H. (1994) The Reference Library System – sharing biological material and experimental data. *Nature* **367:** 489–491.

Chapter 2

Methods for measurement of gene (mRNA) expression

S.J. Blakemore, D.M. Wallace and M.K. Trower

1. Introduction

The term messenger RNA (mRNA) was coined in the early 1960s to describe the function of a class of cellular RNA as the carrier of genetic information from the nucleus to the cytoplasmic protein synthesis machinery (ribosomes), where it is translated into proteins. It has long been recognized that mRNA plays a pivotal role in determining the nature and quantity of proteins produced by cells (*Figure 1*). Indeed the differences in protein content of different cell types are a reflection of differences in the types of mRNA species expressed and of their levels of expression (abundance) during cellular development and maintenance. This was implied by studies using nucleic acid hybridization kinetics where cDNA (complementary DNA copy of mRNA) was found to reanneal more quickly to cDNA generated from the same tissue than to cDNA from a different tissue (Hastie and Bishop, 1976). Once such differences in mRNA populations between tissues/cell types were appreciated, it became important to establish what these differences are. It therefore became apparent that methods for accurate quantification of specific mRNA species in biological samples were required. The original hybridization-based assays were facilitated by the unique selectivity of nucleic acid base pairing (e.g. Northern blotting). Application of such gene-by-gene analysis methods established that some transcripts are abundant in certain tissues whilst absent in others, and that some genes are expressed at relatively consistent levels in many/all tissues. Other techniques for mRNA quantification have since been developed to complement Northern blotting, such as ribonuclease protection assays and quantitative reverse transcription polymerase chain reaction (RT-PCR). However, these are also limited to relatively few genes per assay. Technological developments during the past 5 years have made possible methods for quantification of mRNAs on a genomic scale, where several hundred or thousands of genes are assayed in parallel rather than one gene at a time. The motivation for these developments is the desire to gain insights into the total program of genomic activity underlying a biological change (a typical mammalian cell expresses ~15 000 different species of mRNA). Technological advances based around polymerase chain reaction (PCR), large-scale cDNA library sequencing, and *de novo* nucleic acid synthesis have contributed to the development

Proteomics, edited by S.R. Pennington and M.J. Dunn
© 2001 BIOS Scientific Publishers, Oxford.

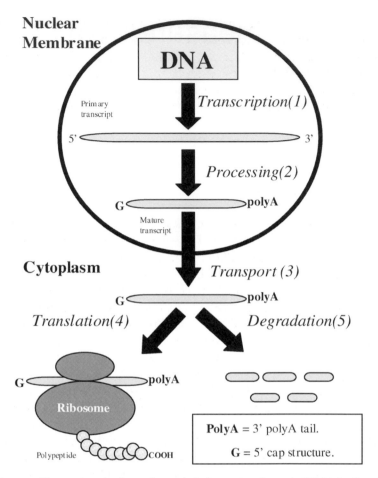

Figure 1. Diagram illustrating the flow of genetic information from the DNA in the cell nucleus to protein synthesis. The pool of translatable mRNA available in the cell cytoplasm defines the nature of the proteins that a cell can produce at any given point in time. The characteristics of this mRNA pool are defined by the rates of the specified cellular processes (1–5).

of a wide range of techniques for mRNA quantification on a (near) genomic scale. Methods such as differential display PCR (DD-PCR), serial analysis of gene expression (SAGE) and DNA array hybridization all have significant benefits over the Northern blot in terms of sensitivity and genes assayed per amount of RNA. Details of these techniques and a critique of their relative strengths and weaknesses are presented in this chapter.

It should be noted that all current techniques for mRNA quantification provide empirical rather than absolute mRNA steady-state level information; however, this is sufficient for detection of changes in levels of mRNA expression.

In the context of a publication on proteomics it should be noted that detection of a difference in abundance of a specific mRNA between two biological samples is not necessarily reflected by an equivalent quantitative difference at the protein level. In prokaryotes it has been known since the 1960s that nearly all regulation of gene expression occurs at the transcriptional level. Although it is also generally true in eukaryotes that an increase (or decrease) of a protein's level follows an increase (or decrease) of its encoding mRNA, the extent and kinetics of protein change are often unpredictable. Therefore when experimental material is eukaryotic, it is prudent not to draw conclusions regarding protein expression based solely on mRNA data. Enzyme-linked immunosorbant assay, Western blot or proteomic studies are required to confirm changes at the protein level.

2. DNA array hybridization

The most commonly used methods for assessing mRNA levels are based on hybridization of a labeled population of nucleic acids (representing an mRNA sample) to an array of individual cDNA sequences, or oligonucleotides, printed as spots onto a solid support. After hybridization the intensity of label associated with each spot represents the expression level for that particular sequence. This intensity can be compared to the intensity of the equivalent spot on an identical gridded array hybridized with material prepared using mRNA isolated from a different source. This provides a measure of differential mRNA expression specific for that spot's DNA sequence (gene). By comparing all equivalent spot intensities between two grids, the changes in expression of many thousands of genes can be monitored simultaneously. The use of distinguishable labels to label the two mRNA populations under study permits cohybridization onto one gridded array (see colour *Figure 2* at the end of this chapter); a measure of differential hybridization is then obtained by comparing intensities of the two labels at each spot. The range of sequences present on an array limits these methods, thus if a particular gene sequence is not represented, then obviously it cannot be assayed. Therefore, the selection of which DNA elements are present on an array is of crucial importance for any study. Such array-based methods of mRNA quantitation are termed 'closed systems'.

Using DNA arrays to measure mRNA abundance is directly analogous to performing multiple reverse Northern blots simultaneously. Instead of RNA being immobilized on a solid support and being hybridized with gene-specific labeled DNA (Northern blot), the gene-specific DNA elements are attached to a solid support and hybridized to labeled material derived from RNA. The power of array technology is derived from the fact that many thousands of DNA elements (genes), printed as spots, can be assayed in parallel. Typically, thousands of DNA elements are robotically printed onto a nylon membrane, or glass slide, at high density. Currently, higher spotting densities are achievable on glass, the benefit of which is an increase in the number of genes assayed per amount of input RNA without any compromise in terms of assay sensitivity. DNA arrays can be classified into two types depending on the chemical nature of the DNA element used to construct the array: either denatured PCR products (derived from a cDNA library) or oligonucleotides. Therefore, array construction requires access to either cDNA clones or sequence information (for design of representative oligonucleotides). Thus, array design is restricted by our current knowledge of the genes which make up an organism's genome (hybridizations are performed in a species-specific manner). For those organisms whose genomes are completely sequenced, e.g. *Saccharomyces cerevisiae*, *Escherichia coli* and *Mycobacterium tuberculosis* (see *Kyoto Encyclopaedia of Genes and Genomes*, http://www.genome.ad.jp/kegg/kegg2.html for a list of completely sequenced genomes), array-based technologies can provide complete genome coverage and are therefore effectively 'open systems'. Indeed results of genome-wide expression profiling experiments using DNA arrays have been published for the species *S. cerevisiae* (Wodicka *et al.*, 1997), *E. coli* (Richmond *et al.*, 1999) and *M. tuberculosis* (Behr *et al.*, 1999). However, for species such as human, mouse and rat, genome-wide arrays will not be available until after completion of their respective genome sequencing projects.

Another major issue when planning DNA array-based experiments is how to convert the RNA into labeled material, and how the hybridization will be performed. Various methods have been employed for generating labeled material using either mRNA or total RNA as starting template. These include first-strand cDNA synthesis alone [with anchored-oligo-(dT) or random hexamer priming], or in combination with a subsequent nucleic acid amplification step using either T7-RNA polymerase or PCR procedures. Radiochemical (e.g. ^{32}P or ^{33}P) or fluorescent (e.g. Cy3 and Cy5) labeled nucleotides are incorporated, the latter being commonly used with glass slide arrays and the former on nylon membranes. Each method has different optimal RNA requirements, for example for first-strand cDNA synthesis ~ 2–5 μg poly(A)RNA, whereas for PCR-based amplification as little as 25 ng total RNA can be sufficient (Gonzalez *et al.*, 1999). As no systematic comparison of the current options has been published, no single 'standard' method can be advocated. Notable concerns about PCR-based amplification include the potential for non-linear amplification of abundant transcripts, and for sequence-specific amplification, which would result in quantitatively sub-optimal labeling. However, recent evidence indicates that perhaps these concerns are unfounded (Endege *et al.*, 1999). Regardless of technique, assay sensitivity appears to be about the same – ~ 1 in 10^5 mRNA molecules discernible when using poly(A)RNA as template (~ 2–5 mRNA copies per cell).

When comparing mRNA samples using only one label, the procedure for identification of differentially expressed genes is relatively simple: labeled RNA (from two or more samples) is applied to identical DNA arrays and the intensities of spots are compared between the resulting images (after normalization to allow for specific activity differences). Another commonly used technique involves using two distinguishable labels (e.g. labeling one mRNA sample using Cy3 and another with Cy5), and applying the two differently labeled samples to one glass slide array. The two fluorescent labels, having distinct characteristic emission wavelengths, can be discriminated from each other, permitting detection of differential hybridization using a single array. The ratio of fluorescence at each spot coordinate gives the result of a sequence-specific competitive hybridization. The principal advantage of using a single array technique over the former method is the elimination of artifacts resulting from subtle quality differences between individual gridded arrays. However, a disadvantage of using single-array dual-hybridization is that the resulting ratios are only valid within a particular data set, for example for a time course the time-point samples would be labeled with Cy5 and each hybridized to a separate array alongside the same Cy3-labeled zero hour mRNA reference sample. As different studies will use different reference samples, the ability to relate results from different studies is compromised, although not impossible.

Using either single or dual labeling approaches the intensity values, and/or ratios, from array studies are stored in databases enabling *in silico* analyses and comparisons between different datasets. One of the major benefits of using defined gene array-based techniques is the relative simplicity of application of multivariate statistical analysis methods to the data (e.g. time course, dose–response, and other multiple sample data). For instance, in the case of a time course, methods for clustering genes whose kinetics of expression are similar can supply insight into coordinate control of gene expression (Iyer *et al.*, 1999). Another example of the power of statistical analysis of multiple gene expression profile data is the novel classification and subclassification of lymphoid tumor types (Golub *et al.*, 1999). Multivariate analysis of expression data has the potential to uncover correlations in expression of genes that would not have been identified by performing binary comparisons of hybridization datasets (e.g. disease vs. normal tissue, or 'treated' vs. 'untreated' cells). To date, published examples of multivariate gene expression analysis have been based mainly on array data (Alizadeh *et al.*, 1999; Fambrough *et al.*, 1999; Golub *et al.*, 1999; Heller *et al.*, 1997; Iyer *et al.*, 1999; Perou *et al.*, 1999), the sole exception being one report using data derived from RT-PCR (Wen *et al.*, 1998). Indeed an advantage of array-based methods is the relative ease with which multivariate statistical analysis can be applied compared with that for data from other platforms, for example SAGE, digital gene expression profiling and DD-PCR.

With the accumulation of large volumes of gene expression data in the databases, it is expected that identification of coordinated regulation of genes will become increasingly accurate and more sensitive. In general terms this will not only provide a more comprehensive appreciation of 'batteries' of gene expression which will lead to a better understanding of their regulation, but also may give clues to their biological functional context. The identification of gene expression profiles associated with similar diseases (e.g. inflammatory disorders), both in common to several disorders or apparently disease-specific, will lead to a more refined understanding of their molecular pathology. However, to facilitate this it will be necessary to employ some forms of 'standardization' to enable cross-comparison of results from different groups. The most appropriate method for generating such 'standardized expression datasets' for all the various platforms remains a source of debate. Recently, both the European Bioinformatics Institute (http://www.ebi.ac.uk/arrayexpress/) and the US National Center for Biotechnology Information (NCBI; Marshall, 1999) have stated their intention to provide a public database to store and facilitate the analysis of array-based hybridization data.

2.1 cDNA arrays

cDNA arrays consist of PCR products, derived from cDNA libraries, robotically spotted onto a solid support. cDNA arrays can be constructed from undefined cDNA libraries (no knowledge of cDNA sequences) or from defined cDNA libraries (DNA sequences previously characterized). There are two major advantages of using defined cDNA libraries: firstly, the ability to check for the presence on an array of 'genes of interest' for a particular study (i.e. positive/negative controls); and, secondly, results analysis is expedited by rapid mapping of hybridization data to gene annotations. The alternative of using undefined cDNA sources requires identification of differentially regulated genes by DNA sequencing of those clones shown to display differential hybridization. This is time-consuming and can be inefficient as the same gene may be repeatedly identified (especially if using a 'standard' redundant cDNA library as array source material). When designing cDNA arrays it is common practice to select 3' cDNA sequences to represent the genes. These are likely to be derived from noncoding sequences which are generally more divergent than coding sequences between gene family members. Selecting such 3' cDNAs reduces the possibility of cross-hybridization between related family members, especially in situations where strong sequence similarities exist.

Establishment of a facility for cDNA array production, hybridization, and for gene expression data analysis requires considerable bioinformatic support. Accurate and reliable computer-based systems are essential at all stages of the process. This includes choice of cDNA sequences for an array, sample tracking to ensure the correct cDNA is located to the correct array spot coordinate, measurement of all spot coordinate hybridization signal intensities, retention of hybridization results in a database (linking the hybridization data to gene annotations), and subsequent data-mining. An example of such a system is ArrayDB, developed by and in use at the NCBI for analyzing cDNA array data (Ermolaeva *et al.*, 1998). Developing such a cDNA array facility is currently beyond the scope of many laboratories because of high costs of robotics and the bioinformatics infrastructure. Therefore, although the original development of cDNA array technology was principally pioneered at the university-based laboratories of Hans Lehrach (Gress *et al.*, 1996; Maier-Ewett *et al.*, 1993; Zehetner and Lehrach, 1994) and Pat Brown (Schena *et al.*, 1995, 1996; Shalon *et al.*, 1996), it is not surprising that the majority of arraying facilities are currently found in biotech companies, large pharmaceutical and a few specialized academic centers. Unfortunately this means that, regardless of the potential strengths of cDNA arrays for identifying differentially expressed genes, the majority of researchers currently cannot easily access the technology. However, this situation should improve, as costs are likely to decrease as more laboratories embrace the technology. Indeed, Pat Brown's laboratory website (http://cmgm.stanford.edu/ pbrown/) already provides information on production of an arrayer at a significantly lower cost than that of an automated DNA sequencer (US$ 23 000). The basic advantage for investigators in having an 'in-house' arraying facility is an opportunity for flexibility in construction of arrays to match local requirements. Alternatively, cDNA array-based studies can be performed without investing in all of the technology, either by purchasing arrays and analysis software from commercial vendors, for example Research Genetics Inc. (http://www.researchgenetics.com/), Genome Systems Inc. (http://www.genomesystems.com/), Clontech (http://www.clontech.com/) or NEN (http://www.nenlifesci.com/), or by collaboration with commercial cDNA array service providers [e.g. Incyte Pharmaceuticals Inc. (http://gem.incyte.com/gem/index.shtml) or Research Genetics Inc.].

The utility of cDNA arrays for generating novel biological information has been demonstrated by the increasing number of publications in this field; examples include cancer gene expression profile studies (DeRisi *et al.*, 1996; Golub *et al.*, 1999; Moch *et al.*, 1999; Perou *et al.*, 1999), identification of a key insulin-resistance gene (Aitman *et al.*, 1999), detection of neural gene expression changes during the circadian cycle (Patten *et al.*, 1998), and the identification of altered lymphocytic gene expression in asthma (Syed *et al.*, 1999). A demonstration of the novel biological insight that this technology can provide is the study on the response of quiescent cultured human fibroblasts to serum (a well-characterized model of cell cycle control) using an 8613 gene array (Iyer *et al.*, 1999). Many of the observed early changes in expression are of genes known to be involved in the proliferative response due to serum (e.g. transcription factors fos/jun). However, at later time points unexpected changes in gene expression typical of a wound-healing response were revealed, for example induction of vascular endothelial growth factor (promotes angiogenesis) and fibroblast growth factor 7 (stimulates re-epithelialization). The authors concluded that the primary effect of exposure of these cells to serum was a wound-healing response, advocating their use as an *in vitro* model for such studies. Therefore, by performing multigene hybridization this study provided a previously unsuspected important insight into the biological response of fibroblasts to serum.

In summary, cDNA arrays can be used to perform transcript profiling in any species (providing an appropriate cDNA library is accessible), enabling the simultaneous monitoring of tens of thousands of gene sequences, facilitating thorough data analysis, and databasing of results. However, cDNA arrays can only provide truly genome wide assays for those species whose whole genome sequence is known. The costs of production of arrays and establishment of bioinformatics tend to be prohibitive for the majority of laboratories, resulting in limited access to the technology.

2.2 Oligonucleotide arrays

In terms of utility and general performance, oligonucleotide arrays are similar to cDNA arrays (see Section 2.1). Both consist of DNA elements arrayed at high density on a solid matrix (e.g. a glass slide), which are used for hybridization-based gene expression profiling. Instead of arraying PCR products prepared from cDNA clones, oligonucleotide arrays are made up of synthetic gene-specific oligodeoxynucleotides. This section will highlight the key strengths and weaknesses between these two forms of DNA array.

The basic principal of 'oligo' arrays is that short oligodeoxynucleotides (usually 20–25-mers) can contain sufficient sequence complexity to selectively hybridize a single transcript. In practice, for any one gene several different component oligo-nucleotide sequences are usually placed on an array. Obviously construction of an oligo array requires prior knowledge of the expressed sequences, limiting their usefulness to those species whose expressed genomes have been extensively sequenced. This is in contrast to cDNA arrays where a completely uncharacterized cDNA library can be used for expression profiling. However, the use of oligos means that there is no need to retain and carefully curate physical collections of cDNA clones and PCR products, which simplifies the logistics of accurate array assembly. Indeed, the use of 'on-chip' oligo synthesis can minimize the risk of array error to its lowest level. Methods are available that facilitate oligo design in order to provide unique sequence-specificity to a single transcript (Wodicka *et al.*, 1997). These procedures offer the optimal choice of sequence based on available genome sequence information to reduce the possibility of artifactual results caused by cross-hybridiza-tion. Effects of cross-hybridization cannot easily be ruled out when using cDNA arrays.

Once oligonucleotide sequences have been chosen to represent the required set of genes there are two basic methods for array production: robotic spotting of presynthesized oligos similar to that for cDNA arrays; or direct photolithographic DNA synthesis on the surface of the array as developed by a US biotechnology company, Affymetrix (http://www.affymetrix.com/; Lockhart *et al.*, 1996). The Affymetrix method can produce arrays of higher spot density (10^6 elements/cm^2; McGall *et al.*, 1996) than robotic spotting of oligonucleotides (3×10^5 elements/cm^2; Yershov *et al.*, 1996). Currently, designing and constructing arrays by the former method is considerably more expensive, limiting its availability to potential users, and involves generating a set of array-specific masks for the photolithography process, causing the technology to be relatively inflexible. However, a recently published report describes a method that may reduce the costs of on-slide oligonucleotide synthesis as it does not require production of array specific masks (Singh-Gasson *et al.*, 1999). One feature of oligonucleotide arrays, which differenti-ates them from cDNA arrays, is that it is possible to include imperfectly matched oligos for the represented gene set, as well as perfectly matched ones. This practice is employed on Affymetrix arrays, which include a single nucleotide mismatch oligo (differing by one central base) alongside each perfect match oligonucleotide. In this way hybridization specificity for each oligo is reported and accommodated, provid-ing an increased level of quantitative accuracy for each gene assayed.

It should be noted that oligonucleotide arrays provide single-stranded templates ready for hybridization, whereas arrayed PCR products must first be denatured to supply single-stranded hybridization templates. It is important that the denaturation procedure is consistent to optimize reproducibility between cDNA arrays – this is not a consideration for oligonucleotide arrays.

The methods used for converting mRNA or total RNA into labeled material are the same as those used for cDNA arrays, the most common being fluorescent labeling engaging some form of linear amplification (e.g. T7-RNA polymerase) to supply sufficient material of high specific activity. Data analysis and databasing issues are similar to those for cDNA arrays (see Section 2.1).

Examples of oligo array applications include simultaneous monitoring of expression of all yeast genes (Holstege *et al.*, 1998; Wodicka *et al.*, 1997), identification of redundancy in mouse receptor tyrosine kinase-activated signaling pathways (Famborough *et al.*, 1999), and analysis of effect of calorific restriction on mouse skeletal muscle aging (Lee *et al.*, 1999).

In summary, oligo arrays offer a technology which has similar attributes to cDNA arrays but which can achieve higher gene-specific hybridization accuracy than cDNA arrays, albeit at a higher cost. The requirement of prior gene sequence knowledge to design oligo arrays will become a lesser consideration as genome sequencing projects mature.

3. Non-DNA array-based methods of mRNA quantification

In contrast to DNA array-based methods for mRNA quantitation, where only genes represented on the array can be assayed, techniques that have the potential to assay all transcripts present within a biological sample are commonly termed 'open systems'. Theoretically, the sensitivity of the assay and the resource available to perform the experiment limit such methods. Experiments using 'open systems' can be performed in any species without the need for access to prior species-specific sequence information.

3.1 Sequencing-based methodologies

First-pass sequencing of cDNA libraries. As part of the Human Genome Project, hundreds of cDNA libraries derived from mRNA isolated from a wide variety of tissues have been constructed and their clones' cDNA inserts sequenced by the Integrated Molecular Analysis of Genomes and their Expression Consortium (IMAGE); (http://www-bio.llnl.gov/bbrp/image/image.html). The first-pass cDNA sequence information (a single-read DNA sequencing usually of both ends of the cloned insert – average read length 200–300 bp) has been deposited in a public database (dbEST section of Genbank collated at the NCBI). Similar proprietary databases have also been created (e.g. Incyte Pharmaceuticals Inc. and Human Genome Sciences Inc., http://www.hgsi.com). The principal aim for this effort has been to achieve rapid identification of reference sequences for all human genes, each first-pass sequence representing an expressed sequence tag (EST) of a gene (as opposed to the complete full-length sequence). Similar projects are underway for mouse and rat at genome centers worldwide. As of November 1999 (release number 115 of Genbank) the number of independent entries in dbEST stood at approximately 1 300 000 human, 513 000 mouse and 115 000 rat. Each sequenced clone of a standard cDNA library represents an independent sampling of the original mRNA population used to generate that library. Therefore, by sequencing thousands of such clones from a cDNA library and then counting the frequency of appearance of a particular gene sequence it is possible to gain an estimate of the relative abundance of each mRNA transcript. As this method involves counting the number of instances each gene is represented by a given EST (or of ESTs known to cluster together), the data produced is digital. Therefore, in theory, the linear range over which accurate mRNA quantitation can be performed is limitless. The sensitivity of this method for accurate quantitation of abundance of a transcript is a direct function of the number of cloned sequences produced from the cDNA library. The strengths of this approach are that it is applicable to any biological sample for which a cDNA library can be prepared, and that it is accessible to any laboratory with a high-throughput DNA sequencing capacity. However, there are several weaknesses. cDNAs that are difficult to clone will not be accurately represented in the cDNA library, resulting in the underestimation of their abundance. Secondly, due to the random sampling approach, the more abundant species will be selected for sequencing most often; therefore many thousand sequencing reactions are required on each cDNA library to accurately assay even moderately abundant transcripts. The NCBI provide web-based tools, xProfiler (http://www.ncbi.nlm.nih.gov//ncicgap/cgapxpsetup.cgi) and Digital Differential Display (http://www.ncbi.nlm.nih.gov/UniGene/ddd.cgi?ORG = Hs), which allow the comparison of sequence data derived from several hundred cDNA libraries in their collection. Statistical methods have also been developed to aid in the interpretation of data resulting from such comparisons (Audic and Claverie, 1997). Put simplistically, these methods dictate that, to detect statistically significant differences in the abundance of transcripts in two cDNA libraries, a given transcript must be sampled (sequenced) more than twice in a given library. Therefore, to detect a statistically significant difference in the expression of a transcript present at one copy in 20 000, at least 60 000 independent sequencing reactions must be carried out from each cDNA library. This quantity of sequencing is costly both in time and money and then is only likely to provide accuracy for moderately abundantly expressed genes (note, low-abundance cellular mRNA transcripts are present at 1:100 000 copies and less). These practical considerations have tended to limit the use of this technique for mRNA quantification and the identification of differentially expressed genes. However, the utility of the quantitative information derived from analyzing the frequency of EST data from cDNA libraries has been demonstrated by the detection of the expression of human prostate-specific genes (Vasmatzis et al., 1998), the identification of differentially expressed genes in PC-12 cells in response to nerve growth factor treatment (e.g. superoxide dismutase and synapsin 2; Lee et al., 1995), and within the human hematopoietic hierarchy (Claudio et al., 1998). In all cases initial experimental findings were confirmed by Northern blot.

In summary, while first-pass sequence data derived from cDNA libraries does have utility in quantitating mRNA abundance, in most instances it would not be the first-choice method due to its relatively low sensitivity and high cost. However, given that information for many cDNA libraries is freely available (dbEST), as are the tools for comparing and interpreting such data, individuals considering performing differential mRNA expression studies should consider the potential application of 'electronic comparisons' of this EST sequence data as a first 'port of call'. Obviously this approach is only valid when sequence data is available from cDNA libraries derived from cellular sources relevant to the investigator. This method can only be applied to standard cDNA library information, and not to normalized cDNA library data, as the normalization process causes radical alteration of the quantitative representation of cDNAs in a cDNA library.

Serial analysis of gene expression – SAGE. Serial analysis of gene expression (SAGE) is a DNA sequencing-based technology for quantifying mRNA abundance first published in 1995 from the laboratory of Bert Voglestein (Velculescu *et al.*, 1995). SAGE overcomes the major shortcoming of gene expression analysis by sequencing cDNA inserts of cloned cDNA libraries – the necessity to sequence tens of thousands of clones' cDNA inserts to provide quantitative accuracy (see above). The fundamentals of SAGE involve isolation of short unique sequence tags (9–14 bases) representing a defined region of each individual transcript, followed by their concatenation, cloning, and then sequencing of the cloned concatenates (*Figure 3*). The frequency of representation of a particular sequence tag within the total number of tags is then a measure of the frequency of its mRNA in the original population. Theoretically, a tag length of 9–14 bases provides sufficient sequence information to unequivocally identify specific mRNA transcripts (Velculescu *et al.*, 1995). For example, if one assumes a random distribution then all possible permutations of 10 bases (4^{10}) yield 1 048 576 possible combinations, which is 10 times greater than the estimated number of genes constituting the human genome (Fields *et al.*, 1994). Therefore, by reducing the DNA sequence to a minimum informative length there is a gain in efficiency over the cDNA library sequencing method: for each 'sequence tag concatenate' clone sequenced, 30–50 fold more gene information is acquired (Bertelsen and Velculescu, 1998).

Just as for sequencing cDNA library clones, SAGE data is digital and the linear range theoretically limitless. Methods have been developed for assessing the significance of differences in sequence tag abundance derived from two biological samples based on simulations (Zhang *et al.*, 1997) and on statistical methods (Audic and Claverie, 1997). Using independent methods, it has been demonstrated that SAGE sequence tag frequencies are an accurate measure of transcript abundance for mRNAs in the less than 0.01% to greater than 0.1% range (Madden *et al.*, 1997; Velculescu *et al.*, 1995, 1997). SAGE applications have included the identification of p53-induced genes prior to apoptosis in a human colorectal cancer cell line (Polyak *et al.*, 1997) and the analysis of gene expression profiles of normal versus cancer cells (Zhang *et al.*, 1997). Further examples of SAGE applications can be found at NCBI (http://www.ncbi.nlm.nih.gov/SAGE/) and Genzyme Oncology (http://www.genzyme.com/prodserv/molecular_oncology/sage/welcome.htm) a bio-technology company providing SAGE as a commercial service.

The 'standard' SAGE method requires a similar amount of RNA to that needed for construction of a cDNA library (2.5–5.0 μg mRNA). This limits application of SAGE to situations where starting biological material is plentiful: cell culture systems and animal models, as opposed to human clinical biopsy material or microdissected samples. However, the recent advent of MicroSAGE (Datson *et al.*,

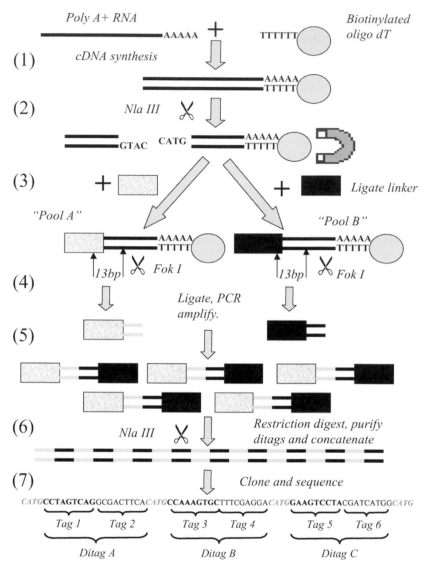

Figure 3. Schematic of serial analysis of gene expression (SAGE) method. Step 1: polyA + RNA is converted to cDNA using a biotinylated-oligodT primer. Step 2: the resulting cDNA is then cleaved by a frequently cutting restriction enzyme, 'anchoring enzyme' (e.g. Nla III, 4bp recognition site, cuts on average every 256 nucleotides). Step 3: the 3′ end of the cDNA is captured by the use of streptavidin-coated magnetic beads. The cDNA pool is then split in two ('pool A' and 'pool B') and separate linkers are ligated to each of the cDNA pools. Step 4: these linkers contain a site for a type IIs restriction enzyme; this 'tagging enzyme' (so named as they cut the cDNA at a site a specific number of nucleotides away from the recognition site for the enzyme, e.g. Fok I cuts 13 bp downstream of its recognition site) is then used to cut the cDNA/linker hybrid and releases a short sequence tag of cDNA attached to the linker. Step 5: the linker-tag molecules are then ligated tail-to-tail and amplified by PCR (via linker sequences). Step 6: linkers are then removed from the ditags by use of the anchoring enzyme; ditags are then purified and ligated together to form concatemers. Step 7: concatamers are then cloned and sequenced to reveal the identity of each tag.

1999) provides a modified SAGE technique which can utilize as little as 5 ng of mRNA.

There are two major drawbacks to SAGE technology. Firstly, as a short sequence tag distinguishes each transcript, high-quality sequence data is essential for its accurate identification. A single base calling error in a 10-nucleotide tag would have far greater impact for transcript identification (by BLAST analysis) than would be observed for a 200 base first-pass cDNA sequence with a 10% error rate. Secondly, the ability to successfully identify the originating transcript for a tag is directly related to the number of sequences (in particular 3′ end sequences) deposited in databases for each species. Therefore, SAGE data interpretation for human material is currently simpler than that for either mouse or rat.

In conclusion, SAGE is broadly applicable to any biological system and, in conjunction with automated DNA sequencing and sufficient bioinformatic support, it is an efficient and accurate method for quantifying mRNA abundance.

3.2 Differential display PCR (DD-PCR)

Differential display PCR is a method for identifying cDNA fragments that are differentially expressed between two biological samples (Liang and Pardee, 1992) (*Figure 4a*). It is based on generating cDNA fragments from mRNA using two oligonucleotide primers, one being complementary to the polyA tail of transcripts (e.g. oligo-d(T)$_{11}$VN), and the other a short random nucleotide sequence (e.g. 10-mer). DD-PCR has the potential to identify all transcripts present in a biological sample when sufficient primer combinations are applied. After cDNA synthesis, the fragments are labeled (radiolabel or fluorescent) during PCR amplification. The products are then separated by electrophoresis on a sequencing gel, and the pattern of amplified cDNA fragments visualized. The intensity of a labeled band reflects the relative abundance of its mRNA transcript within the original mRNA population. Major differences in the cDNA band patterns generated from two biological samples, when using the same set of primers, indicate the presence of differentially expressed transcripts. Cloning and sequencing of the eluted cDNA bands enables the identity of the genes from which these cDNAs originate to be defined. In its early days DD-PCR was reported to produce a significant number of false positives (Liang *et al.*, 1994; McCleland *et al.*, 1995). Obviously this significantly reduced its efficiency. Because of these shortcomings many technical improvements have been proposed, such as optimizing primer composition, increasing PCR primer annealing temperatures, DNAase treatment of RNA prior to reverse transcription, and the use of nondenaturing gels (Bauer *et al.*, 1993; Liang and Pardee, 1995). However, probably the most significant development has been the use of restriction enzyme digestion of double-stranded cDNA (prepared from the mRNA), followed by ligation of an adapter to mediate selective PCR of the extreme 3′ ends of the fragmented cDNAs (Prashar and Weissman, 1996; *Figure 4b*). The use of a selection of restriction enzymes increases the coverage of transcripts sampled for display. These adaptations enable high-stringency PCR to be performed by specific primer annealing at 56°C, resulting in a marked improvement over standard DD-PCR

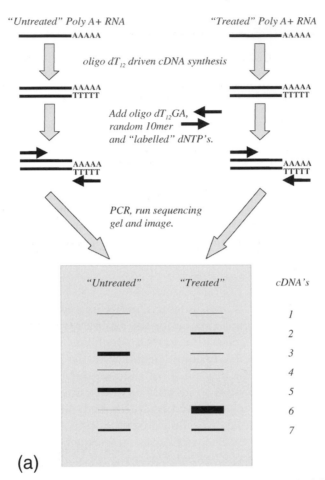

(a)

Figure 4. Schematic of differential display methodology. (a) Standard method for the differential display of reverse transcription PCR products (Liang and Pardee, 1992). RNA from two distinct biological samples (i.e. cell line treated or untreated with modifying agent) is reverse transcribed using oligo-d(T)$_{12}$. The resulting cDNA is then split and subjected to PCR using a series of anchored oligodT primers (in this case d(T)$_{12}$GA), together with a random 10-mer (which primes at different sites in different cDNAs). The material generated in the PCR is visualized by the use of sequencing gel (radioisotopic or fluorescent labels are incorporated during the PCR). The presence/absence of bands with the same electrophoretic mobility indicates the differential expression of the gene from which the cDNA that represents the band is generated.

conditions (nonstringent primer annealing at 40°C). This modified DD-PCR has been claimed to deliver near-quantitative information on gene expression; as it produces single bands per transcript per restriction enzyme it reduces the complexity of the display pattern. Also, the sizes of the bands are predictable for genes of known sequence, permitting provisional gene identification prior to band recovery from the gel (Prashar and Weissman, 1996). Variations on this technique have been developed independently by several biotechnology companies [Curagen Corporation, http://

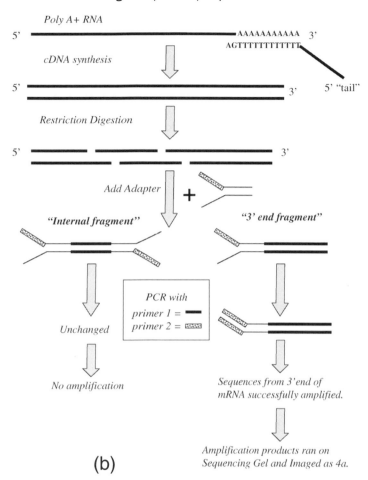

Figure 4 *(continued)* (b) Adaptation of the differential display methodology to improve reproducibility (Prashar and Weissman, 1996). First-strand cDNA synthesis is driven by an anchored dT primer containing a 5′ 'tail' (in bold), that is **CTCGCAATCGGGGCTCGTCG-**$(T)_{12}$**GA**. After a standard second-strand cDNA synthesis the resulting molecules are digested using a restriction enzyme and ligated to a Y-shaped adapter. Selective high-stringency PCR can then be performed to selectively amplify the 3′-end fragments of the cDNA. After amplification PCR products can be run on a sequencing gel and identified as above.

www.curagen.com/ (Shimkets *et al.*, 1999); Digital Gene Technologies Inc., http://www.dgt.com/index.html; and Display Systems Biotech, http://www.display systems.com/] to provide rapid high-throughput DD-PCR as a service.

DD-PCR has been applied in many diverse areas of investigation: for example, identification of novel drug targets (Shiue, 1997; Wang and Feuerstein, 1997), uncovering differentially regulated genes in rheumatoid arthritis (White and Petkovich, 1998), and assessing the effects of environmental stimuli on bacterial gene expression (Fislage, 1998). Indeed, DD-PCR is currently the most widely published technique for the identification of differentially expressed genes. However, this probably reflects the fact that DD-PCR does not require expensive specialized equipment or bioinformatic analysis tools, facilitating its introduction into many laboratories (rather than any intrinsic superiority over other differential expression techniques). A significant advantage of DD-PCR is the relatively small quantity of input RNA required. Originally, the order of several hundred nanograms of RNA was required (Liang and Pardee, 1992). Recently an adaptation of the technique has claimed to be able to perform DD-PCR on RNA derived from a single cell (Renner et al., 1998). Therefore, RNA yield from a biological sample should not preclude application of DD-PCR. However, DD-PCR also has disadvantages. As mentioned earlier, the technique does suffer from false positives. Also, a band identified as differentially expressed may not always be a single molecular species, as more than one mRNA transcript could generate fragments of similar size comigrating on the electrophoresis gel. This can result in difficulties identifying the 'gene' which gave rise to the observed 'band'.

In summary, DD-PCR can be successfully used to identify differentially expressed genes in any tissue from any species from which high-quality RNA can be isolated. The simplicity of this technique and its relative low cost has led to its widespread use. However, results are often more qualitative rather than quantitative and tend to be more error prone than data generated from other techniques such as array data and SAGE.

4. Concluding remarks

Technological advances in the field of molecular biology and genome sequencing have driven the development of methods for mRNA quantitation away from a gene-by-gene approach towards performing genome-wide assays. In the past 5 years several complementary techniques have developed in parallel. Each of these techniques has relative strengths and weaknesses with regards to sensitivity, throughput and genomic breadth of the assay, reliability, cost and ease/power of data analysis. All methods require significant experimental input and subsequent work to validate findings. For this reason there are not currently any published examples where results from two techniques, for example cDNA array and SAGE, have been comprehensively compared. Therefore, it is difficult to provide absolute comparisons of the utility of each technique. However, without doubt the utility of SAGE, array-based hybridization, DD-PCR and electronic DGE for mRNA quantitation has been demonstrated by confirmation of findings by independent methods (RT-PCR or Northern blotting). Other methods used for the identification of differentially expressed genes, such as suppression subtraction hybridization (Diatchenko et al., 1996) and representation difference analysis (Hubank and Schatz, 1994), have not been discussed here as they are primarily designed to identify qualitative differences in mRNA expression (compared to SAGE, arrays etc.) or have not been as widely used as other qualitative methods such as DD-PCR.

References

Aitman, T.J., Glazier, A.M., Wallace, C.A. *et al.* (1999) Identification of *CD36* (*Fat*) as an insulin-resistant gene causing defective fatty acid and glucose metabolism in hypertensive rats. *Nature Genet.* **21**: 76–83.

Alizadeh, A.A., Eisen, M.B., Davis, R.E. *et al.* (2000) Distinct types of diffuse large B-cell lymphoma identified by gene expression profiling. *Nature.* **403**: 503–511.

Audic, S. and Claverie J-M. (1997) The significance of digital gene expression profiles. *Genome Res.* **7**: 986–995.

Bauer, D., Muller, H., Reich, J., Riedel, H., Ahrenkeil, Warthoe, P. and Strauss, M. (1993) Identification of differentially expressed mRNA species by an improved display technique (DDRT-PCR). *Nucleic Acids Res.* **21**: 4272–4280.

Behr, M.A., Wilson, M.A., Gill, W.P., Salamon, H., Schoolnik, G.K., Rane, S. and Small, P.M. (1999) Comparative genomics of BCG vaccines by whole-genome DNA microarray. *Science* **284**: 1520–1523.

Bertelsen, A.H. and Velculescu, V.E. (1998) High-throughput gene expression analysis using SAGE. *Drug Discov. Today* **3**: 152–158.

Claudio, J.O., Liew, C.C., Dempsey, A.A., Cukerman, E., Stewart, A.K., Na, E., Atkins, H.L., Iscove, N.N. and Hawley, R.G. (1998) Identification of sequence-tagged transcripts differentially expressed within the human hematopoietic hierarchy. *Genomics* **50**: 44–52.

Datson, N.A., van der Perk-de Jong, J., van den Berg, M.P., de Kloet, E.R. and Vreugdenhill, E. (1999) MicroSAGE: a modified procedure for the serial analysis of gene expression in limited amounts of tissue. *Nucleic Acids Res.* **27**: 1300–1307.

DeRisi, J., Penland, L., Brown, P.O., Bittner, M.L., Meltzer, P.S., Ray, M., Chen, Y., Su, Y.A. and Trent, J.M. (1996) Use of a cDNA microarray to analyse gene expression patterns in human cancer. *Nature Genet.* **14**: 457–460.

Diatchenko, L., Lau, Y.F.C., Campbell, A.P., Chenchik, A., Moqadam, F., Huang, B., Lukyanov, S., Lukyanov, K., Gurskaya, N., Sverdlov, E.D. and Siebert, P.D. (1996) Suppression subtraction hybridisation – a method for generating differentially regulated or tissue-specific cDNA probes and libraries. *Proc. Natl Acad. Sci. USA* **93**: 6025–6030.

Endege, W.O., Steinmann, K.E., Boardman, L.A., Thibodeau, S.N. and Schlegel, R. (1999) Representative cDNA libraries and their utility in gene expression profiling. *BioTechniques* **26**: 542–550.

Ermolaeva, O., Rastogi, M., Pruitt K.D., Schuler, G., Bittner, M.L., Chen, Y., Simon, R., Meltzer, P., Trent, J.M. and Boguski, M.S. (1998) Data management and analysis for gene expression arrays. *Nature Genet.* **20**: 19–23.

Fambrough, D., McClure, K., Kazlauskas, A. and Lander, E.S. (1999) Diverse signaling pathways activated by growth factor receptors induce broadly overlapping, rather than independent sets of genes. *Cell* **97**: 727–741.

Fields, C., Adams, M.D., White, O. and Venter, J.C. (1994) How many genes in the human genome? *Nature Genet.* **7**: 345–346.

Fislage, R. (1998) Differential display approach to quantitation of environmental stimuli on bacterial gene expression. *Electrophoresis* **19**: 613–616.

Golub, T.R., Slonim, D.K., Tamayo, P., Huard, C., Gaasenbeek, M., Mesirov, J.P., Coller, H., Loh, M.L., Downing, J.R., Caligiuri, M.A., Bloomfield, C.D. and Lander, E.S. (1999) Molecular classification of cancer: class discovery and class prediction by gene expression monitoring. *Science* **286**: 531–537.

Gonzalez, P., Zigler Jr, J.S., Epstein, D.L. and Borras, T. (1999) Identification and isolation of differentially expressed genes from very small tissue samples. *BioTechniques* **26**: 884–892.

Gress, T.M., Muller-Pillasch, F., Geng, M., Zimmerhackl, F., Zehetner, G., Friess, H., Buchler, M., Adler, G. and Lehrach, H. (1996) A pancreatic cancer-specific expression profile. *Oncogene* **13**: 1819–1830.

Gygi, S.P., Rochon, Y., Franza, B.R. and Aebersold, R. (1999) Correlation between protein and mRNA abundance in yeast. *Mol. Cell. Biol.* **19**: 1720–1730.

Hastie, N.D. and Bishop, J.O. (1976) The expression of three abundance classes of messenger mRNA in mouse tissues. *Cell* **9**: 761–774.

Heller, R., Schena, M., Chai, A., Shalon, D., Bedilion, T., Gilmore, J., Woolley, D.E. and Davis, R.W. (1997) Discovery and analysis of inflammatory disease-related genes using cDNA microarrays. *Proc. Natl Acad. Sci. USA* **94**: 2150–2155.

Holstege, F.C.P., Jennings, E.G., Wyrick, J.J., Lee, T.I., Hengartner C.J., Green, M.R., Golub, T.R., Lander, E.S. and Young, R.A. (1998) Dissecting the regulatory circuitry of a eukaryotic genome. *Cell* **95**: 717–728.

Hubank, M. and Schatz, D.G. (1994) Identifying differences in mRNA expression by representational analysis of cDNA. *Nucleic Acids Res.* **22**: 5640–5648.

Iyer, V.R., Eisen, M.B., Ross, D.T., Schuler, G., Moore, T., Lee, J.C.F., Trent, J.M., Staudt, L.M., Hudson Jr, J.H., Boguski, M.S., Lashkari, D., Shalon, D., Botstein, D. and Brown, P.O. (1999) The transcriptional program in the response of human fibroblasts to serum. *Science* **283**: 83–87.

Lee, C-K., Klopp, R.G., Weindruch, R. and Prolla, T.A. (1999) Gene expression profile of aging and its retardation by caloric restriction. *Science* **285**: 1390–1393.

Lee, N.H., Weinstock, K.G., Kirkness, E.F., Earle-Hughes, J.A., Fuldner, R.A., Marmaros, S., Glodek, A., Gocayne, J.D., Adams, M.D., Kerlavage, A.R., Fraser, C.M. and Venter, J.C. (1995) Comparative expressed-sequence tag analysis of differential gene expression profiles in PC-12 cells before and after nerve growth factor treatment. *Proc. Natl Acad. Sci. USA* **92**: 8303–8307.

Liang, P. and Pardee, A.B. (1992) Differential display of eukaryotic messenger RNA by means of the polymerase chain reaction. *Science* **257**: 967–971.

Liang, P. and Pardee, A.B. (1995) Recent advances in differential display. *Curr. Opin. Immunol.* **7**: 274–280.

Liang, P., Zhu, J., Zhang, X., Guo, Z., O'Connell, R.P., Averboukh, L., Wang, F. and Pardee, A.B. (1994) Differential display using one-base anchored oligodT primers. *Nucleic Acids Res.* **22**: 5763–5764.

Lockhart, D.J., Dong, H., Byrne, M.C., Follettie, M.T., Gallo, M.V., Chee, M.S., Mittmann, M., Wang, C., Kobayashi, M., Horton, H. and Brown, E.L. (1996) Expression monitoring by hybridization to high-density oligonucleotide arrays. *Nature Biotechnol.* **14**: 1675–1680.

Madden, S.L., Galella, E.A., Zhu, J., Bertelsen, A.H. and Beaudry, G.A. (1997) SAGE transcript profiles for p53-dependent growth regulation. *Oncogene* **15**: 1079–1085.

Maier-Ewert, S., Maier E., Ahmadi, A., Curtis, J. and Lehrach, H. (1993) An automated approach to generating expressed sequence catalogues. *Nature* **361**: 375–376.

Marshall, E. (1999) Do-it-yourself gene watching. *Science* **286**: 444–447.

McCleland, M., Mathieu-Daude, F. and Welsh, J. (1995) RNA fingerprinting and differential display using arbitrarily primed PCR. *Trends Genet.* **11**: 242–246.

McGall, G., Labadie, J., Brook, P., Wallraff, G., Nguyen, T. and Hinsberg, W. (1996) Light-directed synthesis of high-density oligonucleotide arrays using semiconductor photoresists. *Proc. Natl Acad. Sci. USA* **93**: 13 555–13 560.

Moch, H., Schrami, P., Bubendorf, L., Mirlacher, M., Kononen, J., Gasser, T., Mihatsh, M.J., Kallioniemi, O.P. and Sauter, G. (1999) High-throughput tissue microarray analysis to evaluate genes uncovered by cDNA microarray screening in renal carcinoma. *Am. J. Pathol.* **154**: 981–986.

Patten C., Clayton, C.L., Blakemore, S.J., Trower, M.K., Wallace, D.M. and Hagan, R.M. (1998) Identification of two novel diurnal genes by screening a rat brain cDNA library. *Neuroreport* **9**: 3935–3941.

Perou, C.M., Jeffrey, S.S., van de Rijn, M., Rees, C.A., Eisen, M.B., Ross, D.T., Pergamenshikov, A., Williams, C.F., Lee, J.C.F., Lashkari, D., Shalon, D., Brown, P.O. and Botstein, D. (1999) Distinctive gene expression patterns in human mammary epithelial cells and breast cancers. *Proc. Natl Acad. Sci. USA* **16**: 9212–9217.

Polyak, K., Xia, Y., Zweier, J.L., Kinzler, K.W. and Vogelstein, B. (1997) A model for p53 induced apoptosis. *Nature* **389**: 300–305.

Prashar, Y. and Weissman, S. M. (1996) Analysis of differential gene expression by display of 3′ end restriction fragments of cDNA's. *Proc. Natl Acad. Sci. USA* **93**: 659–663.

Renner, C., Trumper, L., Pfitzenmeier, J-P., Loftin, U., Gerlach, G., Stehle, I., Wadle, A. and

Pfreundschuh, M. (1998) Differential mRNA display at the single cell level. *BioTechniques* **24:** 720–724.

Richmond, C.S., Glasner, J.D., Mau, R., Jin, H. and Blattner, F.R. (1999) Genome-wide expression profiling in *Escherichia coli* K-12. *Nucleic Acids Res.* **27:** 3821–3835.

Schena, M., Shalon, D., Davis, R.W. and Brown, P.O. (1995) Quantitative monitoring of gene expression patterns with a complementary DNA microarray. *Science* **270:** 467–470.

Schena, M., Shalon, D., Heller, R., Chai, A., Brown, P.O. and Davis, R.W. (1996) Parallel human genome analysis: microarray-based expression monitoring of 1000 genes. *Proc. Natl Acad. Sci. USA* **93:** 10 614–10 619.

Shalon, D., Smith, S.J. and Brown P.O. (1996) A DNA microarray system for analyzing complex DNA samples using two-color fluorescent probe hybridisation. *Genome Res.* **6:** 639–645.

Shimkets, R.A., Lowe, D.G., Tsu-Ning Tai, J., Sehl, P., Jin, H., Yang, R., Predki, P.F., Rothberg, B.E.G., Murtha, M.T., Roth, M.E., Shenoy, S.G., Windemuth, A., Simpson, J.W., Simons, J.W., Daley, M.P., Gold, S.A., McKenna, M.P., Hillan, K., Went, G.T. and Rothberg, J.A. (1999) Gene expression analysis by transcript profiling coupled to a gene database query. *Nature Biotechnol.* **17:** 798–803.

Shiue, L. (1997) Identification of candidate genes for drug discovery by differential display. *Drug Devl. Res.* **41:** 142–159.

Singh-Gasson, S., Green, R.D., Yue, Y., Nelson, C., Blattner, F., Sussman, M.R. and Cerrina, F. (1999) Maskless fabrication of light-directed oligonucleotide microarrays using a digital micromirror array. *Nature Biotechnol.* **17:** 974–978.

Syed, F., Blakemore, S.J., Wallace, D.M., Trower, M.K., Johnson, M., Markham, A.F. and Morrison, J.F.J. (1999) CCR7 (EBI 1) receptor down-regulation in asthma: differential gene expression in human CD4+ T lymphocyte. *Q. J. Med.* **92:** 463–471.

Vasmatzis, G., Essand, M., Brinkmann, U., Lee, B. and Pastan, I. (1998) Discovery of three genes specifically expressed in human prostate by expressed sequence tag database analysis. *Proc. Natl Acad. Sci. USA* **95:** 300–304.

Velculescu, V.E., Zhang, L., Vogelstein, B. and Kinzler, K.W. (1995) Serial analysis of gene expression. *Science* **270:** 484–487.

Velculescu, V.E., Zhang, L., Zhou, W., Vogelstein, J., Basrai, M.A., Bassett, D.E., Hieter, P., Vogelstein, B. and Kinzler, K.W. (1997) Characterization of the yeast transcriptome. *Cell* **88:** 243–251.

Wang X. and Feuerstein G.Z. (1997) The use of mRNA differential display for discovery of novel therapeutic targets in cardiovascular disease. *Cardiovasc. Res.* **35:** 414–421.

Wen, X., Fuhrman, S., Michaels, G.S., Carr, D.B., Smith, S., Barker, J.L. and Somogyi, R. (1998) Large-scale gene expression mapping of central nervous system development. *Proc. Natl Acad. Sci. USA* **95:** 334–339.

White, J.A. and Petkovich, M. (1998) Identification and cloning of RA-regulated genes by mRNA-differential display. *Methods Mol. Biol.* **89:** 389–404.

Wodicka, L., Dong, H., Mittmann, M., Ho., M.H. and Lockhart, D.J. (1997) Genome-wide expression monitoring in *Saccharomyces cerevisiae*. *Nature Biotechnol.* **15:** 1359–1367.

Yershov, K., Barsky, V., Belgovisky, A., Kirillov, E., Kreindlin, E., Ivanov, I., Parinov, S., Guschin, D., Drobishev, A., Dubiley, S. and Mirzabekov, A. (1996) DNA analysis and diagnostics on oligonucleotide microchips. *Proc. Natl Acad. Sci. USA* **93:** 4913–4198.

Zehetner, G. and Lehrach, H. (1994) The Reference Library System – sharing biological material and experimental data. *Nature* **367:** 489–491.

Zhang, L., Zhou, W., Velculescu, V.E., Kern, S.E., Hruban, R.H., Hamilton, S.R., Vogelstein, B. and Kinzler, K.W. (1997) Gene expression profiles in normal and cancer cells. *Science* **276:** 1268–1272.

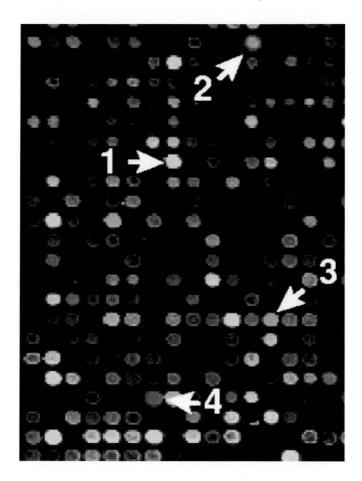

Chapter 2, Figure 2. Typical image of a glass microarray after cohybridization of two mRNA populations distinguished by separate labels. A section of a microarray containing human cDNAs hybridized with RNA-derived material from both serum-deprived (Cy3) and serum-stimulated (Cy5) fibroblasts. Yellow spots represent genes whose expression does not vary significantly between the two samples (1). mRNAs which are more abundant in the serum-stimulated fibroblasts appear red (2 and 4), whereas those mRNAs which are more abundant in serum-deprived fibroblasts appear green (3). (Reprinted with permission from Iyer *et al.*, The transcriptional program in the response of human fibroblasts to serum, *Science*, volume 283, copyright 1999 American Association for the Advancement of Sciences).

Chapter 3

Two-dimensional polyacrylamide gel electrophoresis for proteome analysis

Michael J. Dunn and Angelika Görg

1. Introduction

The first requirement for proteome analysis is the separation, visualization and analysis of the complex mixtures containing as many as several thousand proteins obtained from whole cells, tissues or organisms. Recently, progress has been made in the development of alternative methods of protein separation for proteomics (*see* Chapter 10), such as the use of chip-based technologies (Merchant and Weinberger, 2000; Nelson *et al.*, 2000), the direct analysis of protein complexes using mass spectrometry (Link *et al.*, 1999), the use of affinity tags (Gygi *et al.*, 1999; Rigaut *et al.*, 1999), and large-scale yeast two-hybrid screening (Uetz *et al.*, 2000). However, two-dimensional polyacrylamide gel electrophoresis (2-DE) remains the core technology of choice for separating complex protein mixtures in the majority of proteome projects. This is due to its unrivalled power to separate simultaneously thousands of proteins, the subsequent high-sensitivity visualization of the resulting 2-D separations that are amenable to quantitative computer analysis to detect differentially regulated proteins, and the relative ease with which proteins from 2-DE gels can be identified and characterized using highly sensitive microchemical methods. Recent developments that will be described in this chapter have resulted in a current 2-DE method that combines increased resolving power and high reproducibility with relative simplicity of use.

2. Brief history of 2-DE

The first 2-DE separations are attributed to the work of Smithies and Poulik (1956) describing a 2-D combination of paper and starch gel electrophoresis for the separation of serum proteins. Subsequent developments in electrophoretic technology, such as the use of polyacrylamide as a support medium and the use of polyacrylamide concentration gradients were rapidly applied to 2-D separations (reviewed in Dunn, 1987). In particular, the application of isoelectric focusing (IEF) techniques to 2-D separations made it possible for the first dimension separation to

Proteomics, edited by S.R. Pennington and M.J. Dunn
© 2001 BIOS Scientific Publishers, Oxford.

be based on the charge properties of the proteins. The coupling of IEF with sodium dodecyl sulfate polyacrylamide gel electrophoresis (SDS-PAGE) in the second dimension resulted in a 2-D method that separated proteins according to two independent parameters (i.e. charge and size). This methodology was then adapted to a wide range of samples with differing solubility properties by the use of urea and nonionic detergents for IEF. Thus, by 1975 a 2-DE system had evolved that could be applied to the analysis of protein mixtures of whole cells and tissues (Iborra and Buhler, 1976; Klose, 1975; Scheele, 1975).

Based on these developments, O'Farrell in 1975 described a method of 2-DE optimized for the separation of the proteins of *Escherichia coli*. This method used a combination of IEF in cylindrical 4%T, 5%C polyacrylamide gels containing 8M urea and 2% (w/v) Nonidet P-40 (NP-40) with the discontinuous SDS-PAGE system of Laemmli (1970). This orthogonal combination of charge separation (isoelectric point, pI) with size separation (relative molecular mass, M_r) resulted in the sample proteins being distributed across the two-dimensional gel profile. This technique has formed the basis for most developments in 2-DE over the last 25 years and several thousand papers have been published using this methodology.

3. 2-DE with immobilized pH gradients

For proteome analysis, it is essential that 2-DE should generate reproducible high-resolution protein separations. However, until recently this was a major problem largely because of the nature of the synthetic carrier ampholytes (SCA) used to generate the pH gradients required for IEF. SCA are produced by a complex synthetic process that is difficult to control reproducibly. This results in considerable batch to batch variability and limits the reproducibility of 2-DE protein separations. Perhaps more importantly, SCA are relatively small molecules that are not fixed within the IEF gel. As a consequence the electroendosmotic flow of water that occurs during IEF results in migration of the SCA molecules towards the cathode. This process, known as cathodic drift, results in pH gradient instability and is exacerbated using tube IEF gels due to the negatively charged groups present on the walls of the glass capillaries. In practice, pH gradients using the O'Farrell method of 2-DE rarely extend far beyond pH 7, with the resultant loss of basic proteins. This problem was recognized by O'Farrell, who developed an alternative procedure, known as non-equilibrium pH gradient electrophoresis (NEPHGE), for the 2-DE separation of basic proteins (O'Farrell *et al.*, 1977). In this method, separation occurs on the basis of protein mobility in the presence of a rapidly forming pH gradient, but reproducibility is extremely difficult to control. Fortunately, this problem has been overcome with the development of immobilized pH gradient (IPG) IEF (Bjellqvist *et al.*, 1982) based on the use of the Immobiline reagents (Amersham Pharmacia Biotech). Early attempts to apply the IPG technology to 2-DE separations encountered many problems, but these have been overcome (Görg *et al.*, 1988, 2000) so that IPG IEF is the current method of choice for the first dimension of 2-DE for most proteomic applications. This method of 2-DE, known as IPG-DALT, will be discussed in detail in this chapter. A recent article describes in detail the use of SCA IEF for 2-DE (Monribot and Boucherie, 2000).

4. Sample preparation

No single method of sample preparation can be applied universally due to the diverse nature of samples that are analyzed by 2-DE. However, some general considerations can be mentioned. It is essential to minimize protein modifications that might result in artifactual spots on 2-DE gel profiles. In particular, samples containing urea must not be heated as this can introduce considerable charge heterogeneity due to carbamylation of the proteins by isocyanate formed in the decomposition of urea. In addition, proteases present within samples can readily result in artifactual spots, so that samples should be subjected to minimal handling and kept cold at all times. Cocktails of protease inhibitors can be added, but it should be remembered that such reagents could modify proteins leading to charge artifacts.

4.1 Soluble samples

Soluble, liquid samples, such as serum, plasma, urine, cerebrospinal fluid (CSF) and aqueous extracts of cells and tissues, can often be analyzed by 2-DE with minimal pretreatment. Samples with a relatively high protein concentration, such as plasma and serum, can be analyzed directly by 2-DE after dilution with the appropriate sample solubilization buffer (*see* section 5). However, the high abundance of proteins such as albumin and immunoglobulins obscure many of the minor components on 2-D profiles. This problem can be overcome by depleting these proteins (Dunn, 1987; Lollo *et al.*, 1999), but there is always the possibility that this can result in the nonspecific removal of other protein components.

Other types of liquid sample may have a relatively low protein concentration or contain high concentrations of salts that can interfere with the protein separation during IEF. Such samples can be desalted by dialysis or liquid chromatography prior to 2-DE. The samples then require concentration by methods such as lyophilization, dialysis against polyethylene glycol, or precipitation with TCA or acetone. TCA/ acetone precipitation has been found to be very useful for the inactivation of proteases to minimize protein degradation, the removal of interfering compounds, and for the enrichment of very alkaline proteins such as ribosomal proteins from total cell lysates (Görg *et al.*, 1997).

4.2 Tissue samples

Solid tissue samples are usually disrupted in the presence of solubilization buffer (*see* section 5). The best approach is to break up the tissue while it is still frozen, preferably at liquid nitrogen temperature. Small tissue specimens that have been wrapped in aluminum foil and frozen in liquid nitrogen can be crushed between two cooled blocks or ground under liquid nitrogen using a pestle and mortar. Large tissue pieces can be processed by homogenization in solubilization buffer using a rotating-blade type homogenizer, but heating and foaming must be minimized.

A particular problem with the analysis of tissue samples is the heterogeneous nature of the sample. Many different types of cell can be present in the tissue sample and in samples from disease it may not be possible to separately collect diseased and nondiseased areas of tissue. There are several approaches to this problem that have been used for 2-DE analysis. Using positive or negative selection, usually based on an immunoaffinity method, it is possible to enrich particular populations of cells. These can either be subjected to 2-DE analysis directly or the numbers of cells can be increased by short-term culture prior to 2-DE. Alternatively, microdissection techniques can be employed to obtain pure populations of cells from tissue sections. Of particular interest is the use of laser capture microdissection (LCM) (reviewed by Simone *et al.*, 1998). In this technique, cell selection is achieved by activation of a transfer film placed in contact with a tissue section by a laser beam that is focused on the selected area of tissue using an inverted microscope. The precise area of film targeted by the laser bonds to the tissue beneath it and these cells are then lifted free of surrounding tissue. This technique is readily applicable to the analysis of nucleic acids through the availability of the polymerase chain reaction (PCR) to amplify the small amount of material present in such specimens. The absence of a comparable method for the amplification of proteins is clearly a major challenge for the application of LCM to isolate cells for proteomic analysis. Nevertheless, Banks *et al.* (1999) demonstrated the use of LCM for proteomic analysis. In this study, dissection of epithelial tissue from a sample of normal human cervix resulted in enrichment of some proteins compared with analysis of the whole tissue.

4.3 Cells

Cells grown *in vitro* in suspension culture or samples of circulating cells (e.g. erythrocytes, lymphocytes) should ideally be harvested by centrifugation, washed in phosphate-buffered saline (PBS) and solubilized in sample buffer (*see* section 5). For cells cultured on solid substrates, the culture medium should first be removed and the cell layer then washed with PBS or an isotonic sucrose solution. The latter is a particularly efficient way of minimizing salts that can interfere with the IEF dimension. The cells can then be harvested by scraping, but the use of proteolytic enzymes should be avoided. Alternatively, the cells can be lysed directly while still attached to the substrate by the addition of a small volume of solubilization buffer. If the cells contain high levels of nucleic acids leading to high sample viscosity, the samples should be treated with a protease-free DNase/RNase mixture (Garrels, 1979).

4.4 Sample fractionation

Owing to the high dynamic range and diversity of the proteins expressed in eukaryotic tissues, it is sometimes necessary to perform a fractionation step to reduce the complexity of the sample and enrich for low copy number proteins. Prefractionation of proteins can be achieved by methods such sub-cellular fractionation, electrophoresis in the liquid phase, adsorption chromatography, and selective precipitation (reviewed by Corthals *et al.*, 2000). An alternative approach is the sequential extraction of proteins from a cell or tissue on the basis of their solubility in a series of buffers with increasingly powerful solubilizing properties (Cordwell *et al.*, 2000; Molloy *et al.*, 1998).

5. Solubilization

The ideal solubilization procedure for 2-DE would result in the disruption of all noncovalently bound protein complexes and aggregates into a solution of individual polypeptides. If this is not achieved, persistent protein complexes in the sample are likely to result in new spots in the 2-D profile, with a concomitant reduction in the intensity of those spots representing the single polypeptides. In addition, the method of solubilization must allow the removal of substances, such as salts, lipids, polysaccharides and nucleic acids that can interfere with the 2-DE separation. Finally, the sample proteins must remain soluble during the 2-DE process. For the foregoing reasons sample solubility is one of the most critical factors for successful protein separation by 2-DE.

The most popular method for protein solubilization for 2-DE remains that originally described by O'Farrell (1975), using a mixture of 9.5 M urea, 4% w/v of the nonionic detergent NP-40, 1% w/v of the reducing agent dithiothreitol (DTT) and 2% w/v of SCA of the appropriate pH range (so-called 'lysis buffer'). While this method works well for many types of sample, it is not universally applicable, with membrane proteins representing a particular challenge (Santoni *et al.*, 2000). The zwitterionic detergent, CHAPS (3[(cholamidopropyl)dimethylammonio]-1-propane sulfonate), has been found to be effective for solubilization of membrane proteins (Perdew *et al.*, 1983), particularly when used at a concentration of 4% w/v in combination with a mixture of 2 M thiourea and 8 M urea (Molloy *et al.*, 1998; Rabilloud *et al.*, 1997). Thiourea is a much stronger denaturant than urea, but cannot be used alone as it is weakly soluble in water. However, it is more soluble in concentrated solution of urea, so that urea–thiourea mixtures exhibit improved solubilizing power. Linear sulphobetaine detergents, such as SB 3–10 or SB 3–12, are also effective solubilizing agents, but these are not compatible with high concentrations of urea (Rabilloud *et al.*, 1997). This can be overcome by using these reagents at 2% w/v in combination with 5 M urea, 2 M thiourea and 2% CHAPS (Herbert, 1999; Rabilloud *et al.*, 1997).

The detergent sodium dodecyl sulfate (SDS) is able to disrupt most noncovalent protein interactions and is a very effective agent for the solubilization of membrane proteins. However, its anionic character precludes its inclusion in IEF gels. However, SDS can be used as a pre-solubilization procedure for samples prior to 2-DE. In this approach, the sample is initially solubilized in 1% w/v SDS and then diluted with the normal solubilization solution (e.g. lysis buffer). The aim here is to displace the SDS from the proteins and replace it with nonionic or zwitterionic detergent, thereby maintaining the proteins in a soluble state. The ratio of protein to the nonionic or zwitterionic detergent (1:3) and the ratio of SDS to the nonionic or zwitterionic detergent (1:8) must both be carefully controlled for effective solubilization while minimising the deleterious effects of SDS on IEF (Dunn, 1987).

The presence of nucleic acids can be problematical during IEF. This is due to an increase in the viscosity of the sample and in some cases formation of complexes with the sample proteins, leading to artifactual migration and streaking. If problems of this type are suspected, it is best to degrade the nucleic acid by the addition of a suitable pure (i.e. protease free) endonuclease to the sample solubilization solution. An alternative method is to utilize the ability of SCA to form complexes with nucleic acids and then remove the complexes by ultracentrifugation (Rabilloud et al., 1986).

6. Reduction

Protein disulphide bonds are normally reduced with free-thiol containing reagents such as DTT or β-mercaptoethanol. However, reagents such as DTT are charged so that they migrate out of the gel during IEF, leading to reoxidation of the sample proteins that can result in loss of sample solubility during electrophoresis. It has recently been reported that replacing the thiol containing reducing agents with a noncharged reducing agent such as tributyl phosphine (TBP) can greatly increase protein solubility during the IEF dimension and result in increased transfer to the second dimension gel (Herbert et al., 1998; Herbert, 1999).

7. The first dimension: IEF with IPG

IPG IEF gels are prepared using Immobilines (Amersham Pharmacia Biotech), a series of eight acrylamide derivatives with the structure CH_2=CH—CO—NH—R, where R contains either a carboxyl or tertiary amino group. These form a series of buffers with different pK values distributed throughout the pH 3–10 range. The appropriate IPG reagents, calculated according to published recipes, are added to the mixture used for gel polymerization. Thus, during polymerization, the buffering groups that will form the pH gradient are covalently attached via vinyl bonds to the polyacrylamide backbone. IPG generated in this way are, therefore, immune to the effects of electroendosmosis, so that they provide the opportunity to carry out IEF separations which are extremely stable, allowing the true equilibrium state to be attained. Initial attempts to implement the IPG technology to 2-D separations encountered several problems (reviewed by Görg et al., 1988). Fortunately these problems have been solved and IPG IEF has become the method of choice for the first dimension separation of 2-DE (reviewed by Görg et al., 2000).

7.1 IPG gel preparation

The method (Görg et al., 1988) is shown schematically in Figure 1. Briefly, IPG slab gels of the desired pH range are cast on GelBond PAGfilm (Figure 1b) according to the extensive library of published recipes (reviewed by Görg et al., 2000). After polymerization, the gels are extensively washed with deionized water (6×10 min), immersed in 2% w/v glycerol (30 min), and dried at room temperature in a dust-free cabinet, and then covered with a plastic film. If the gels are not used immediately, they can be stored for up to 1 year at $-20°C$. Prior to use, the required number of gel strips (3–5 mm wide) for 2-DE are cut off of the IPG gel slab using a paper cutter (*Figure 1c*). Alternatively, ranges of ready-made IPG strips are available commercially from Amersham Pharmacia Biotech (Immobiline DryPlate or Immobiline DryStrip) and Bio-Rad (ReadyStrip). IPG strips of any desired length can be used, but it should be remembered that, in general, the larger the separation area of a 2-D gel, the greater the number of proteins that can be resolved. Strips of 18 or 24 cm (Görg *et al.*, 1999) are usually employed for high-resolution separations, while shorter strips (4, 7 or 11 cm) are used for rapid screening applications. A large selection of both wide and narrow range pH gradients is available for IPG IEF and this topic is discussed in section 7.4.

7.2 Rehydration of IPG strips

For use in 2-DE, the strips may be rehydrated in a reswelling cassette (*Figure 1d*) in a solution containing 8 M urea (or, alternatively, 2 M thiourea and 5 M urea), 0.5% to 4% nonionic (e.g. NP-40, Triton X-100) or zwitterionic (CHAPS) detergent, 15 mM DTT and 0.5% SCA of the appropriate pH range. The strips are then placed directly on the surface of the cooling plate of a horizontal flat-bed electrophoresis apparatus (e.g. Multiphor, Amersham Pharmacia Biotech) (*Figure 1g*). A convenient alternative is to use the special strip tray available from Pharmacia (*Figure 1h*). This tray is fitted with a corrugated plastic plate that contains grooves allowing easy alignment of the IPG strips. In addition, the tray is fitted with bars carrying the electrodes and a bar fitted with sample cups allowing application of samples at any desired point on the gel surface. When required, this tray is filled with silicon oil that protects the gel from the effects of the atmosphere during IEF. Horizontal streaking can often be observed at the basic end of 2-DE protein profiles, particularly when basic IPG gradients are used for the first dimension. This problem can be resolved by applying an extra electrode strip soaked in 20 mM DTT on the surface of the IPG strip alongside the cathodic electrode strip (Görg *et al.*, 1995). This has the advantage that the DTT within the gel, which migrates towards the anode during IEF, is replenished by the DTT released from the strip at the anode. An alternative approach is to use the noncharged reducing agent, TBP, which does not migrate during IEF and has been found to greatly improve protein solubility during IEF, especially for wool proteins (Herbert *et al.*, 1998; Herbert, 1999).

7.3 Sample application and running conditions

Samples are usually applied into silicon rubber frames or special sample cups (*Figure 1g, h*) placed either at the anodic or cathodic end of the IPG strips, the optimal position being determined empirically for each type of sample. The initial voltage should be limited to 150 V for 30 min to allow maximal sample entry and then progressively increased until 3500 V is attained. The time required for the run

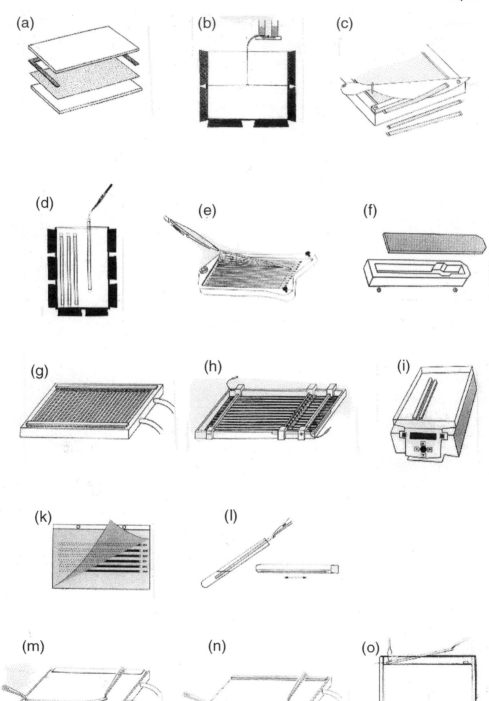

depends on several factors, including the type of sample, the amount of protein applied, the length of the IPG strips used, and the pH gradient being used. The IEF run should be performed at 20°C, as at lower temperatures there is a risk of urea crystallization and different temperatures have been found to result in alterations in the relative positions of some proteins on the final 2-D patterns (Görg et al., 1991). Some typical running conditions are given in *Table 1*.

This method of sample application can result in protein precipitation and this effect is more pronounced when high protein loadings (1 mg or more) are used. This problem can be overcome by reswelling the IPG strips directly in the solution containing the protein sample to be analyzed (Rabilloud et al., 1994; Sanchez et al., 1997). Very high protein loads (> 10 mg) have been successfully separated using this method, but there can be a selective loss of high molecular weight, very basic and membrane proteins. Recently a new integrated instrument, named the IPGphor (Amersham Pharmacia Biotech), has been developed to simplify the IPG IEF dimension 2-DE. This instrument features a strip holder that provides for rehydration of individual IPG strips with or without sample, optional separate sample loading, and subsequent IEF, all without handling the strip after it is placed in the ceramic strip holder (*Figure 1f*). The instrument can accommodate up to 12 individual strip holders and incorporates Peltier solid-state cooling and a programmable 8000 V, 1.5 mA power supply (*Figure 1i*). Some typical running conditions for IPG IEF using the IPGphor are given in *Table 2*. For improved entry of high molecular weight proteins, low voltage is applied during reswelling. Equipment facilitating sample application by in-gel rehydration and for carrying out the IPG IEF dimension of 2-DE is also available from Genomic Solutions (IPGpHaser) and Bio-Rad (Protean IEF cell).

7.4 Choice of pH gradient for IPG IEF

A wide-range, linear pH 3.5–10 gradient is often useful for the initial analysis of a new type of sample. However, for many samples this can result in loss of resolution in the region pH 4 to 7, in which the pI values of many proteins occur. This problem can be overcome to some extent with the use of a nonlinear pH 3.5–10 IPG IEF gel (Bjellqvist et al., 1993), in which the pH 4–7 region contains a much flatter gradient than in the pH 7–10.5 region. This allows good separation in the pH 4–7 region while

---

Figure 1. Procedure of IPG-Dalt according to Görg et al. (2000). (a) Assembly of the polymerization cassette prior to casting IPG and SDS gels on plastic backing (Glass plates, GelBond PAGfilm, 0.5 mm thick U-frame). (b) Casting of IPG- and/or SDS pore gradient gels. (c) Cutting of washed and dried IPG slab gels (or Immobiline DryPlates) into individual IPG strips. Rehydration of individual IPG gel strips (d) in a vertical rehydration cassette, (e) in the reswelling tray, and (f) in the IPGphor strip holder. IEF in individual IPG gel strips (g) placed directly on the cooling plate of the IEF unit, (h) in the DryStrip kit, and (i) on the IPGphor. (k) Storage of the IPG strips after IEF. (l) Equilibration of IPG gel strips prior to SDS-PAGE. Transfer of the equilibrated IPG gel strip onto the surface (m) of a laboratory-made horizontal SDS gel alongside the cathodic electrode wick or (n) onto a ready-made horizontal SDS gel along the cathodic buffer strip. (o) Loading of the equilibrated IPG gel strip onto a vertical SDS gel.

Table 1. Running conditions using the Multiphor according to Görg *et al.* (2000)

Gel length	180 mm
Temperature	20°C
Current max.	0.05 mA per IPG strip
Power max.	0.2 W per IPG strip
Voltage max.	3500 V

I Analytical IEF
Initial IEF

Cup loading (20–50 μL)	In-gel rehydration (350 μL)
150 V, 1 h	150 V, 1 h
300 V, 1–3 h	300 V, 1–3 h
600 V, 1 h	

IEF to the steady state at 3500 V

1–1.5 pH units	4 pH units	7 pH units
e.g. IPG 5–6 ..... 24 h	IPG 4–8 .... 10 h	IPG 3–10 L ..... 6 h
e.g. IPG 4–5.5 .. 20 h	IPG 6–10 .... 10 h	IPG 3–10 NL ... 6 h

3 pH units	5–6 pH units	8–9 pH units
IPG 4–7 .... 12 h	IPG 4–9 .... 8 h	IPG 3–12 .... 6 h
IPG 6–9 .... 12 h	IPG 6–12 .... 8 h	IPG 4–12 .... 8 h

II Extended separation distances (240 mm)
IEF to the steady-state at 3500 V
IPG 3–12 .... 　8 h
IPG 4–12 .... 12 h
IPG 5–6 .... 40 h

III Micropreparative IEF
Initial IEF

Cup loading (100 μL)	In-gel rehydration (350 μL)
50 V, 12–16 h	50 V, 12–16 h
300 V, 1 h	300 V, 1 h

IEF to the steady state at 3500 V
Focusing time of analytical IEF plus approximately 50%

still resolving the majority of the more basic species (*Figure 2*). However, use of a pH 4–7 IPG IEF gel will result in even better protein separation within this range (*Figure 3*). Commercial IPG strips are available for these pH ranges (Amersham Pharmacia Biotech, Bio-Rad). Laboratory-made pH 4–9 IPG gels also give good coverage of the majority of cellular proteins from many types of sample. These gradients are compatible with cup-loading as well as with in-gel rehydration, and

Table 2. Running conditions using the IPGphor according to Görg *et al.* (2000)

Gel length	180 mm
Temperature	20°C
Current max.	0.05 mA per IPG strip
Voltage max.	8000 V

I Analytical IEF

Reswelling	• 30 V, 12–16 h
Initial IEF	• 200 V, 1 h
	• 500 V, 1 h
IEF to the steady-state	• 1000 V, 1 h
	• Gradient from 1000 V to 8000 V within 30 min
	• 8000 V to the steady-state, depending on the pH interval used:

1–1.5 pH units	4 pH units	7 pH units
e.g. IPG 5–6 .... 8 h	IPG 4–8 .... 4 h	IPG 3–10 L 3h
e.g. IPG 4–5.5.. 8 h		IPG 3–10 NL .... 3 h

3 pH units	5–6 pH units	8–9 pH units
IPG 4–7 .... 4 h	IPG 4–9 .... 4 h	IPG 3–12 .... 3 h
		IPG 4–12 .... 3 h

II Micropreparative IEF

Reswelling	• 30 V, 12–16 h
IEF to the steady state	• Focusing time of analytical IEF + additional 50% (approx.)

they work well on the Multiphor and on the IPGphor systems. They are suitable for both analytical (sample loading 50–100 µg) and micropreparative runs (sample loading up to 1 mg). Typical running conditions are given in *Tables 1* and *2*.

With complex samples such as eukaryotic cell extracts, 2-DE on a single wide-range pH gradient reveals only a small percentage of the whole proteome because of insufficient spatial resolution and the difficulty of visualizing low copy number proteins in the presence of the more abundant species. One approach to overcoming the problem of the high dynamic range and diversity of the proteins expressed in eukaryotic tissues is sample prefractionation (*see* section 4.4). A powerful alternative is to use multiple, overlapping narrow range IPGs spanning 1–1.5 pH units; an approach that has become known as 'zoom gels' (Wildgruber *et al.*, 2000), 'composite gels' or 'subproteomics' (Cordwell *et al.*, 2000). In addition, extended separation distances up to 24 cm can also be used (Görg *et al.*, 1999). Narrow range IPGs between pH 4 and 7 work with both in-gel rehydration and cup-loading, and can be run on either the Multiphor or IPGphor systems (*Figure 4*). These gels are ideal for micropreparative purposes where protein loadings of several milligrams can be used. These gels require extended focusing times (*see Tables 1* and *2*).

Strongly alkaline proteins such as ribosomal and nuclear proteins with closely related pIs between 10.5 and 11.8 can be separated using narrow range pH 10–12 or pH 9–12 IPGs (*Figure 5*; Görg *et al.*, 1997). In order to obtain highly reproducible

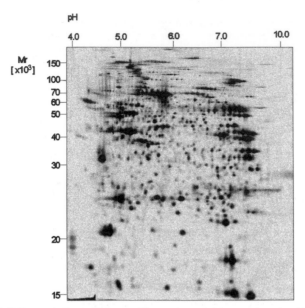

Figure 2. A 2-DE separation of 80 μg of heart (ventricle) proteins. An 18 cm nonlinear pH 3–10 IPG IEF gel was used in the first dimension. The second dimension was a 21 cm 12% SDS-PAGE gel. The proteins were detected by silver staining. The scale at the top indicates the nonlinear pH range of the first-dimension IPG strip from which the apparent pIs of the separated proteins can be estimated. The M_r scale can be used to estimate the molecular weights of the separated proteins.

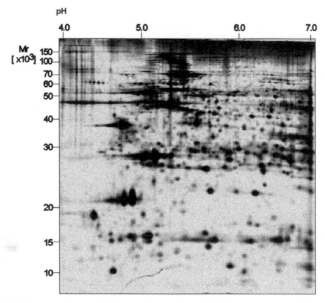

Figure 3. A 2-DE separation of 80 μg of heart (ventricle) proteins. An 18 cm linear pH 4–7 IPG IEF gel was used in the first dimension. The second dimension was a 21 cm 12% SDS-PAGE gel. The proteins were detected by silver staining. The scale at the top indicates the linear pH range of the first-dimension IPG strip from which the apparent pIs of the separated proteins can be estimated. The M_r scale can be used to estimate the molecular weights of the separated proteins.

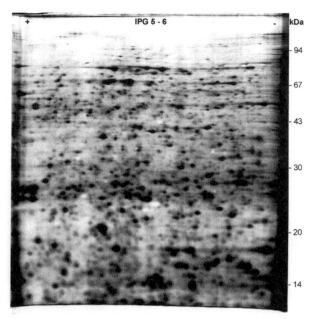

Figure 4. Micropreparative IPG-Dalt of mouse liver proteins. First dimension: IPG 5–6 (ready-made). Separation distance: 180 mm. Sample application by in-gel rehydration (500 μg protein). IEF was performed on the IPGphor. Running conditions: 12 h at 8000 V max. (for details *see* Table 1). Second dimension: vertical SDS-PAGE (13%T constant). Silver stain.

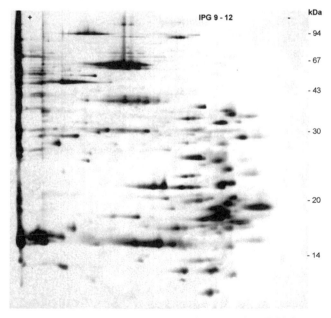

Figure 5. Analytical IPG-Dalt of very alkaline mouse liver proteins (TCA/acetone precipitation) according to Görg *et al.* (1997). First dimension: IPG 9–12 (8 h at 5000 V max.). Separation distance: 180 mm. Sample application by cup-loading near the anode. Second dimension: vertical SDS-PAGE (13%T constant). Silver stain.

2-D patterns, different optimization steps with respect to pH engineering and gel composition, such as the substitution of *N*,*N*-dimethylacrylamide (DMAA) for acrylamide and the addition of isopropanol to the IPG rehydration solution, were necessary in order to suppress the reverse electroendosmotic flow which gives rise to highly streaky 2-D profiles (Görg *et al.*, 1997). Sample cup-loading at the anode and IEF under silicon oil is essential.

Theoretical 2-D profiles calculated from sequenced genomes (Link *et al.*, 1997) not only indicate that the majority of proteins of a total cell lysate possess pIs between pH 4 and pH 9, but also that a considerable number of proteins with pI values up to pH 12 can be present. In classical 2-DE using SCA these proteins can only be separated by NEPHGE (*see* section 3), but these proteins can be readily resolved using IPG IEF. This can be accomplished using the narrow range pH 10–12 or pH 9–12 IPGs described above. However, in order to obtain an overview of the 'total' cellular or tissue proteome, wide IPGs covering the range pH 3–12 and pH 4–12 have been generated (Görg *et al.*, 1998, 1999). Excellent 2-D profiles of cell and tissue extracts and TCA/acetone precipitated proteins for the visualization of basic proteins with pI values exceeding pH 10 that are usually absent in lysis buffer extracts (Görg *et al.*, 1998, 1999), as well as the Triton X-100 insoluble cell fraction of *Mycoplasma pneumoniae* (Hermann *et al.*, submitted) have been obtained. More-over, a pH 4–12 IPG with a flattened region between pH 9 and 12 was found to give excellent separation of very basic proteins such as ribosomal proteins (Görg *et al.*, 1998). These wide gradients can be run under standard conditions without isopro-panol, as the strong water transport to the cathode (reverse electroendosmosis) characteristic of narrow IPGs exceeding pH 10 is negligible. In addition, the entry and focusing of high molecular weight proteins is significantly improved (*Figure 6*).

Extended separation distances (> 30 cm in each dimension) for maximum resolution of complex protein patterns have been described for the classical O'Farrell system (Klose, 1999). However, using the tube IEF gel technology, this is achieved at the expense of ease of gel handling and processing. In contrast, handling of long IPG gel strips cast on a plastic backing does not require special precautions. For example, 24 cm long IPG gel strips covering the ranges pH 3–12 and pH 4–12 have been successfully used (*Figure 6*; Görg *et al.*, 1999). Furthermore, using 24 cm long IPG strips covering a narrow pH range (e.g. pH 5–6), highly resolved 2-D profiles can be obtained (*Figure 7*).

7.5 *Optimization of focusing time*

In theory, the optimum focusing time required to achieve the best quality and reproducibility is the time needed for the IEF separation pattern to achieve the steady state (Görg *et al.*, 1988, 1997). If the focusing time is too short, this will result in horizontal and vertical streaking. However, over-focusing should be avoided. Although in contrast to the classical O'Farrell method, this does not result in migration of proteins towards the cathode (cathodic drift), it does result in excess water exudation at the surface of the IPG gel due to active water transport (electroendosmosis). This leads to distorted protein patterns, horizontal streaks at

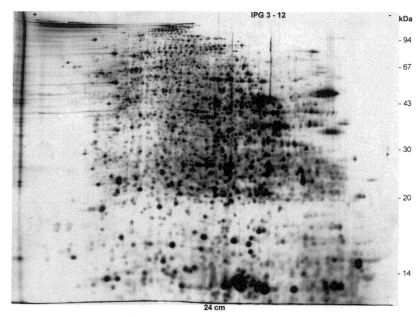

Figure 6. Wide IPG with extended separation distance according to Görg *et al.* (1999). Analytical IPG-Dalt of mouse liver proteins (Lysis buffer extract). First dimension: IPG 3–12. Separation distance: 240 mm. Sample application: cup-loading near the anode. Running conditions: 6 h at 3500 V max. (*see* Table 1). Second dimension: vertical SDS-PAGE (13%T constant). Silver stain.

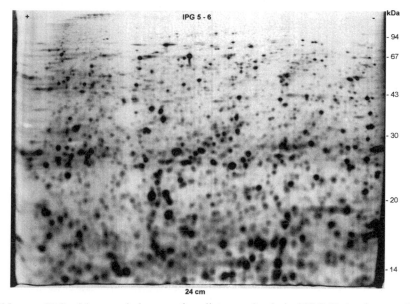

Figure 7. Narrow IPG with extended separation distance. Analytical IPG-Dalt of mouse liver proteins (Lysis buffer extract) according to Görg *et al.* (2000). First dimension: IPG 5–6 (laboratory-made). Separation distance: 240 mm. Sample application by cup-loading at the anode. Running conditions: 40 h at 3500 V max. (*see* Table 1). Second dimension: vertical SDS-PAGE (13%T constant). Silver stain.

the basic end of the gel, and loss of proteins. The optimum focusing time must be established empirically for each combination of protein sample, protein loading and the particular pH range and length of IPG gel strip used. As a guideline, optimum focusing times determined for a number of different wide and narrow pH range IPGs are given in *Tables 1* and *2*.

8. Equilibration between dimensions

After the IPG IEF dimension, strips can be used immediately for the second dimension. Alternatively, strips can be stored between two sheets of plastic film at $-80°C$ for periods of several months. Prior to the second-dimension separation, it is essential that the IEF gels are equilibrated to allow the separated proteins to interact fully with SDS so that they will migrate properly during SDS-PAGE (*Figure 11*). The recommended protocol is to incubate the IPG IEF gel strips for 15 min in 50 mM Tris buffer, pH 8.8 containing 2% w/v SDS, 1% w/v DTT, 6 M urea and 30% w/v glycerol. The urea and glycerol are used to reduce electroendosmotic effects which otherwise result in reduced protein transfer from the first to the second dimension (Görg *et al.*, 1988). This is followed by a further 15 min equilibration in the same solution containing 5% w/v iodoacetamide in place of DTT. The latter step is used to alkylate any free DTT, as otherwise this migrates through the second-dimension SDS-PAGE gel, resulting in an artifact known as point-streaking that can be observed after silver staining. An alternative procedure, allowing equilibration to be achieved in a single step, is to replace the DTT in the equilibration buffer with 5 mM TBP, which is uncharged and so does not migrate during SDS-PAGE (Herbert *et al.*, 1998). After equilibration, the IPG strips should be drained along the edge on filter paper for 1 min to remove excess liquid before application to the second dimension SDS-PAGE gel.

9. The second dimension: SDS-PAGE

The discontinuous buffer system of Laemmli (1970) is most commonly used for the second dimension of 2-DE. Gels of either a single polyacrylamide concentration or containing a linear or nonlinear polyacrylamide concentration gradient, to extend the range over which proteins of different molecular mass can be effectively separated, may be used. The IPG gel strips are easy to handle compared with the fragile tube IEF gels used in the classical O'Farrell procedure as they are bound to a flexible plastic support. After equilibration, the IPG strips are applied directly to the surface of the second-dimension SDS-PAGE gels, which can be of either a vertical (*Figure 1m, n*) or horizontal (*Figure 1o*) format.

Horizontal SDS-PAGE can be carried out on either ready-made (ExcelGel, Amersham Pharmacia Biotech) or laboratory-made slab gels cast on GelBond PAGfilm (Amersham Pharmacia Biotech) by the procedure described by Görg *et al.* (1988) for ultrathin pore gradient gels. It is essential to use a 6%T stacking gel and either a homogeneous (e.g. 12%T) or gradient (e.g. 12–15%T) resolving gel can be used. The procedure is described in detail in Görg *et al.* (2000). The horizontal second-dimension system has the advantage that the gels are attached to a plastic support, thereby preventing alterations in gel size during the staining procedure. In addition, spot sharpness can be superior to that obtained using vertical systems due to the decreased gel thickness (typically 0.5 mm compared with 1 or 1.5 mm) that allows the application of higher voltages and consequent reduction in protein diffusion due to the reduced running time.

Large-scale proteome analysis usually requires the simultaneous electrophoresis of batches of second-dimension SDS-PAGE gels in order to maximize the reproducibility of 2-DE protein profiles. In addition, the gel format must be compatible with computer-driven robotic spot cutting devices. These requirements are most easily met using multiple vertical second-dimension SDS-PAGE gels, such as provided by the DALT (Amersham Pharmacia Biotech; Anderson and Anderson, 1978a,b), Investigator (Genomic Solutions; Patton *et al.*, 1990) or Protean II xi (Bio-Rad) systems. It is not necessary to use stacking gels with the vertical format as the protein zones within the IPG strips are already concentrated and the non-restrictive (low polyacrylamide concentration) IEF gel can be considered to act as a stacking gel (Dunn, 1993).

10. Resolution

The resolving capacity of 2-DE gels is dependent on the separation length in both dimensions and is usually considered to be proportional to the total gel area available for the separation. Using 'standard' 18 cm long IPG IEF gels in combination with 20 cm long second-dimension SDS-PAGE gels, around 2000 proteins from complex mixtures, such as whole cell and tissue lysates, can be readily resolved. Only a few hundred proteins can be separated using mini-gel formats, but these are much quicker to run and can be useful for rapid screening purposes. For maximal resolution of very complex mixtures, very large format gels can be used with the potential to separate as many as 5000 to 10000 proteins from whole cell lysates (Klose, 1999). However, there may be distinct advantages in using multiple, overlapping narrow range 'zoom' IPG gels to generate composite 2-DE protein maps of complex samples (*see* section 7.4).

11. Reproducibility of 2-DE

Until recently reproducibility was a major problem limiting the more widespread application of 2-DE. Using the classical tube gel technique of O'Farrell, it was often difficult to obtain reproducible separations of a particular type of sample even within a single laboratory, while comparison of 2-DE separation patterns generated in different laboratories was often considered to be impossible. The use of dedicated equipment for 2-DE such as the ISO-DALT and Investigator systems helps in this regard as it allows the simultaneous electrophoresis of large numbers (between 5 and 20) of 2-DE gels under reproducibly controlled conditions. More importantly, inter-laboratory studies of various types of sample (heart, barley, yeast) have unequi-vocally demonstrated that 2-DE using IPG IEF results in 2-D protein separations with very high spatial and quantitative reproducibility (Blomberg *et al.*, 1995; Corbett *et al.*, 1994).

12. Outlook

The reproducible high-resolution separation of complex protein mixtures is crucial for successful proteomics. Although recent progress has been made in the develop-ment of alternative methods of protein separation for proteomics (*see* section 1 and Chapter 10), there is still no generally applicable method that can replace 2-DE in its ability to simultaneously separate and display several thousand proteins from complex mixtures. In comparison with the classical O'Farrell method of 2-DE based on the use of SCA-generated pH gradients, 2-DE using IPG IEF has proved to be extremely flexible with respect to the requirements of proteome analysis. In particular, the development of basic IPGs up to pH 12 has facilitated the analysis of very alkaline proteins, the introduction of overlapping narrow-range IPGs permits higher resolution separations and the analysis of less abundant proteins, and the availability of ready-made IPG strips and integrated running devices such as the IPGphor have gone some way towards the goal of automation. Although automa-tion of 2-DE is perceived to be an important issue in proteomics, in reality it may be impossible to devise a fully automated system that is capable of performing the whole process. In fact, one of the major bottlenecks in the proteomic process is the analysis and datamining of the 2-D gel images. It is possible that developments in automation and efficiency in this area will have the greatest impact on proteomic projects (Lopez, 2000). Limitations remain in the analysis of hydrophobic and/or membrane proteins, as well as the lack of sensitive and reliable methods for protein quantification, although recent developments in fluorescent dye technology have gone some way to addressing this problem (Patton, 2000). In conclusion, although by no means perfect, 2-DE nevertheless remains the core technology of choice for separating complex protein mixtures in proteomic projects and it seems set to remain with us for some time to come.

Acknowledgments

The proteomic research in MJD's laboratory is supported by the British Heart Foundation, the Arthritis Research Campaign, the Biotechnology and Biological Sciences Research Council, EUREKA (DTI), and Proteome Sciences plc.

References

Anderson, N.G. and Anderson, N.L. (1978) Analytical techniques for cell fractions. XXI. Two-dimensional analysis of serum and tissue proteins: multiple isoelectric focusing. *Anal. Biochem.* **85**: 331–340.

Anderson, N.G. and Anderson, N.L. (1978) Analytical techniques for cell fractions. XXII. Two-dimensional analysis of serum and tissue proteins: multiple gradient-slab gel electrophoresis. *Anal. Biochem.* **85**: 341–354.

Banks, R.E., Dunn, M.J., Forbes, M.A., Stanley, A., Pappin, D., Naven, T., Gough, M., Harnden, P. and Selby, P.J. (1999) The potential use of laser capture microdissection to selectively obtain distinct populations of cells for proteomic analysis – preliminary findings. *Electrophoresis* **20**: 689–700.

Bjellqvist, B., Ek, K., Righetti, P.G., Gianazza, E., Görg, A., Westermeier, R. and Postel, W. (1982) Isoelectric focusing in immobilized pH gradients: principle, methodology and some applications. *J. Biochem. Biophys. Methods* **6**: 317–339.

Bjellqvist, B., Sanchez, J.C., Pasquali, C., Ravier, F., Paquet, N., Frutiger, S., Hughes, G.J. and Hochstrasser, D.F. (1993) A nonlinear wide-range immobilised pH gradient for two-dimensional electrophoresis and its definition in a relevant pH scale. *Electrophoresis* **14**: 1357–1365.

Blomberg, A., Blomberg, L., Norbeck, J. *et al.* (1995) Interlaboratory reproducibility of yeast protein patterns analyzed by immobilized pH gradient two-dimensional gel electrophoresis. *Electrophoresis* **16**: 1935–1945.

Corbett, J.M., Dunn, M.J., Posch, A. and Görg, A. (1994) Positional reproducibility of protein spots in two-dimensional polyacrylamide gel electrophoresis using immobilised pH gradient isoelectric focusing in the first dimension: an interlaboratory comparison. *Electrophoresis* **15**: 1205–1211.

Cordwell, S.J., Nouwens, A.S., Verrils, N.M., Basseal, D.J. and Walsh, B.J. (2000) Sub-proteomics based upon protein cellular location and relative solubilities in conjunction with composite two-dimensional gels. *Electrophoresis* **21**: 1094–1103.

Corthals, G.L., Wasinger, V.C., Hochstrasser, D.F. and Sanchez, J.C. (2000) The dynamic range of protein expression: A challenge for proteomic research. *Electrophoresis* **21**: 1104–1115.

Dunn, M.J. (1987) Two-dimensional polyacrylamide gel electrophoresis. In: *Advances in Electrophoresis* (ed. Chrambach, A., Dunn, M.J. and Radola, B.J.), Vol. 1, VCH, Weinheim, pp. 1–109.

Dunn, M.J. (1993) *Gel Electrophoresis: Proteins.* BIOS Scientific Publishers, Oxford.

Garrels, J.I. (1979) Two dimensional gel electrophoresis and computer analysis of proteins synthesized by clonal cell lines. *J. Biol. Chem.* **254**: 7961–7977

Garvik, B.M. and Yates, J.R. 3rd. (1999) Direct analysis of protein complexes using mass spectrometry. *Nature Biotechnol.* **17**: 676–682.

Görg, A., Postel, W. and Günther, S. (1988) The current state of two-dimensional electrophoresis with immobilized pH gradients. *Electrophoresis* **9**: 531–546.

Görg, A., Postel, W., Friedrich, C., Kuick, R., Strahler, J.R. and Hanash, S.M. (1991) Temperature-dependent spot positional variability in two-dimensional polypetide patterns. *Electrophoresis* **12**: 653–658.

Görg, A., Boguth, G., Obermaier, C., Posch, A. and Weiss, W. (1995) Two-dimensional polyacrylamide gel electrophoresis with immobilized pH gradients in the first dimension (IPG-Dalt): the state of the art and the controversy of vertical versus horizontal systems. *Electrophoresis* **16**: 1079–1086.

Görg, A., Obermaier, C., Boguth, G., Csordas, A., Diaz, J.J. and Madjar, J.J. (1997) Very alkaline immobilized pH gradients for two-dimensional electrophoresis of ribosomal and nuclear proteins. *Electrophoresis* **18**: 328–337.

Görg, A., Boguth, G., Obermaier, C. and Weiss, W. (1998) Two-dimensional electrophoresis of proteins in an immobilized pH 4–12 gradient. *Electrophoresis* **19**: 1516–1519.

Görg, A., Obermaier, C., Boguth, G. and Weiss, W. (1999) Recent developments in two-dimensional gel electrophoresis with immobilized pH gradients: wide pH gradients up to pH 12, longer separation distances and simplified procedures. *Electrophoresis* **20**: 712–717.

Görg, A., Obermaier, C., Boguth, G., Harder, A., Scheibe, B., Wildgruber, R. and Weiss, W. (2000)

The current state of two-dimensional electrophoresis with immobilized pH gradients. *Electrophoresis* **21**: 1037–1053.

Gygi, S.P., Rist, B., Gerber, S.A., Turecek, F., Gelb, M.H. and Aebersold, R. (1999) Quantitative analysis of complex protein mixtures using isotope-coded affinity tags. *Nature Biotechnol.* **17**: 994–999.

Herbert, B., Molloy, M.P., Gooley, A.A., Walsh, B.J., Bryson, W.G. and Williams, K.L. (1998) Improved protein solubility in two-dimensional electrophoresis using tributyl phosphine as a reducing agent. *Electrophoresis* **19**: 845–851.

Herbert, B. (1999) Advances in protein solubilisation for two-dimensional electrophoresis. *Electrophoresis* **20**: 660–663.

Iborra, F., Buhler, J.M. (1976) Protein subunit mapping. A sensitive high-resolution method. *Anal. Biochem.* **74**: 503–511.

Klose, J. (1975) Protein mapping by combined isoelectric focusing and electrophoresis of mouse tissues. A novel approach to testing for induced point mutations in mammals. *Humangenetik* **26**: 231–243.

Klose, J. (1999) Genotypes and phenotypes. *Electrophoresis* **20**: 643–652.

Laemmli, U.K. (1970) Cleavage of structural proteins during the assembly of the head of bacteriophage T4. *Nature* **22**: 680–685.

Link, A.J., Robison, K. and Church, G.M. (1997) Comparing the predicted and observed properties of proteins encoded in the genome of *Escherichia coli* K-12. *Electrophoresis* **18**: 1259–1313.

Link, A.J., Eng, J., Schieltz, D.M., Carmack, E., Mize, G.J., Morris, D.R., Garvik, B.M. and Yates, J.R. 3rd. (1999) Direct analysis of protein complexes using mass spectrometry. *Nature Biotechnol.* **17**: 676–682.

Lollo, B.A., Harvey, S., Liao, J., Stevens, A.C, Wagenknecht, R., Sayen, R., Whaley, J. and Sajjadi, F.G. (1999) Improved two-dimensional gel electrophoresis representation of serum proteins by using ProtoClear. *Electrophoresis* **20**: 854–859.

Lopez, M.F. (2000) Better approaches to finding the needle in a haystack: Optimizing proteome analysis through automation. *Electrophoresis* **21**: 1082–1093.

Merchant, M. and Weinberger, S.R. (2000) Recent advances in surface enhanced laser-desorption/ionization time-of-flight mass spectrometry. *Electrophoresis* **21**: 1165–1177.

Molloy, M.P., Herbert, B., Walsh, B.J., Tyler, M.I., Traini, M., Sanchez, J.C., Hochstrasser, D.F., Williams, K.L. and Gooley, A.A. (1998) Extraction of membrane proteins by differential solubilization for separation using two-dimensional electrophoresis. *Electrophoresis* **19**: 837–844.

Monribot, C. and Boucherie, H. (2000) Two-dimensional electrophoresis with carrier ampholytes. In: *Proteome Research: Two-Dimensional Gel Electrophoresis and Identification Methods* (ed. Rabilloud, T.), Springer, Berlin, pp. 31–55.

Nelson, R.W., Nedelkov, D. and Tubbs, K.A. (2000) Biosensor chip mass spectrometry: a chip-based approach. *Electrophoresis* **21**: 1155–1163.

O'Farrell, P.H. (1975) High resolution two-dimensional electrophoresis of proteins. *J. Biol. Chem.* **250**: 4007–4021.

O'Farrell, P.Z., Goodman, H.M. and O'Farrell, P.H. (1977) High resolution two-dimensional electrophoresis of basic as well as acidic proteins. *Cell* **12**: 1133–1141.

Patton, W.F., Pluskal, M.G., Skea, W.M., Buecker, J.L., Lopez, M.F., Zimmermann, R., Belanger, L.M. and Hatch, P.D. (1990) Development of a dedicated two-dimensional gel electrophoresis system that provides optimal pattern reproducibility and polypeptide resolution. *BioTechniques* **8**: 518–527.

Patton, W.F. (2000) A thousand points of light: The application of fluorescence detection technologies to two-dimensional gel electrophoresis and proteomics. *Electrophoresis* **21**: 1123–1144.

Perdew, G.H., Schaup, H.W. and Selivonchick, D.P. (1983) The use of a zwitterionic detergent in two-dimensional gel electrophoresis of trout liver microsomes. *Anal. Biochem.* **135**: 453–455.

Rabilloud, T., Hubert, M. and Tarroux, P. (1986) Procedures for two-dimensional electrophoretic analysis of nuclear proteins. *J. Chromatogr.* **351**: 77–89.

Rabilloud, T., Valette, C. and Lawrence, J.J. (1994) Sample application by in-gel rehydration improves the resolution of two-dimensional electrophoresis with immobilized pH gradients in the first dimension. *Electrophoresis* **15**: 1552–1558.

Rabilloud, T., Adessi, C., Giraudel, A. and Lunardi, J. (1997) Improvement of the solubilization of proteins in two-dimensional electrophoresis with immobilized pH gradients. *Electrophoresis* **18**: 307–316.

Rigaut, G., Shevchenko, A., Rutz, B., Wilm, M., Mann, M. and Seraphin, B. (1999) A generic protein purification method for protein complex characterization and proteome exploration. *Nature Biotechnol.* **17**: 1030–1032.

Sanchez, J.C., Rouge, V., Pisteur, M., Ravier, F., Tonella, L., Moosmayer, M., Wilkins, M.R. and Hochstrasser, D.F. (1997) Improved and simplified in-gel sample application using reswelling of dry immobilized pH gradients. *Electrophoresis* **18**: 324–327.

Santoni, V., Molloy, M. and Rabilloud, T. (2000) Membrane proteins and proteomics: Un amour impossible? *Electrophoresis* **21**: 1054–1070.

Scheele, G.A. (1975) Two-dimensional gel analysis of soluble proteins. Characterization of guinea pig exocrine pancreatic proteins. *J. Biol. Chem.* **250**: 5375–5385.

Simone, N.L., Bonner, R.F., Gillespie, J.W., Emmert-Buck, M.R. and Liotta, L.A. (1998) Laser-capture microdissection: opening the microscopic frontier to molecular analysis. *Trends Genetics* **14**: 272–276.

Smithies, O. and Poulik, M.D. (1956) Two-dimensional electrophoresis of serum proteins. *Nature* **177**: 1033.

Uetz, P., Giot, L., Cagney, G. *et al.* (2000) A comprehensive analysis of protein–protein interactions in *Saccharomyces cerevisiae*. *Nature* **403**: 623–627.

Wildgruber, R., Harder, A., Obermaier, C., Boguth, G., Weiss, W., Fet. S.J., Larsen, P.M. and Görg, A. (2000) Towards higher resolution: 2D-Electrophoresis of *Saccharomyces cerevisiae* proteins using overlapping narrow IPGs. *Electrophoresis* **21**: (in press).

Chapter 4

Detecting proteins in polyacrylamide gels and on electroblot membranes

Wayne F. Patton

1. Introduction

The primary objective of this review is to describe modern total protein visualization technologies applicable to gel electrophoresis and electroblotting, with particular emphasis on their suitability for proteomics investigations. An overview of milestones in the development of these techniques is given in *Table 1*. Many excellent review articles and comparative studies discussing various methods of staining proteins in gels and on membranes have been published and should be consulted by interested readers (Brush, 1998; Dunn, 1999; Fernandez-Patron *et al.*, 1998; Fowler, 1994; Li *et al.*, 1989; Merril, 1987; Neuhoff *et al.*, 1985; Patton *et al.*, 1999a,b; Rabilloud, 1990; Rabilloud *et al.*, 1994; Wirth and Romano, 1995).

2. Organic dyes and silver stains

Coomassie blue and silver staining are the most widely utilized methods for routine detection and quantitation of proteins separated by polyacrylamide gel electrophoresis (reviewed in Brush, 1998; Merril, 1987; Wirth and Romano, 1995). Increases in the sensitivity of sequencing techniques and mass spectrometry have transformed the role of protein stains from simple visualization reagents to key components of integrated protein microchemical characterization procedures. While commonly used in proteome studies, the limitations of Coomassie blue and silver-staining methods are increasingly being encountered as high-sensitivity protein analytical methods are interfaced with postelectrophoretic identification technologies.

2.1 Organic dyes

Conventional Coomassie blue staining of polyacrylamide gels in an aqueous solution containing methanol and acetic acid has enjoyed widespread popularity since its introduction in the early 1960s (Fazekas de St. Groth *et al.*, 1963; Meyer and Lamberts, 1965). At least 600 variations of the Coomassie blue staining procedure

Proteomics, edited by S.R. Pennington and M.J. Dunn
© 2001 BIOS Scientific Publishers, Oxford.

Table 1. A brief 50 year history of stains used to visualize proteins after gel electrophoresis and electroblotting

Year	Investigator	Type of stain	Milestone
1950	Durrum	Organic dye	Bromophenol blue staining on filter paper
1952	Grassman and Hannig	Organic dye	Amido black staining on filter paper
1963	Fazakas de St. Groth *et al.*	Organic dye	Coomassie blue R-250 staining on cellulose acetate membranes, agar or starch gels
1965	Meyer and Lamberts	Organic dye	Coomassie blue R-250 staining in polyacrylamide gels
1967	Chrambach *et al.*	Organic dye	Background-free staining in polyacrylamide gels with Coomassie blue R-250 dye in trichloroacetic acid
1969	Hartman and Udenfriend	Organic fluorophore	Noncovalent staining of hydrophobic sites on proteins with anilinonaphthalene sulfonate
1971	Talbot and Yphantis	Organic fluorophore	Covalent derivatization of proteins with dansyl chloride prior to gel electrophoresis
1972	Diezel *et al.*	Organic dye	Improved, background-free staining of gels with Coomassie blue G-250 dye in trichloroacetic acid. Recognized that the dye is converted to a colloid
1972	Kerenyi and Gallyas	Silver	Potassium ferrocyanide-based histochemical silver staining of proteins in agarose gels
1974	Wallace *et al.*	Reverse stain	Chilling of SDS gels to produce a detergent precipitate and clear protein zones
1978	Graham *et al.*	Metal chelate	Ferrous-bathophenanthroline disulfonate staining in gels
1979	Switzer *et al.*	Silver	Silver staining in polyacrylamide gels using successive alkaline and acidic silvering steps
1980	Hager and Burgess	Reverse stain	Negative staining in gels with KCl
1980	Oakley *et al.*	Silver	Histochemically based silver staining in polyacrylamide gels (alkaline/silver diamine method)
1981	Merril *et al.*	Silver	Non-diamine, chemically based silver staining in polyacrylamide gels (acidic/silver nitrate method)

Year	Author	Type	Description
1982	Zapolski et al.	Metal chelate	Radioactive, ferrous-bathophenanthroline disulfonate staining in gels
1983	Hancock and Tsang	Colloidal dispersion	India ink stain for electroblotted proteins (precursor to colloidal gold)
1985	Neuhoff et al.	Organic dye	Demonstrated colloidal nature of colloidal Coomassie blue stain and generalized formulations to a variety of acid/alcohol mixtures
1985	Rohringer and Holden	Colloidal dispersion	Colloidal gold stain for electroblotted proteins
1985	Moeremans et al.	Colloidal dispersion	Sensitive colloidal gold, iron and silver stain for electroblotted proteins
1987	Lee et al.	Reverse stain	Negative staining in gels with copper chloride
1988	Dzandu et al.	Reverse stain	Zinc chloride negative stain in gels
1988	Jackson et al.	Organic fluorophore	Covalent derivatization of proteins in isoelectric focusing gels using MDPF, followed by second dimension separation
1991	Daban et al.	Organic fluorophore	Fluorescent staining of protein–SDS complexes in polyacrylamide gels with Nile red dye
1992	Ortiz et al.	Reverse stain	Sensitive, zinc chloride/imidazole negative stain
1994	Patton et al.	Metal chelate	Colorimetric, ferrozine-ferrous metal chelate stain (nitrocellulose, PVDF)
1996	Steinberg et al.	Organic fluorophore	Sensitive, fluorescent staining of protein–SDS complexes with SYPRO Red and Orange dye
1997	Lim et al.	Fluorescent metal chelate	Luminescent europium-bathophenanthroline disulfonate stain (nitrocellulose, PVDF, polyacrylamide gels). SYPRO Rose dye
1997	Unlu et al.	Organic fluorophore	Covalent derivatization of two protein samples using propyl-Cy3 and methyl-Cy5 dyes, followed by simultaneous visualization on a single 2-D gel
1999	Berggren et al.	Fluorescent metal chelate	Sensitive, luminescent ruthenium-based metal chelate stain (nitrocellulose, PVDF, polyacrylamide gels). SYPRO Ruby dye

have been examined with respect to linearity of staining response and sensitivity of detection in polyacrylamide gels (Neuhoff *et al.*, 1985). Coomassie blue dye is capable of detecting as little as 30–100 ng of protein, but this is considerably less sensitive than silver staining or fluorescence detection (Brush, 1998).

Perhaps the most significant improvement in the Coomassie blue staining procedure has been the evolution of colloidal Coomassie blue staining for background-free detection of proteins in polyacrylamide gels (Chrambach *et al.*, 1967; Diezel *et al.*, 1972; Neuhoff *et al.*, 1985, 1988). Related dyes such as acid violet 17, serva violet 49 and fast green FCF also form colloids in strongly acidic solutions and stain proteins in polyacrylamide gels with low background (Neuhoff *et al.*, 1990; Patestos *et al.*, 1988). It is thought that during staining an equilibrium is established between the colloidal dye and freely dispersed dye in solution. The small amounts of free dye in solution penetrate the gel matrix and preferentially stain the proteins while the colloidal dye is excluded, thus preventing background staining (Brush, 1998; Neuhoff *et al.*, 1988). The limit of detection for colloidal Coomassie blue stain is approximately 8–10 ng of protein (Brush, 1998). Staining is usually performed in concentrated trichloroacetic acid, perchloric acid or phosphoric acid, often in combination with methanol or ethanol. Studies of protein modifications by mass spectrometry indicate that staining proteins in Coomassie blue dye solutions containing trichloroacetic acid and alcohol leads to irreversible, acid-catalyzed esterification of glutamic acid side-chain carboxyl groups (Haebel *et al.*, 1998). This can complicate interpretation of peptide mapping data from mass spectrometry, but algorithms can be added to analysis software to account for this modification.

Amido black was one of the earliest organic dyes used to visualize proteins after electrophoresis (Grassman and Hannig, 1952). The dye is now largely relegated to medium sensitivity, colorimetric detection of electroblotted proteins on polyvinylidene difluoride (PVDF) and nitrocellulose membranes (Dunn, 1999). While either amido black or Coomassie blue dye may be used to visualize electroblotted proteins, Coomassie blue dye typically stains nitrocellulose membranes with a very high background. However, a modified procedure for low background staining of proteins on nitrocellulose membranes may allow effective utilization of Coomassie blue dye for this application (Metkar *et al.*, 1995). Both amido black and Coomassie blue dyes are suitable for detection of proteins on PVDF membranes. While compatible with downstream microchemical characterization, amido black and Coomassie blue staining of electroblotted proteins on membrane supports often interferes with subsequent immunodetection and photographic documentation procedures (Eynard and Lauriere, 1998; Pryor *et al.*, 1992). One solution to this problem is to destain the proteins prior to immunodetection (Pryor *et al.*, 1992). A method has been described for simultaneous visualization of specific proteins using alkaline phosphatase-conjugated second antibodies and colorimetric substrates such as fast red/naphthol AS MX phosphate or 5-bromo-4-chloro-indolyl phosphate/ nitroblue tetrazolium, followed by total protein detection with Coomassie blue dye (Zeindl-Eberhart *et al.*, 1997). Using the procedure, large amounts of protein are typically applied to a two-dimensional gel (400 μg), blotting must be performed on PVDF membranes and the target protein visualized should be fairly abundant. At present, this-two color visualization procedure must be tested on a case-by-case basis to determine whether it is suitable for detecting a protein of interest.

2.2 Silver stains

The origins of silver staining can be traced back to photography and histology techniques developed in the 1800s (see Merril, 1987; Merril *et al.*, 1984; Rabilloud, 1990). Kerenyi and Gallyas (1972) first devised a general silver stain for the detection of proteins separated by electrophoresis, immunoelectrophoresis and immunodiffusion; however, the stain was optimized for agarose and not polyacrylamide gels. Detection sensitivity was reported to be 10–20 times higher than amido black stain. The method was later improved and shown to stain DNA as well as protein with high sensitivity in agarose gels (Peats, 1984). Silver staining of proteins in polyacrylamide gels was introduced 7 years after the first agarose gel method (Switzer *et al.*, 1979). The newer method was reported to be at least 100 times more sensitive than Coomassie blue staining and is generally credited with launching silver staining as a revolutionary protein detection technique. The linear dynamic range of typical silver-staining procedures extends over a 40-fold range of concentration and is thus comparable to Coomassie blue dye, which has a 20-fold range of linearity (Dunn, 1987). With the introduction of silver staining, proteins could be detected in the nanogram range instead of the microgram range (but see earlier figures for sensitivity of colloidal Coomassie blue).

More than 100 different silver-staining procedures have been published (see Rabilloud, 1992). The enormous number of these procedures devised attests to the fact that no single method combines maximal convenience, speed and sensitivity (Rabilloud, 1992, 1999). A drawback to many silver-staining methods is that certain classes of proteins, such as calcium-binding proteins and glycoproteins, stain rather poorly (Jay *et al.*, 1990; Schleicher and Watterson, 1983; Wirth and Romano, 1995). With respect to glycoproteins, an inverse relationship between the intensity of silver staining and the proportion of the molecular weight that is composed of carbohydrate has been noted (Jay *et al.*, 1990). Prestaining proteins with cationic dyes prior to silver staining substantially improves detection of glycoproteins (Dubray and Bezard, 1982; Jay *et al.*, 1990).

Silver stains allow detection of low nanogram amounts of protein, but they also interact with cysteine residues and utilize glutaraldehyde and formaldehyde, which alkylate α- and ε-amino groups of proteins. This prevents further analysis of proteins from gels by Edman-based protein sequencing, although mass spectroscopic analysis has been successfully applied for protein identification by removing glutaraldehyde from the stain (Christiansen and Houen, 1992; Shevchenko *et al.*, 1996; Wilm *et al.*, 1996). This modification of the procedure is accompanied by decreases in stain sensitivity and uniformity as well as increases in gel background. Using the modified silver staining procedure, the advantage of higher detection sensitivity compared with Coomassie blue or zinc-imidazole staining appears to be nullified by inferior sequence coverage in peptide mass fingerprinting experiments (Scheler *et al.*, 1998). Low sequence coverage of silver-stained proteins is thought to arise from their modification by formaldehyde during the staining process (Scheler *et al.*, 1998). Unfortunately, formaldehyde cannot be omitted from present silver-staining protocols.

Two general forms of silver staining, alkaline/silver diamine and acidic/silver nitrate, are most commonly used (Rabilloud, 1999). The acidic/silver nitrate methods are based on photographic procedures and rely upon gel impregnation with silver ions at acidic pH, followed by reduction to elemental metallic silver at alkaline pH using formaldehyde (Wirth and Romano, 1995). The acidic/silver nitrate procedures are reputed to stain basic proteins with slightly lower sensitivity and acidic proteins with higher sensitivity than alkaline/silver diamine methods (Rabilloud, 1999; Rabilloud *et al.*, 1994). The acidic/silver nitrate methods are also somewhat more susceptible to metachromatic staining artifacts, but this can be greatly minimized when the mechanically durable polyacrylamide gel matrix Duracryl (Genomic Solutions, Ann Arbor, MI) is used in place of standard polyacrylamide (Patton *et al.*, 1992). The acidic/silver nitrate methods are suitable for protein quantitation, provided image development is limited to about 10 min (Patton, 1995).

The alkaline/silver diamine methods are based on histological procedures and use ammonium hydroxide to form soluble silver–diamine complexes (Wirth and Romano, 1995). Proteins are subsequently visualized by reduction of free silver ions with formaldehyde in an acidified developer. Optimal results are achieved with the alkaline/silver diamine methods when gels contain the crosslinker piperazine diacrylamide instead of N, N' methylenebisacrylamide (Hochstrasser *et al.*, 1988; Rabilloud, 1999). Gels utilizing the Tris tricine or bicine buffer chemistry, such as certain commercially available, long-shelf-life, precast gels and gels routinely used for the resolution of low-molecular-weight proteins, are poorly stained using the alkaline/silver diamine methods. Proteins separated using immobilized pH gradient gels are also poorly stained using alkaline/silver diamine methods (Rabilloud *et al.*, 1994). In addition, polyacrylamide gels with plastic backings often produce a silver mirroring artifact that greatly compromises visualization of proteins (Rabilloud, 1999). Due to the volatile nature of ammonia, the alkaline/silver diamine methods are susceptible to run-to-run variability (Rabilloud, 1999) and are less suited for quantitative analysis due to their nonstoichiometric staining properties (Rodriguez *et al.*, 1993).

3. Reverse stains

One disadvantage of standard Coomassie blue dye procedures is the incorporation of a protein fixation step during staining. Silver stains also employ a fixation step and, in addition, a sensitization step. These steps can reduce total protein yield after elution from gels, leaving lower amounts of material for subsequent microchemical characterization. Reverse stains were developed specifically to improve protein recovery from polyacrylamide gels. They cannot be utilized for the staining of transfer membranes and have not routinely been employed for rigorous quantification of proteins. The stains produce a semi-opaque background on the gel surface, while proteins are detected as transparent zones when gels are viewed on a black background or with proper back lighting (Wirth and Romano, 1995). Staining is quite rapid, usually requiring 5–15 min to complete, and the biological activity of proteins is often preserved. Protein mobilization is achieved by elution after chelation of the metal ions with agents such as ethylenediaminetetraacetic acid (EDTA) or Tris/glycine transfer buffer (Castellanos-Serra et al., 1996; Lee et al., 1987). Reverse stains are suitable for visualization of proteins, passive elution of intact proteins from gels and their analysis by mass spectrometry (Cohen and Chait, 1997).

3.1 Metal salt reverse stains

A number of reverse-stain methods for visualization of proteins in SDS–polyacryl-amide gels have been described over the years, including methods using low temperature (Wallace et al., 1974), cationic detergent (Takagi et al., 1977), sodium acetate (Higgins and Dahmus, 1979), potassium chloride (Hager and Burgess, 1980), silver (Merril et al., 1984), colloidal gold (Casero et al., 1985), copper chloride (Lee et al., 1987), zinc chloride (Dzandu et al., 1988), cobalt acetate (Dzandu et al., 1988), nickel chloride (Dzandu et al., 1988), zinc-imidazole (Ortiz et al., 1992) and methyl trichloroacetate (Candiano et al., 1996). In a patented approach that is analogous to the zinc-imidazole procedure, a planar dye such as rhodamine, acridine orange, eosin or fluorescein is complexed with a metal salt such as potassium dichromate, barium chromate, magnesium chloride or calcium chloride, producing a highly sensitive negative stain in which the gel is markedly stained a visible color and the separated proteins remain transparent (Berube, 1990).

The three reverse-stain methods that have been used most extensively are the potassium chloride, copper chloride and zinc chloride procedures. Copper chloride staining is slightly more sensitive, while potassium chloride staining is significantly less sensitive than that of Coomassie blue (Adams and Weaver, 1990). Copper chloride staining is reported to visualize cytochrome c, histones and other basic proteins with lower sensitivity than more acidic proteins (Vanflerteren and Peeters, 1990). Zinc chloride staining is the most sensitive reverse stain developed to date (Fernandez-Patron et al., 1998).

3.2 Zinc-imidazole stain

Nucleic acids, glycolipids and proteins display similar staining properties with some formulations of zinc-imidazole stain (Fernandez-Patron et al., 1998). In the presence of imidazole, free or weakly bound zinc ions are readily precipitated as zinc-imidazole, while tightly bound ions associated with the biopolymers are refractory to precipitation. This leads to clear zones on a semi-opaque background that typify reverse-stain procedures. Attention to the duration of staining is necessary as over-staining of gels can obscure bands and on occasions complex patterns of negative and positive staining are obtained using zinc-imidazole (Fernandez-Patron et al., 1998). The limit of detection for the zinc-imidazole stain is 10–20 ng per band for protein separated by sodium dodecyl sulfate (SDS)–polyacrylamide gel electrophoresis, 1–10 ng per band for lipopolysaccharides separated by SDS–polyacrylamide gel electrophoresis and 10–100 ng per band for small-sized oligonucleotides (28 bp to 1.3 kbp) separated on urea-denaturing polyacrylamide gels (Fernandez-Patron *et al.*, 1998). The limit of detection for bovine serum albumin separated on native polyacrylamide gels is somewhat poorer at 40–80 ng per band.

Zinc-imidazole stained gels can be dried with no alteration of the staining pattern (Ferreras *et al.*, 1993). Proteins can be evaluated by densitometry, although the linear dynamic range for protein detection is restricted to 10–100 ng (Ferreras *et al.*, 1993; Matsui *et al.*, 1997). Zinc-imidazole stained gels can also be electroblotted by addition of a chelating agent to the transfer buffer (Fernandez-Patron *et al.*, 1995). The electroblotted proteins may be further characterized by Edman-based microsequencing (Fernandez-Patron *et al.*, 1995). Proteins stained with zinc-imidazole are also compatible with in-gel tryptic digestion followed by matrix-assisted laser desorption ionization time-of-flight (MALDI-TOF) mass spectrometry for determination of peptide masses and sequences (Matsui *et al.*, 1997).

4. Colloidal dispersion stains

Colloidal dispersion stains contain fine particles with high affinity for protein. The stains were developed to address the need for sensitive detection of proteins electroblotted, dot-blotted or slot-blotted to nitrocellulose or PVDF membranes. The stains are not generally used for detection of proteins in polyacrylamide gels, although a colloidal gold reverse-stain procedure appears in the literature (Casero *et al.*, 1985). In addition, colloidal palladium or platinum particles with an organic shell affixed to their surfaces have been added to protein mixtures, subjected to polyacrylamide gel electrophoresis and visualized with a silver enhancement step (Powell, 1998). On transfer membranes, the best colloidal dispersion stain, colloidal gold stain, is nearly as sensitive as silver staining in polyacrylamide gels.

4.1 India ink stain

Diluted India ink from fountain pens was the first successful colloidal dispersion stain developed for visualization of electroblotted proteins (Hancock and Tsang, 1983). The principal advantage of the stain is that it is simple to perform and inexpensive. Fountain pen inks are proprietary formulations containing metalized, acidic dyes (sulfonates, chlorosulfonates or sulfonamides) as well as basic dye salts (Rohde *et al.*, 1998). In addition, fine pigment dispersions of carbon along with copper-phthalocyanide, glycol, glycerol, surfactants, antioxidants and viscosity adjusters are added to formulations of water-soluble inks (Rohde *et al.*, 1998). Analysis of the black fountain pen ink commonly used to stain electroblots (Pelikan, Hannover, Germany) using capillary electrophoresis with ultraviolet/visible absorbance and laser-induced fluorescence reveals at least seven distinct dye species in the formulation (Rohde *et al.*, 1998). Although Hancock and Tsang claimed that Pelikan fountain pen ink is the most sensitive ink for protein staining, others have obtained better results with Higgins black ink no. 893 (Faber-Castell Corporation, Newark, NJ) or Speedball black ink no. 3211 (Hunt Manufacturing Company, Statesville, NC; Hughes *et al.*, 1988; Lek *et al.*, 1995).

Direct, colorimetric or autoradiographic immunostaining of proteins has been achieved prior to or after India ink staining (Glenney, 1986; Mobbs *et al.*, 1989; Ono and Tuan, 1990). If India ink staining is performed prior to immunodetection, staining must be limited to 2–3 h to avoid interference in subsequent binding steps with an antibody. Better results have been achieved combining chemiluminescence immunodetection with India ink staining (Eynard and Lauriere, 1998). While Mobbs utilized India ink in combination with immunodetection, membranes utilized for Edman-based sequencing were invariably stained with Coomassie blue dye (Mobbs *et al.*, 1989).

India ink displays relatively high protein-to-protein staining variability compared with the organic dyes and the colloidal metal stains (Li *et al.*, 1989). For example, a number of proteins, including β-lactoglobulin, trypsin inhibitor, bacitracin, C-hordeins, thyrotropin-releasing hormone, ovalbumin, ovomucoid, carbonic anhydrase, ferritin and insulin, stain poorly with India ink (Dunn, 1999; Eynard and Lauriere, 1998; Li *et al.*, 1989; Rohringer and Holden, 1985). The need to include Tween-20 to reduce background when staining membranes with India ink is also a significant disadvantage that is shared with the other colloidal dispersion stains. The detergent partially removes proteins from transfer membrane supports (Li *et al.*, 1988). This and other problems associated with incorporation of Tween-20 in stains are discussed in greater detail in Section 4.2. Finally, fountain pen inks are designed for writing devices and do not undergo any quality control analyses with respect to protein staining capability. Staining with India ink requires in-house optimization with respect to the type of membrane, membrane lot, brand of India ink and ink lot (Hughes *et al.*, 1988).

4.2 Colloidal metal stains

Shortly after the development of India ink as a stain for electroblotted proteins, other particle dispersions were perfected for the same application (Moeremans *et al.*, 1985, 1986; Rohringer and Holden, 1985). Sensitive colloidal gold (AuroDye), iron (FerriDye), and silver stains were developed at Janssen Pharmaceuticals (Beerse, Belgium). FerriDye, which contained toxic cacodylic acid and generated cyanide gas, was later discontinued. Colloidal silver stain was never commercialized by the company. A direct comparison of the common colloidal dispersion stains available in the 1980s indicated that the detection sensitivity of colloidal gold stain (Aurodye) is superior to India ink or colloidal iron stain (FerriDye), but the useful dynamic range of colloidal gold stain is very limited (1–20 ng; Hunter and Hunter, 1987; Li *et al.*, 1989). Log transformation of protein concentrations extends the useful dynamic range of colloidal gold stain to about 100 ng (Li *et al.*, 1989).

Colloidal gold staining can be utilized with conventional, colorimetric immunoblotting procedures, as well as with chemiluminescent immunodetection (Chevallet *et al.*, 1997; Egger and Bienz, 1987; Schapira and Keir, 1988). However, the colloidal gold stain pattern tends to fade somewhat during chemiluminescent procedures owing to oxidation of the metal. Also, since all proteins in the pattern become weakly chemiluminescent, distinguishing between weak reactivity and nonspecific background may be problematic. While removal of colloidal gold from electro-blotted proteins is not feasible, the stain can be decolorized by oxidation with potassium dichromate in aqueous, 10% HCl solution (Nelson, 1993). The stain's intensity can also be enhanced using a silver intensification step (Moeremans et al., 1984).

Although colloidal gold staining offers exceptional sensitivity, as with other colloidal dispersion stains, it requires pretreatment of the membrane with a blocking agent such as Tween 20, followed by a 2–18 h incubation in a detergent-stabilized colloid solution (Moeremans et al., 1985). As alluded to earlier, nonionic surfactants partially remove proteins from transfer membranes, resulting in lower overall yields of material for later characterization (Flanagan and Yost, 1984; Li et al., 1988; Miranda et al., 1993). Fixation with agents such as potassium hydroxide and glutaraldehyde or baking of membranes at 100°C reduce protein losses due to detergent extraction, but often render the proteins unsuitable for other applications (Li et al., 1988). Detergents can also compromise the collection of high-quality spectra by MALDI-TOF and electrospray ionization (ESI) mass spectrometry, as background ion signals from detergent clusters can dominate mass spectra and detergents can suppress signals from protein (Gharahdaghi et al., 1996; Loo et al., 1994). For instance, Tween-20 is known to generate polymer background (mass-to-charge ratio of 500–1200), severe signal suppression and form adducts with proteins in ESI-MS (Loo et al., 1994). No reports demonstrating the successful use of colloidal gold for such mass spectrometry applications appear in the literature. In addition, colloidal gold stain consistently interferes with protein sequence analysis by Edman degradation, as demonstrated by poor initial and repetitive sequencing yields (Christiansen and Houen, 1992). Decolorizing the gold stain does not improve sequencing yields (Nelson, 1993).

The incompatibility of colloidal gold stain with downstream protein microchemical methods has spurred interest in the development of alternative colloidal stains. A 'colloidal' silver stain that does not contain detergents may be substituted for colloidal gold stain in the detection of electroblotted proteins that are to be analyzed by liquid chromatography–electrospray ionization tandem mass spectrometry (van Oostveen et al., 1997). The stain is not actually colloidal, but is a slurry of particles. Our own experiences with the silver slurry is that it offers very good sensitivity for carbonic anhydrase and bovine serum albumin, the test proteins described in the original publication, but stains other proteins present in commercially available molecular weight standards rather poorly. Another suspension for staining proteins on electroblots utilizes an alkaline copper iodide solution (Root and Reisler, 1989; Root and Wang, 1993). As with colloidal gold staining, the copper stain may be intensified using a silver enhancement step (Root and Wang, 1993). The stain is easily removed using EDTA and the destained blot can subsequently be used for colorimetric immunodetection (Root and Reisler, 1989). However, compatibility with other modern microchemical methods has not yet been established.

5. Organic fluorophore stains

Fluorescent detection of proteins after electrophoresis is gaining popularity, particularly with laboratories engaging in large-scale proteome research. Several of the newer fluorescent stains are suitable for incorporation into integrated proteomics platforms for global analysis of protein expression. The linear dynamic range of detection using fluorescent stains is usually superior to colorimetric stains.

5.1 Covalently bound fluorophores

A variety of methods for fluorescent detection of proteins in gels and on transfer membranes have been devised utilizing fluorescamine, fluorescein isothiocyanate, monobromobimane, 2-methoxy-2,4-diphenyl-3(2H)furanone (MDPF), dichloro-triazynlamino-fluorescein or dansyl chloride (Houston and Peddie, 1989; Jackson et al., 1988; Ragland et al., 1974; Szewczyk and Summers, 1987, 1992; Talbot and Yphantis, 1971; Urwin and Jackson, 1993; Vandekerckhove et al., 1985; Vera and Rivas, 1988). More recently, propyl-Cy3 and methyl-Cy5 dyes have been utilized to fluorescently tag two different protein samples and simultaneously visualize differences between the samples on the same 2-D gel (Unlu et al., 1997).

Drawbacks to the cited methods include the need to covalently modify proteins, altered mobilities of labeled proteins and rapid decay of signal due to photobleaching of fluorescent probes. Sensitivity varies considerably from protein to protein depending upon the number of functional groups accessible to modification by the fluorophore. Addition of the fluorophore may also decrease protein solubility (Unlu et al., 1997). This can be compensated for by labeling a small percentage of the functional groups in a sample, but slight molecular weight differences between labeled and unlabeled protein could lead to errors in post-separation protein identification by Edman degradation or mass spectroscopy. When performing 2-DE or isoelectric focusing (IEF), prederivatization with fluorescent molecules may lead to aberrant protein migration (Urwin and Jackson, 1993). In general, the total number of amino acid residues derivatized in the sample is usually quite small and thus of little consequence with respect to amino acid analysis, Edman-based sequencing or peptide-mapping studies (Alba and Daban, 1998; Stephens, 1975; Unlu et al., 1997).

A protein detection method for transfer membranes has been developed that uses a nonenzymatic, chemiluminescent reaction comprising bis (2,4,6-trichlorophenyl) oxylate and hydrogen peroxide for visualization of electroblotted or dot-blotted proteins covalently labeled with MDPF (Alba and Daban, 1997). Direct fluorescence detection of MDPF-labeled proteins on PVDF membranes using a 300 nm UV transilluminator was subsequently found to be simpler, less time-consuming and more sensitive than the original chemiluminescent method (Alba and Daban, 1998). The direct fluorescence method allows detection of 5–10 ng of protein and is compatible with chemiluminescent immunodetection. However, proteins labeled with MPDF must be viewed using wet PVDF membranes, as the fluorescence signal decreases 500-fold upon drying. Currently, there is no information regarding the suitability of the MDPF-labeling procedure for visualization of proteins electro-blotted to nitrocellulose membranes. If proteins are to be used for N-terminal sequencing, staining must be performed with a suboptimal dye concentration for a short period of time so that all primary amines are not consumed in the reaction. Under these conditions the fluorescent stain is no more sensitive than standard Coomassie blue staining procedures (Alba and Daban, 1998).

5.2 Noncovalently bound fluorophores

A number of fluorophores such as 1-anilino-8-naphthalene sulfonate (ANS), bis (8-toluidino-1-naphthalenesulfonate) (bis-ANS), 9-diethylamino-5H-benzo[α]phenox-azine-5-one (Nile red), SYPRO Orange and SYPRO Red dyes interact with proteins noncovalently (Aragay et al., 1985; Bermudez et al., 1994; Daban and Aragay, 1984; Daban et al., 1991a,b; Hartman and Undenfriend, 1969; Horowitz and Bowman, 1987; Pina et al., 1985; Steinberg et al., 1996a,b, 1997). These fluorophores are virtually nonfluorescent in aqueous solution, but become fluorescent in nonpolar solvents or upon association with SDS–protein complexes. SDS binds to protein with a fairly constant stoichiometry of 1.4 to 1, suggesting that protein quantitation based upon an interaction with the detergent should be more reliable than one based upon interactions with primary amines.

ANS can be excited with a 300 nm transilluminator and emits at about 480 nm. The dye has been used to detect proteins in SDS–polyacrylamide gels, but sensitivity is typically below that of Coomassie blue staining (Aragay et al., 1985; Daban and Aragay, 1984; Pina et al., 1985). The related dye, bis-ANS, in combination with 2 M KCl provides slightly better sensitivity, with a limit of detection in the vicinity of 100 ng of protein (Horowitz and Bowman, 1987). Nile red dye can be excited at 300 or 540 nm and emits at approximately 640 nm (Daban et al., 1991a), and the detection sensitivity of 20–100 ng for Nile red staining is slightly better than that obtained with Coomassie blue. Furthermore, the dye is compatible with standard Edman sequen-cing and immunodetection procedures (Alba and Daban, 1998; Bermudez et al., 1994; Daban et al., 1991b).

SYPRO Red and SYPRO Orange dyes can detect proteins in SDS–polyacryla-mide gels using a simple, one-step staining procedure that requires 30–60 min to complete. As little as 2–10 ng of protein can be detected, rivaling the sensitivity of standard silver-staining techniques. Documentation of stained gels can readily be achieved by photography on a standard laboratory 300 nm UV transilluminator. Alternatively, the dyes can be quantified with commercially available CCD camera-based image analysis workstations or laser scanners, providing a linear dynamic range of three orders of magnitude. SYPRO Red, Orange and Ruby dyes are the only commercially available electrophoresis stains to match the dynamic range of gene expression levels measured in yeast as determined by serial analysis of gene expression (SAGE) (0.3–200 transcript copies per cell; Velculescu et al., 1997). The SYPRO dyes are quite resistant to photobleaching. After staining in Tris/glycine electroblot buffer, proteins can also be transferred to membranes and specific proteins identified by immunostaining or Edman-based protein sequencing (Hamby, 1996, 1997; Steinberg *et al.*, 1996b). Finally, SYPRO Orange and Red dyes may be utilized for the detection of picomolar quantities of protein using SDS–capillary gel electrophoresis (Harvey *et al.*, 1998). Compared with the SYPRO dyes, the detection sensitivity with Nile red is poorer, staining is not stable, and the dye is prone to precipitation in aqueous solution, consequently making it more difficult to handle. The photolabile nature of Nile red staining makes quantitation of proteins by image analysis difficult.

6. Metal chelate stains

The metal chelate stains are a relatively new family of protein visualization reagents developed to be compatible with modern proteomics research. They have been designed specifically for compatibility with commonly used microchemical characterization procedures. The stains do not contain extraneous chemicals such as glutaraldehyde, formaldehyde or Tween-20, which are known to compromise these procedures. Luminescent metal chelate stains are easily incorporated into integrated proteomics platforms that utilize automated gel stainers, image analysis workstations, robotic spot excision instruments, protein digestion work stations and mass spectrometers. The most sensitive luminescent stain, SYPRO Ruby dye, is more sensitive and exhibits a broader linear dynamic range than the best alkaline silver diamine and acidic/silver nitrate staining procedures.

6.1 Colorimetric metal chelate stains

The earliest report employing a metal chelate for protein detection after gel electrophoresis utilized the pink bathophenanthroline disulfonate/ferrous complex (Graham *et al.*, 1978). The limit of detection of the method was reported as roughly 600 ng. With Coomassie blue staining already routinely practised, the metal chelate stain never achieved widespread acceptance, as it offered no clear advantage over established methods. The introduction of radioactive ^{59}Fe into the bathophenanthroline disulfonate complex allowed detection of 10–25 ng of protein, but by today's standards this is only as sensitive as colloidal Coomassie blue staining (Gersten *et al.*, 1984; Zapolski *et al.*, 1982). Hazards associated with working with radioactivity and the increased burden of license application for an infrequently utilized radioisotope precluded widespread use of this method. The metal chelate, copper phthalocyanine tetrasulfonate, stains polyacrylamide gels with a sensitivity that is comparable to standard Coomassie blue staining (Bickar and Reid, 1992). The dye is also useful for staining proteins immobilized on nitrocellulose transfer membranes, with a detection limit of 10–20 ng of protein.

Although relatively insensitive, amido black, Ponceau S, and Coomassie blue are the stains most commonly used for detection of electroblotted proteins since they are also compatible with downstream microchemical characterization techniques. Our group addressed this sensitivity problem by devising a number of metal chelate stains for detection of electroblotted proteins (Lim *et al.*, 1996; Patton *et al.*, 1994, 1999a,b;

Shojaee *et al.*, 1996). Noting the success of Chung in detecting iron-binding proteins in polyacrylamide gels with ferrene S, we reasoned that if exogenously added ferrous ions were attached to the proteins followed by exposure to an iron indicator such as ferrene S or ferrozine, total protein profiles should be visualized on electroblots (Chung, 1985; Patton *et al.*, 1994). As we optimized the procedure, it was simplified by mixing the ferrous ion with the chelator prior to incubation with the electro-blotted proteins, but at the time of publication we were still convinced that binding to the proteins was primarily mediated by the metal ion itself (*figure 1* in Patton *et al.*, 1994). Only after screening large numbers of metal chelates in an effort to develop alternative metal chelate stains did we realize that binding to proteins is mediated by the sulfonate groups on the metal chelate (Shojaee *et al.*, 1996). Thus, the metal chelate stains rightly belong in the same family as the bathophenanthroline disulfonate/ferrous stain developed in 1978.

Ferrozine/ferrous and pyrogallol red/molybdate stains are about as sensitive as amido black stain, generally detecting about 20–50 ng of protein. The advantages of the ferrozine/ferrous stain is that it is very easily destained by increasing solution pH and easily intensified by subsequent incubation with ferrocyanide/ferric solution (Lim *et al.*, 1996; Patton *et al.*, 1994). After intensification, the limit of detection of the stain is about 5–10 ng of protein (Patton *et al.*, 1994). The metal chelate stains are compatible with Edman-based sequencing, mass spectrometry and immunoblotting (Patton *et al.*, 1994).

6.2 Luminescent metal chelate stains

Dissatisfaction with the detection sensitivity obtained using colorimetric dyes prompted us to investigate luminescent metal chelate stains (Lim *et al.*, 1997; Patton *et al.*, 1999a,b). The measurement of light emission is intrinsically more sensitive than measurement of light absorbance, as absorption is limited by the molar extinction coefficient of the colored complex (Gossling, 1990). Thus, luminescent protein stains utilizing metal chelates complexed to certain transition metal ions, such as europium, terbium, ruthenium and rhenium, offer greater sensitivity compared to their colorimetric predecessors.

Bathophenanthroline disulfonate/europium (SYPRO Rose Protein Blot stain, Molecular Probes, Eugene, OR) is a medium-sensitivity stain that routinely detects 2–4 ng/mm^2 of applied protein as determined by slot blotting, which translates to about 15–30 ng of gel applied protein in routine electroblotting applications. The stain can be visualized using 300 nm UV epi-illumination and emits at 590 and 615 nm. The stain is readily removed from proteins by incubating at neutral to mildly alkaline pH. Compatibility with immunoblotting, lectin blotting, Edman sequencing and mass spectrometry make SYPRO Rose stain an excellent routine protein detection reagent that is nearly as simple to use as ponceau S stain (Lim *et al.*, 1997). Limitations of the dye are that it stains nucleic acids as well as proteins, staining gels requires a destain step much like conventional Coomassie blue staining, and the dye cannot be utilized with many laser-based gel scanners since it is not excited by visible light.

SYPRO Ruby dye is a proprietary ruthenium-based metal chelate stain developed to address the limitations of SYPRO Rose dye. SYPRO Ruby Protein Blot stain (Molecular Probes, Eugene, OR) permanently stains electroblotted proteins on nitrocellulose and PVDF membranes with a detection sensitivity of 0.25–1 ng of protein/mm^2 in slot-blotting applications. Approximately 2–8 ng of protein can routinely be detected by electroblotting, which side-by-side comparisons demonstrate is as sensitive as colloidal gold stain (AuroDye, Amersham Pharmacia Biotech, Arlington Heights, IL). While colloidal gold staining requires 2–4 h, SYPRO Ruby dye staining is complete in 15 min. The linear dynamic range of SYPRO Ruby Protein Blot stain is vastly superior to colloidal gold stain, extending over a 1000-fold range. The dye can be excited using a standard 300 nm UV transilluminator or using imaging systems equipped with 450, 473, 488 or even 532 nm lasers. Unlike colloidal gold stain, SYPRO Ruby stain does not interfere with mass spectrometry or immunodetection procedures, and unlike SYPRO Rose dye, it does not stain nucleic acids.

SYPRO Ruby 2-DE and IEF stains allow one-step, low background staining of proteins in polyacrylamide gels without resorting to lengthy destaining steps. The linear dynamic range of these dyes extends over three orders of magnitude, thus surpassing silver and Coomassie blue stains in performance. An evaluation of 11 protein standards ranging in isoelectric point from 3.5 to 9.3 indicates that SYPRO Ruby IEF gel stain is 3–30 times more sensitive than highly sensitive silver stains (Steinberg et al., 2000). Proteins that stain poorly with silver stain techniques are often readily detected by SYPRO Ruby dye (*Figure 1*). Although more sensitive than SYPRO Orange and Red dyes, optimal staining is somewhat slower, requiring about 4 h. Similar to colloidal Coomassie blue stain, but unlike silver stain, SYPRO Ruby dye stains are end-point stains. Thus, staining times are not critical and staining can be performed overnight without gels overdeveloping.

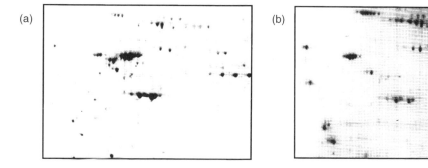

Figure 1. Region of 2-DE gel stained with (a) SYPRO Ruby protein gel stain (Molecular Probes, Eugene, OR), or (b) acidic/silver nitrate staining kit (Genomic Solutions, Ann Arbor, MI). Approximately 5 μg of rat fibroblast lysate protein was loaded on the first-dimension isoelectric focusing gel. The gray scale values of the SYPRO Ruby dye stained gel have been inverted for easier comparison with the silver-stained gel.

Table 2. Properties of an ideal general protein stain; the seven Ss of superior staining

---

- *Safety*: a stain should not contain hazardous materials such as radioisotopes or highly toxic chemicals such as cyanide or cacodylic acid. This ensures safety of personnel and eliminates time-consuming monitoring, clean-up and disposal issues
- *Sensitivity*: a stain should detect sub-nanogram amounts of protein. The linear dynamic range of the stain should be broad, minimally extending over three orders of magnitude
- *Simplicity*: staining should require simple incubation of the gel or blot in a single solution. Development time should not be critical as the stain should reach an end-point and not overdevelop
- *Specificity*: a stain should specifically detect proteins, not lipids, saccharides or nucleic acids. The stain should uniformly detect different classes of proteins such as calcium-binding proteins, lipoproteins, glycoproteins and phosphoproteins
- *Speed*: a stain should rapidly detect proteins and no destaining step should be required. Longer incubations are often convenient, such as overnight staining, as long as there is no requirement to monitor development or change solutions
- *Stability*: stain solutions should be chemically stable, have a long shelf-life, and not require fresh preparation prior to use. The stain solution should be stable upon storage at room temperature. Stained gels should be stable for days to months without deterioration of the signal
- *Synergy*: a stain should be compatible with a wide range of electrophoretic/separations technologies (carrier ampholyte IEF, immobilized pH gradient IEF, Tris-glycine SDS gels, Tris-tricine pre-cast SDS gels, nondenaturing gels, 2-DE gels, PVDF membranes, nitrocellulose membranes, TLC plates). The stain should not cause irreversible covalent modification of proteins. The stains should not contain extraneous chemicals such as glutaraldehyde, formaldehyde or Tween-20 that commonly interfere with Edman sequencing and mass spectrometry

---

7. Conclusions

As the analysis of biological processes at the DNA and mRNA level is rapidly being completed through large-scale genome sequencing projects and the implementation of gene arrays on chips, attention is beginning to turn to the dynamics of protein expression. Proteome research endeavors to examine the total protein complement encoded by a particular genome and to address biological problems that cannot be answered by examining nucleic acid sequence alone (De Francesco, 1999). With the development of proteomics into a high-throughput approach for the study of global protein regulation, new demands are being placed on protein visualization methods (*Table 2*). The older protein stains such as Coomassie blue and silver are beginning to fall short of expectations. Fortunately, newer stains such as Zinc-imidazole, and the noncovalent fluorescent stains (SYPRO dyes) are available to meet the growing demands of modern protein microcharacterization technologies.

Acknowledgments

The author is grateful to Kiera Berggren, Thomas Steinberg, Zhenjun Diwu, Victoria Singer and Richard Haugland, all of Molecular Probes, Inc. for their contributions and efforts leading to the development of SYPRO Ruby dye. I thank Mark Lim and Negin Shojaee of Boston University, as well as Kiera Berggren of Molecular Probes Inc., for their contributions and efforts leading to the development of SYPRO Rose protein blot stain.

References

Adams, L. and Weaver, K. (1990) Detection and recovery of proteins from gels following zinc chloride staining. *Appl. Theor. Electrophoresis* **1**: 279–282.

Alba, F. and Daban, J. (1997) Nonenzymatic chemiluminescent detection and quantitation of total protein on Western and slot blots allowing subsequent immunodetection and sequencing. *Electrophoresis* **18**: 1960–1966.

Alba, F. and Daban, J. (1998) Rapid fluorescent monitoring of total protein patterns on sodium dodecyl sulfate–polyacrylamide gels and Western blots before immunodetection and sequencing. *Electrophoresis* **19**: 2407–2411.

Aragay, A., Diaz, P. and Daban, J. (1985) A fluorescent method for the rapid staining and quantitation of proteins in sodium dodecyl sulfate–polyacrylamide gel electrophoresis. *Electrophoresis* **6**: 527–531.

Bermudez, A., Daban, J., Garcia, J. and Mendez, E. (1994) Direct blotting, sequencing and immunodetection of proteins after five-minute staining of SDS and SDS-treated IEF gels with Nile red. *BioTechniques* **16**: 621–624.

Berggren, K., Steinberg, T., Lauber, W., Carroll, J., Lopez, M., Chernokalskaya, E., Zieske, L., Diwu, Z., Haugland, R. and Patton, W. (1999) A luminescent ruthenium complex for ultrasensitive detection of proteins immobilized on membrane supports. *Anal. biochem.* **276**: 129–143.

Berube, G. (1990) Visualization of proteins on electrophoresis gels using planar dyes. US patent no. 4,946,749, 7 August.

Bickar, D. and Reid, P. (1992) A high-affinity protein stain for western blots, tissue prints, and electrophoretic gels. *Anal. Biochem.* **203**: 109–115.

Brush, M. (1998) Dye hard; protein gel staining products. *Scientist* **12**: 16–22.

Candiano, G., Porotto, M., Lanciotti, M. and Ghiggeri, G. (1996) Negative staining of proteins in polyacrylamide gels with methyl trichloroacetate. *Anal. Biochem.* **243**: 245–248.

Casero, P., Del Campo, G. and Righetti, P. (1985) Negative Aurodye for polyacrylamide gels: the impossible stain. *Electrophoresis* **6**: 362–372.

Castellanos-Serra, L., Fernandez-Patron, C., Hardy, E. and Huerta, V. (1996) A procedure for protein elution from reverse-stained polyacrylamide gels applicable at the low picomole level: an alternative route to the preparation of low abundance proteins for microanalysis. *Electrophoresis* **17**: 1564–1572.

Chevallet, M., Procaccio, V. and Rabilloud, T. (1997) A nonradioactive double detection method for the assignment of spots in two-dimensional blots. *Anal. Biochem.* **251**: 69–72.

Chrambach, A., Reisfeld, R., Wyckhoff, M. and Zaccari, J. (1967) A procedure for rapid and sensitive staining of protein fractionated by polyacrylamide gel electrophoresis. *Anal. Biochem.* **20**: 150–154.

Christiansen, J. and Houen, G. (1992) Comparison of different staining methods for polyvinylidene difluoride membranes. *Electrophoresis* **13**: 179–183.

Chung, M. (1985) A specific iron stain for iron-binding proteins in polyacrylamide gels: application to transferrin and lactoferrin. *Anal. Biochem.* **148**: 498–502.

Cohen, S. and Chait, B. (1997) Mass spectrometry of whole proteins eluted from sodium dodecyl sulfate-polyacrylamide gel electrophoresis gels. *Anal. Biochem.* **247**: 257–267.

Daban, J. and Aragay, A. (1984) Rapid fluorescent staining of histones in sodium dodecyl sulfate-polyacrylamide gels. *Anal. Biochem.* **138**: 223–228.

Daban, J., Samso, M. and Bartolome, S. (1991a) Use of Nile red as a fluorescent probe for the study of the hydrophobic properties of protein–sodium dodecyl sulfate complexes in solution. *Anal. Biochem.* **199**: 162–168.

Daban, J., Bartolome, S. and Samso, M. (1991b) Use of the hydrophobic probe Nile red for the fluorescent staining of protein bands in sodium dodecyl sulfate–polyacrylamide gels. *Anal. Biochem.* **199**: 169–174.

De Francesco, L. (1999) One step beyond; going beyond genomics with proteomics and two-dimensional gel technology. *Scientist* **13**(1): 16–18.

Diezel, W., Kopperschlager, G. and Hofman, E. (1972) An improved procedure for protein staining in polyacrylamide gels with a new type of Coomassie brilliant blue. *Anal. Biochem.* **48:** 617–620.

Dubray, G. and Bezard, G. (1982) A highly sensitive periodic acid-silver stain for 1,2-diol groups of glycoproteins and polysaccharides in polyacrylamide gels. *Anal. Biochem.* **119:** 325–329.

Dunn, M. (1987) Two-dimensional polyacrylamide gel electrophoresis. In: *Advances in Electrophoresis*, Vol. 1 (eds A. Chrambach, M. Dunn and B. Radola). VCH Publishers, New York, pp. 4–109.

Dunn, M. (1999) Detection of total proteins on Western blots of 2-D polyacrylamide gels. In: *Methods in Molecular Biology*, Vol. 112, *2-D Proteome Analysis Protocols* (ed. A. Link). Humana Press, Totowa, NJ, pp. 319–329.

Durrum, E. (1950) A microelectrophoretic and microionophoretic technique. *J. Am. Chem. Soc.* **72:** 2943–2948.

Dzandu, J., Johnson, J. and Wise, G. (1988) Sodium dodecyl sulfate-gel electrophoresis: staining of polypeptides using heavy metal salts. *Anal. Biochem.* **174:** 157–167.

Egger, D. and Bienz, K. (1987) Colloidal gold staining and immunoprobing of proteins on the same nitrocellulose blot. *Anal. Biochem.* **166:** 413–417.

Eynard, L. and Lauriere, M. (1998) The combination of Indian ink with chemiluminescence detection allows precise identification of antigens on blots: application to the study of glycosylated barley storage proteins. *Electrophoresis* **19:** 1394–1396.

Fazekas de St. Groth, S., Webster, R. and Datyner, A. (1963) Two new staining procedures for quantitative estimation of proteins on electrophoretic strips. *Biochim. Biophys. Acta* **71:** 377–391.

Fernandez-Patron, C., Calero, M., Garcia, J., Madrazo, J., Musacchio, A., Soriano, F., Estrada, R., Frank, R., Castellanos-Serra, L. and Mendez, E. (1995) Protein reverse staining: high-efficiency microanalysis of unmodified proteins detected on electrophoresis gels. *Anal. Biochem.* **224:** 203–211.

Fernandez-Patron, C., Castellanos-Serra, L., Hardy, E., Guerra, M., Estevez, E., Mehl, E. and Frank, R. (1998) Understanding the mechanism of the zinc-ion stains of biomacromolecules in electrophoresis gels: generalization of the reverse-staining technique. *Electrophoresis* **19:** 2398–2406.

Ferreras, M., Gavilanes, J. and Garcia-Segura, J. (1993) A permanent Zn^{2+} reverse staining method for the detection and quantification of proteins in polyacrylamide gels. *Anal. Biochem.* **213:** 206–212.

Flanagan, S. and Yost, B. (1984) Calmodulin-binding proteins: visualization by ^{125}I-calmodulin overlay on blots quenched with Tween 20 or bovine serum albumin and poly(ethylene oxide). *Anal. Biochem.* **140:** 510–519

Fowler, S. (1994) The detection of proteins on blots using gold or immunogold. *Methods Mol. Biol.* **32:** 239–255.

Gersten, D., Zapolski, E. and Ledley, R. (1984) Radioactive staining of gels to identify proteins. US Patent no. 4,459,356, 10 July.

Gharahdaghi, F., Kirchner, M., Fernandez, J. and Mische, S. (1996) Peptide-mass profiles of polyvinylidene difluoride-bound proteins by matrix-assisted laser desorption/ionization time-of-flight mass spectrometry in the presence of nonionic detergents.. *Anal. Biochem.* **233:** 94–99.

Glenney, J. (1986) Antibody probing of western blots which have been stained with India ink. *Anal. Biochem.* **156:** 315–319.

Gossling, J. (1990) A decade of development in immunoassay methodology. *Clin. Chem.* **36:** 1408–1427.

Graham, G., Nairn, R. and Bates, G. (1978) Polyacrylamide gel staining with Fe^{2+}-bathophenanthroline sulfonate. *Anal. Biochem.* **88:** 434–441.

Grassman, W. and Hannig, K. (1952) Ein quantitatives verfahren zur analyse der serumproteine durch papierelektrophorese. *Hoppe-Seylev's Z. Physiol. Chem.* **290:** 1–27.

Haebel, S., Albrecht, T., Sparbier, K., Walden, P., Korner, R. and Steup, M. (1998) Electrophoresis-related protein modification: alkylation of carboxy residues revealed by mass spectrometry. *Electrophoresis* **19:** 679–686.

Hager, D. and Burgess, R. (1980) Elution of proteins from sodium dodecyl sulfate–polyacrylamide gels, removal of sodium dodecyl sulfate, and renaturation of enzymatic activity: results with sigma subunit of *Escherichia coli* RNA polymerase, wheat germ DNA topoisomerase, and other enzymes. *Anal. Biochem.* **109**: 76–86.

Hamby, R. (1996) A sensitive fluorescent gel stain. *Am. Biotechnol. Lab.* **14**: 12.

Hamby, R. (1997) A fluorescent gel stain for detection of proteins following 1-D, 2-D and native polyacrylamide gel electrophoresis. *Biotechnol. Int.* **1**: 339–343.

Hancock, K. and Tsang, V. (1983) India ink staining of proteins on nitrocellulose paper. *Anal. Biochem.* **133**: 157–162.

Hartman, B. and Udenfriend, S. (1969) A method for immediate visualization of proteins in acrylamide gels and its use for preparation of antibodies to enzymes. *Anal. Biochem.* **30**: 391–394.

Harvey, M., Bandilla, D. and Banks, P. (1998) Subnanomolar detection limit for sodium dodecyl sulfate–capillary gel electrophoresis using a fluorogenic, noncovalent dye. *Electrophoresis* **19**: 2169–2174.

Higgins, R. and Dahmus, M. (1979) Rapid visualization of protein bands in preparative SDS–polyacrylamide gels. *Anal. Biochem.* **93**: 257–260.

Hochstrasser, D., Patchornik, A. and Merril, C. (1988) Development of polyacrylamide gels that improve the separation of proteins and their detection by silver staining. *Anal. Biochem.* **173**: 412–423.

Horowitz, P. and Bowman, S. (1987) Ion-enhanced fluorescence staining of sodium dodecyl sulfate–polyacrylamide gels using bis(8-*p*-toluidino-1-naphthalenesulfonate). *Anal. Biochem.* **165**: 430–434.

Houston, B. and Peddie, D. (1989) A method for detecting proteins immobilized on nitrocellulose membranes by in situ derivatization with fluorescein isothiocyanate. *Anal. Biochem.* **177**: 263–267.

Hughes, J., Mack, K. and Hamparian, V. (1988) India ink staining of proteins on nylon and hydrophobic membranes. *Anal. Biochem.* **173**: 18–25.

Hunter, J. and Hunter, S. (1987) Quantification of proteins in the low nanogram range by staining with the colloidal gold stain AuroDye. *Anal. Biochem.* **164**: 430–433.

Jackson, P., Urwin, V. and Mackay, C. (1988) Rapid imaging, using a cooled charge-coupled-device, of fluorescent two-dimensional polyacrylamide gels produced by labelling proteins in the first-dimensional isoelectric focusing gel with fluorophore 2-methoxy-2,4-diphenyl-3(2H)fur-anone. *Electrophoresis* **9**: 330–339.

Jay, G., Culp, D. and Jahnke, M. (1990) Silver staining of extensively glycosylated proteins on sodium dodecyl sulfate–polyacrylamide gels: enhancement by carbohydrate-binding dyes. *Anal. Biochem.* **185**: 324–330.

Kerenyi, L. and Gallyas, F. (1972) A highly sensitive method for demonstrating proteins in electrophoretic, immunoelectrophoretic and immunodiffusion preparations. *Clin. Chem. Acta* **38**: 465–467.

Lee, C., Levin, A. and Branton, D. (1987) Copper staining: a five-minute protein stain for sodium dodecyl sulfate–polyacrylamide gels. *Anal. Biochem.* **166**: 308–312.

Lek, L., Yang, E., Wang, D. and Cheng, L. (1995) Comparison of the sensitivity of different India inks staining of electro-blotted proteins on filter membranes. *J. Biochem. Biophys. Methods* **30**: 9–20.

Li, K., Geraerts, W., van Elk, R. and Joose, J. (1988) Fixation increases sensitivity of India ink staining of proteins and peptides on nitrocellulose paper. *Anal. Biochem.* **174**: 97–100.

Li, K., Geraerts, W., van Elk, R. and Joose, J. (1989) Quantification of proteins in the subnanogram and nanogram range: comparison of the AuroDye, FerriDye and India ink staining methods. *Anal. Biochem.* **182**: 44–47.

Lim, M., Patton, W., Shojaee, N. and Shepro, D. (1996) Solid-phase metal chelate assay for quantifying total protein: resistance to chemical interference. *BioTechniques* **21**: 888–897.

Lim, M., Patton, W., Shojaee, N., Lopez, M., Spofford, K. and Shepro, D. (1997) A luminescent europium complex for the sensitive detection of proteins and nucleic acids immobilized on membrane supports. *Anal. Biochem.* **245**: 184–195.

Loo, R., Dales, N. and Andrews, P. (1994) Surfactant effects on protein structure examined by electrospray ionization mass spectrometry. *Protein Sci.* **3**: 1975–1983.

Matsui, N., Smith, D., Clausner, K., Fichmann, J., Andrews, L., Sullivan, C., Burlingame, A. and

Epstein, L. (1997) Immobilized pH gradient two-dimensional gel electrophoresis and mass spectrometric identification of cytokine-regulated proteins in ME-180 cervical carcinoma cells. *Electrophoresis* **18:** 409–417.

Merril, C. (1987) Detection of proteins separated by electrophoresis. In: *Advances in Electrophoresis*, Vol. 1 (eds A. Chrambach, M. Dunn and B. Radola). VCH Publishers, New York, pp. 111–140.

Merril, C., Goldman, D., Sedman, S. and Ebert, M. (1981) Ultrasensitive stain for proteins in polyacrylamide gels shows regional variation in cerebrospinal fluid proteins. *Science* **211:** 1437–1438.

Merril, C., Goldman, D. and Van Keuren, M. (1984) Gel protein stains: silver stain. *Methods Enzymol.* **104:** 441–447.

Metkar, S., Mahajan, S. and Sainis, J. (1995) Modified procedure for nonspecific protein staining on nitrocellulose paper using Coomassie Brilliant Blue R-250. *Anal. Biochem.* **227:** 389–391.

Meyer, T. and Lamberts, B. (1965) Use of Coomassie brilliant blue R250 for the electrophoresis of microgram quantities of paratoid saliva proteins on acrylamide-gel strips. *Biochim. Biophys. Acta* **107:** 144–145.

Miranda, P., Brandelli, A. and Tezon, J. (1993) Instantaneous blocking for immunoblots. *Anal. Biochem.* **209:** 376–377.

Mobbs, C., Berman, J., Marquardt, M. and Pfaff, D. (1989) Comprehensive polypeptide analysis of microdissected rat brain areas; combining 2-dimensional gel electrophoresis with 2-dimensional HPLC and immunoanalysis and sequencing procedures. *J. Neurosci. Methods* **29:** 5–15.

Moeremans, M., Daneels, G., Van Dijck, A., Langanger, G. and De Mey, J. (1984) Sensitive visualization of antigen–antibody reactions in dot and blot immune overlay assays with immunogold and immunogold/silver staining. *J. Immunol. Methods* **74:** 353–360.

Moeremans, M., Daneels, G. and De May, J. (1985) Sensitive colloidal (gold or silver) staining of protein blots on nitrocellulose membranes. *Anal. Biochem.* **145:** 315–321.

Moeremans, M., De Raeymaeker, M., Daneels, G. and De Mey, J. (1986) Ferridye: colloidal iron binding followed by Perls' reaction for the staining of proteins transferred from sodium dodecyl sulfate gels to nitrocellulose and positively charged nylon membranes. *Anal. Biochem.* **153:** 18–22.

Nelson, T. (1993) Destaining of nitrocellulose blots after staining with silver or colloidal gold. *Anal. Biochem.* **214:** 325–328.

Neuhoff, V., Stamm, R. and Eibl, H. (1985) Clear background and highly sensitive protein staining with Coomassie Blue dyes in polyacrylamide gels. *Electrophoresis* **6:** 427–448.

Neuhoff, V., Arold, N., Taube, D. and Ehrhardt, W. (1988) Improved staining of proteins in polyacrylamide gels including isoelectric focusing gels with clear background at nanogram sensitivity using Coomassie Brilliant Blue G-250 and R-250. *Electrophoresis* **9:** 255–262.

Neuhoff, V., Stamm, R., Pardowitz, I., Arold, N., Ehrhardt, W. and Taube, D. (1990) Essential problems in quantification of proteins following colloidal staining with Coomassie Brilliant Blue dyes in polyacrylamide gels, and their solution. *Electrophoresis* **11:** 101–117.

Oakley, B., Kirsch, D. and Morris, N. (1980) Simplified ultrasensitive silver stain for detecting proteins in polyacrylamide gels. *Anal. Biochem.* **105:** 361–363.

Ono, T. and Tuan, R. (1990) Double staining of immunoblot using enzyme histochemistry and India ink. *Anal. Biochem.* **187:** 324–327.

Ortiz, M., Calero, M., Fernandez-Patron, C., Patron, C., Castellanos, L. and Mendez, E. (1992) Imidazole-SDS-Zn reverse staining of proteins in gels containing or not SDS and microsequence of individual unmodified electroblotted proteins. *FEBS Lett.* **296:** 300–304.

Patestos, N., Fauth, M. and Radola, B. (1988) Fast and sensitive protein staining with colloidal Acid Violet 17 followed by isoelectric focusing in carrier ampholyte generated and immobilized pH gradients. *Electrophoresis* **9:** 488–496.

Patton, W. (1995) Biologist's perspective on analytical imaging systems as applied to protein gel electrophoresis. *J. Chromatogr. A* **698:** 55–87.

Patton, W., Lopez, M., Barry, P. and Skea, W. (1992) A mechanically strong matrix for protein electrophoresis with enhanced silver staining properties. *BioTechniques* **12:** 580–585.

Patton, W., Lam, L., Su, Q., Lui, M., Erdjument-Bromage, H. and Tempst, P. (1994) Metal chelates as reversible stains for detection of electroblotted proteins: application to protein microsequencing and immunoblotting. *Anal. Biochem.* **220:** 324–335.

Patton, W., Lim, M. and Shepro, D. (1999a) Protein detection using reversible metal chelate stains. In: *Methods in Molecular Biology*, Vol. 112, *2-D Proteome Analysis Protocols* (ed. A. Link). Humana Press, Totowa, NJ, pp. 331–339.

Patton, W., Lim, M. and Shepro, D. (1999b) Image acquisition in 2-D electrophoresis. In: *Methods in Molecular Biology*, Vol. 112, *2-D Proteome Analysis Protocols* (ed. A. Link). Humana Press, Totowa, NJ, pp. 353–362.

Peats, S. (1984) Quantitation of protein and DNA in silver-stained agarose gels. *Anal. Biochem.* **140:** 178–182.

Pina, B., Aragay, A., Suau, P. and Daban, J. (1985) Fluorescent properties of histone-1-anilinonaphthalene 8-sulfonate complexes in the presence of denaturant agents: application to the rapid staining of histones in urea and Triton–urea–polyacrylamide gels. *Anal. Biochem.* **146:** 431–433.

Powell, R. (1998) Small organometallic probes. US patent no. 5,728,590, 17 March.

Pryor, J., Xu, W. and Hamilton, D. (1992) Immunodetection after complete destaining of Coomassie blue-stained proteins on immobilon-PVDF membrane. *Anal. Biochem.* **202:** 100–104.

Rabilloud, T. (1990) Mechanisms of protein silver staining in polyacrylamide gels: a 10-year synthesis. *Electrophoresis* **11:** 785–794.

Rabilloud, T. (1992) A comparison between low background silver diammine and silver nitrate protein stains. *Electrophoresis* **13:** 429–439.

Rabilloud, T. (1999) Silver staining of 2-D electrophoresis gels. In: *Methods in Molecular Biology*, Vol. 112, *2-D Proteome Analysis Protocols* (ed. A. Link). Humana Press, Totowa, NJ, pp. 297–305.

Rabilloud, T., Vuillard, L., Gilly, C. and Lawrence, J. (1994) Silver staining of proteins in polyacrylamide gels: a general overview. *Cell. Mol. Biol.* **40:** 57–75.

Ragland, W., Pace, J. and Kemper, D. (1974) Fluorimetric scanning of fluorescamine-labeled proteins in polyacrylamide gels. *Anal. Biochem.* **59:** 24–33.

Rodriguez, L., Gersten, D., Ramagli, L. and Johnston, D. (1993) Towards stoichiometric silver staining of proteins resolved in complex two-dimensional electrophoresis gels: real-time analysis of pattern development. *Electrophoresis* **14:** 628–637.

Rohde, E., Vogt, C. and Heineman, W. (1998) The analysis of fountain pen inks by capillary electrophoresis with ultraviolet/visible absorbance and laser-induced fluorescence detection. *Electrophoresis* **19:** 31–41.

Rohringer, R. and Holden, D. (1985) Protein blotting: detection of proteins with colloidal gold, and of glycoproteins and lectins with biotin-conjugated and enzyme probes. *Anal. Biochem.* **144:** 118–127.

Root, D. and Reisler, E. (1989) Copper iodide staining of protein blots on nitrocellulose membranes. *Anal. Biochem.* **181:** 250–253.

Root, D. and Wang, K. (1993) Silver-enhanced copper staining of protein blots. *Anal. Biochem.* **209:** 15–19.

Schapira, A. and Keir, G. (1988) Two-dimensional protein mapping by gold stain and immunoblotting. *Anal. Biochem.* **169:** 167–171.

Scheler, C., Lamer, S., Pan, Z., Li, X., Salnikow, J. and Jungblut, P. (1998) Peptide mass fingerprint sequence coverage from differently stained proteins on two-dimensional electrophoresis patterns by matrix assisted laser desorption/ionization-mass spectrometry (MALDI-MS). *Electrophoresis* **19:** 918–927.

Schleicher, M. and Watterson, D. (1983) Analysis of differences between Coomassie blue stain and silver stain procedures in polyacrylamide gels: conditions for the detection of calmodulin and troponin C. *Anal. Biochem.* **131:** 312–317.

Shevchenko, A., Wilm, M., Vorm, O. and Mann, M. (1996) Mass spectrometric sequencing of proteins from silver-stained polyacrylamide gels. *Anal. Chem.* **68:** 850–858.

Shojaee, N., Patton, W., Lim, M. and Shepro, D. (1996) Pyrogallol red-molybdate: a reversible, metal chelate stain for detection of proteins immobilized on membrane supports. *Electrophoresis* **17:** 687–693.

Steinberg, T., Jones, L., Haugland, R. and Singer, V. (1996a) SYPRO Orange and SYPRO Red protein gel stains: one-step fluorescent staining of denaturing gels for detection of nanogram levels of protein. *Anal. Biochem.* **239**: 223–237.

Steinberg, T., Haugland, R. and Singer, V. (1996b) Applications of SYPRO Orange and SYPRO Red protein gel stains. *Anal. Biochem.* **239**: 238–245.

Steinberg, T., White, H. and Singer, V. (1997) Optimal filter combinations for photographing SYPRO Orange or Red dye-stained gels. *Anal. Biochem.* **248**: 168–172.

Steinberg, T., Chernokalskaya, E., Berggren, K., Lopez, M., Diwu, Z., Haugland, R. and Patton, W. (2000) Ultrasensitive fluorescence protein detection in isoelectric focusing gels using a ruthenium metal chelate stain. *Electrophoresis* **21**: 486–496.

Stephens, R. (1975) High-resolution preparative SDS–polyacrylamide gel electrophoresis: fluorescent visualization and electrophoretic elution-concentration of protein bands. *Anal. Biochem.* **65**: 369–379.

Switzer, R., Merril, C. and Shifrin, S. (1979) A highly sensitive silver stain for detecting proteins and peptides in polyacrylamide gels. *Anal. Biochem.* **98**: 231–237.

Szewczyk, B. and Summers, D. (1987) Fluorescent staining of proteins transferred to nitrocellulose allowing for subsequent probing with antisera. *Anal. Biochem.* **164**: 303–306.

Szewczyk, B. and Summers, D. (1992) Fluorescent protein staining on nitrocellulose with subsequent immunodetection of antigen. In: *Methods in Molecular Biology*, vol. 10: *Immunochemical Protocols* (ed. M. Manson). Humana Press, Totowa, NJ.

Takagi, T., Kubo, K. and Isemura, T. (1977) Simple visualization of protein bands in SDS–polyacrylamide gel electrophoresis by the insoluble complex formation between SDS and a cationic surfactant. *Anal. Biochem.* **79**: 104–109.

Talbot, D. and Yphantis, D. (1971) Fluorescent monitoring of SDS gel electrophoresis. *Anal. Biochem.* **44**: 246–253.

Unlu, M., Morgan, M. and Minden, J. (1997) Difference gel electrophoresis: a single gel method for detecting changes in protein extracts. *Electrophoresis* **18**: 2071–2077.

Urwin, V. and Jackson, P. (1993) Two-dimensional polyacrylamide gel electrophoresis of proteins labeled with the fluorophore monobromobimane prior to first-dimensional isoelectric focusing: imaging of the fluorescent protein spot patterns using a cooled charge-coupled device. *Anal. Biochem.* **209**: 57–62.

Vandekerckhove, J., Bawn, G., Puype, M., Van Damme, J. and Van Montagu, M. (1985) Protein-blotting on Polybrene-coated glass-fiber sheets. A basis for acid hydrolysis and gas-phase sequencing of picomole quantities of protein previously separated on sodium dodecyl sulfate/polyacrylamide gel. *Eur. J. Biochem.* **152**: 9–19.

Vanfleteren, J. and Peeters, K. (1990) Chromatographic recovery of polypeptides from copper-stained sodium dodecyl sulfate polyacrylamide gels. *J. Biochem. Biophys. Methods* **20**: 227–235.

van Oostveen, I., Ducret, A. and Aebersold, R. (1997) Colloidal silver staining of electroblotted proteins for high sensitivity peptide mapping by liquid chromatography–electrospray ionization tandem mass spectrometry. *Anal. Biochem.* **247**: 310–318.

Velculescu, V., Zhang, L., Zhou, W., Vogelstein, J., Basrai, M., Bassett, D., Hieter, P., Vogelstein, B. and Kinzler, K. (1997) Characterization of the yeast transcriptome. *Cell* **88**: 243–251.

Vera, J. and Rivas, C. (1988) Fluorescent labeling of nitrocellulose-bound proteins at the nanogram level without changes in immunoreactivity. *Anal. Biochem.* **173**: 399–404.

Wallace, R., Yu, P., Dieckert, J. and Dieckert, J. (1974) Visualization of protein–SDS complexes in polyacrylamide gels by chilling. *Anal. Biochem.* **59**: 24.

Wilm, M., Shevchenko, A., Houthaeve, T., Breit, S., Schweigerer, L., Fotsis, T. and Mann, M. (1996) Femtomole sequencing of proteins from polyacrylamide gels by nano-electrospray mass spectrometry. *Nature* **379**: 466–469.

Wirth, P. and Romano, A. (1995) Staining methods in gel electrophoresis, including the use of multiple detection methods. *J. Chromatogr. A* **698**: 123–143.

Zapolski, E., Gersten, D. and Ledley, R. (1982) [59Fe]Ferrous bathophenanthroline sulfonate: a radioactive stain for labeling proteins in situ in polyacrylamide gels. *Anal. Biochem.* **123**: 325–328.

Zeindl-Eberhart, E., Jungblut, P. and Rabes, H. (1997) A new method to assign immunodetected spots in the complex two-dimensional electrophoresis pattern. *Electrophoresis* **18**: 799–801.

Chapter 5

Mass spectrometry-based methods for protein identification and phosphorylation site analysis

Scott D. Patterson, Ruedi Aebersold and David R. Goodlett

1. Introduction

Mass spectrometry has been a significant driving technology in the formation of the field of proteomics. Advances in mass spectrometry have occurred through the invention of instruments capable of relatively soft ionization of proteins/peptides and other biological molecules with very fine sensitivity. The increased accessibility of mass spectrometers to a range of scientists has further expanded their utility. Computer algorithms have been developed which use mass spectrometric data (either peptide or fragment ions masses) in correlative-based approaches to identify proteins in sequence databases. When these two advances are combined with the public availability of ever increasing amounts of nucleotide sequence information from a number of species (including human), it is apparent why the discovery/ identification component of proteomics has been accelerated. This chapter will describe the mass spectrometers employed for such analyses, the computational approaches employed for protein identification using mass spectrometry-derived data, and also the use of mass spectrometry to identify sites of post-translational modification, specifically phosphorylation. The chapter will not focus on describing in detail the protocols employed to generate the mass spectrometric data, as this has been covered in several recent reviews and book chapters (Aebersold and Patterson, 1998; Courchesne and Patterson, 1998; Lamond and Mann, 1997; Moritz et al., 1995).

1.1 Protein identification approaches

Many aspects of science are driven by technological advances, and protein identification provides an ideal example. Edman sequencing had been the mainstay of protein identification since the mid 1980s, when automated sequencers began to become available (Hewick et al., 1981). Edman sequencing employs step-wise chemical degradation of a protein or peptide from its N-terminus and the subsequent identification of the released and derivatized amino acids by correlation of their retention time with that of a series of standards, separated by reversed-phase HPLC

Proteomics, edited by S.R. Pennington and M.J. Dunn
© 2001 BIOS Scientific Publishers, Oxford.

(RP-HPLC) with UV detection. Many proteins are 'blocked' to Edman chemistry due to modification of the *N*-terminal amino group and hence yield no data when the intact protein is analyzed (Brown, 1979; Brown and Roberts, 1976). Gel electro-phoresis is the preferred method of choice for protein separation and, when two-dimensional gel electrophoresis (2-DE) is employed, extremely high-resolution separation of complex mixtures is achievable. Introduction of electroblotting of gel-separated proteins to a membrane (polyvinylidine difluoride, PVDF) which is compatible with Edman sequencing chemistry provided a direct link between these two methods (Aebersold *et al.*, 1986; Matsudaira, 1987; Pluskal *et al.*, 1986). By the end of the 1980s methods had been developed for the chemical or enzymatic digestion of low quantities of gel-separated proteins blotted to various membranes, and separation/purification of the resulting peptides in preparation for Edman sequencing (Aebersold *et al.*, 1986, 1987; Vandekerckhove *et al.*, 1985). If the intact protein was blocked to Edman sequencing, this approach provided the opportunity to generate amino acid sequences from internal peptides. Furthermore, internal sequencing provided significantly more sequence coverage than *N*-terminal sequen-cing alone. It also facilitated gene cloning efforts employing polymerase chain reaction (PCR)-based approaches by providing amino acid sequences from different regions of the protein, which could be used to design degenerate oligonucleotide primers (Tempst *et al.*, 1990). During the early 1990s efforts were directed towards improving both the methodologies (e.g. both membranes and digestion protocols) for identification of gel-separated proteins involving either the capture or release of proteins and peptides (reviewed in Aebersold and Leavitt, 1990; Patterson, 1994), and increased sensitivity of the Edman sequencers. Edman sequencing generates amino acid sequences *de novo*, therefore there is no requirement for correlation of the experimentally derived data to amino acid sequence databases to assist in the identification process, as is the case with the mass spectrometry-based approaches which are the subject of this chapter. Hence, Edman sequencing continues to play an important role in the identification of proteins from species with poorly character-ized genomes.

The advent of the Human Genome Project, with its aim of sequencing the entire human genome, triggered sequencing efforts in a number of other species. Through both public and private sector efforts the genomes of a number of organisms of clinical and basic research relevance have been entirely sequenced. For example, the entire genomes of 21 bacteria [for review see Fraser and Fleischmann (1997) and The Institute for Genomics Research (TIGR) Microbial Database web site: http://www.tigr.org/tdb/mdb/mdb.html], yeast (*Saccharomyces cerevisiae*), round worm (*Caenorhabidopsis elegans*) and the fruit fly (*Drosophila melanogaster*) have been completed. There are already over two million human and over one million murine expressed sequence tags (ESTs, partial single-pass cDNA sequences of approxi-mately 300–500 bp derived from reverse translation of mRNA isolated from differentiated cells or tissues) publicly available, and both public and private efforts have at least three-fold coverage of the entire human genome sequence already. Algorithms have been written that provide a rapid means of protein identification to be achieved by correlation of primary structural data (measured by the mass spectrometer) with predicted sequences (translations of nucleotide sequences) in these databases. This new approach to protein identification (established during the mid 1990s) was accompanied by a new wave of technological development which built upon the methods designed for purification/separation of peptides from gel-separated proteins for subsequent Edman analysis (Lamond and Mann, 1997; Patterson and Aebersold, 1995). Improvements in the mass spectrometers and their interfaces [e.g. nanospray (Wilm and Mann, 1994, 1996)] were also made during this period and continue to be made [e.g. microfabricated devices (Figeys *et al.*, 1998); for a review see Figeys and Aebersold (1998)]. These developments have converged into a mature, robust, sensitive and rapid technology which has dramatically advanced the ability to identify proteins and constitutes the basis of the emerging field of proteomics.

2. Background to mass spectrometry

The identification of proteins by correlation with sequence databases relies on the availability of constraining parameters which distinguish specific matches from all the other sequences in the database. Among the parameters (protein amino acid composition, stretches of amino acid sequence, protein and peptide masses, and peptide fragment masses) which have been used either by themselves or in combination, the accurate mass of a molecule is particularly attractive because it is highly constraining and can be determined with great accuracy, rapidly and sensitively by mass spectrometry (MS). MS is limited to the mass measurement of those molecules that can be ionized and for which the ions can be transferred to a vacuum. MS became more compatible with the analysis of biopolymers, proteins, nucleic acids and carbohydrates, with the introduction in the late 1980s of two 'soft' ionization methods: electrospray ionization (ESI; Fenn *et al.*, 1989) and matrix-assisted laser desorption/ionization (MALDI; Karas and Hillenkamp, 1988). These methods are referred to as 'soft' because they do not (in the most part) degrade the molecule during the ionization process. Thus the mass analysis of large [> 10 000 mass units (u)] biologically derived polymers was realized.

Mass spectrometers can be described as consisting of three components: (1) an ionization source; (2) a mass analyzer; and (3) a detector. They measure the mass-to-charge ratios (m/z) of ions under vacuum, and various combinations of ionization sources and mass analyzers (and detectors) have been constructed. The three commercially available types of instruments which are most commonly used for protein and peptide analysis are the MALDI time-of-flight (TOF) mass spectrometer, the ESI-triple quadrupole (TQ) mass spectrometer and the ESI ion trap (IT) mass spectrometer. These three types of instrument will be described further.

2.1 MALDI-MS

Linear MALDI-MS. Probably the simplest type of mass spectrometer conceptually and in design is the MALDI-MS. Molecules (analytes) are ionized by MALDI and their m/z ratio measured in a time-of-flight mass analyzer (see *Figure 1*). The

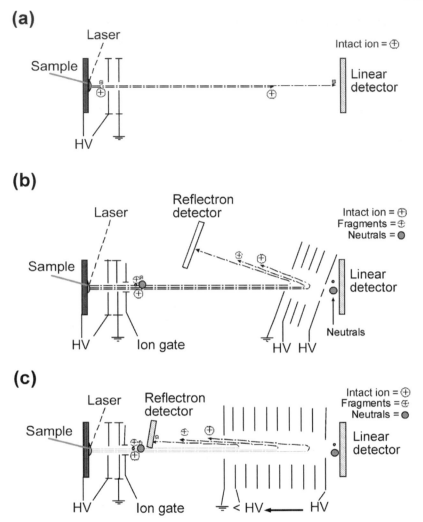

Figure 1. Matrix-assisted laser desorption/ionization mass spectrometers. Schematics of (a) MALDI-MS, (b) MALDI-MS with reflectron and (c) MALDI-MS with curved field reflectron. See text for details.

accuracy of the measurement is dependent on the m/z, with accuracies of $\sim \pm 0.01$–0.05% up to ~ 25 ku, and ± 0.05–0.3% up to 300 ku being achieved. MALDI is also able to be performed when the analyte of interest is dissolved in some biological buffers such as phosphate, Tris, which can also include small amounts of urea, non-ionic detergents, and some alkali metal salts.

Analysis of proteins and peptides by MALDI-MS involves a series of steps:

(i) Analytes of interest are mixed with a matrix compound (small aromatic molecules). The matrix is usually dissolved in an acidic organic solvent and added to the analyte at a ~ 1000-fold molar excess. Analyte and matrix are then placed on a metallic slide or probe. Evaporation of the solvent in air results in the formation of analyte : matrix cocrystals.

(ii)   The probe containing the analyte : matrix cocrystal is inserted into the vacuum chamber (10^{-5}–10^{-8} Torr) of the mass spectrometer.

(iii)   A high voltage (e.g. $+20$–$30\,kV$ to generate positive ions) is applied to the metallic slide/probe and a short laser pulse is directed at the dried sample.

(iv)   The desorption event is initiated by absorption of energy at the wavelength of the laser (UV at 337 nm; IR at 2.94 μm) by the matrix crystals, resulting in emission of this absorbed energy as heat. This rapid heating in turn results in sublimation of the matrix crystals and expansion of the matrix and analyte molecules into the gas-phase of the mass spectrometer.

(v)   Ionization occurs by protonation/deprotonation, cation attachment/cation detachment, or oxidation/reduction in the gas phase. The ions generated are repelled from the surface (held at high voltage, $+20$–$30\,kV$) and accelerated towards a series of lenses held at close to ground voltage which direct the ions into the field-free drift region (50–300 cm) of the time-of-flight mass analyzer. The end of the field-free drift region is fitted with a detector which records the arrival of the ions drifting through the tube.

(vi)   The MALDI process delivers a packet of ions to the mass analyzer with essentially the same final kinetic energy. At constant kinetic energy the velocity of the ions is inversely related to their m/z ratio. Therefore, the arrival time of the ions at the linear detector at the other end of the field-free drift tube reflects the m/z ratio of the detected ions.

(vii) The flight time of ions through the field-free drift tube is recorded by triggering the detector with the laser pulse. Hence the time it takes for specific ions to reach the detector following each ionization event is recorded and the m/z value is calculated (Carr and Annan, 1997).

The smaller the ion the faster it will fly, hence the m/z ratios can be practically calculated from the flight time using compounds of known mass and charge state as calibrants (usually a matrix ion and a peptide or protein which bracket the ion of interest). However, the relationship between the kinetic energy of the ion (E), its mass (m) and its velocity (v) can be expressed as $E = \frac{1}{2}\,mv^2$. Therefore, for a singly charged ion the flight times (a result of their velocities) of the ions will be inversely proportional to the square root of their molecular mass.

An inherent problem with the MALDI process is the small spread of kinetic energies that occurs during ionization. The small difference in kinetic energies imparted on the ions generated by a single laser pulse reduces the resolving power of MALDI-MS. Typical mass resolutions (defined as $m/\Delta m$) for peptides are $\sim$40–800 FWHM (full-width, half maximum) and for proteins $\sim$50–400 FWHM. At this resolution the isotope distribution even of peptides cannot be determined. The initial energy spread is mass dependent with peak broadening increasing with increasing ion mass. Two approaches have been developed to increase MALDI-MS mass resolution. They are based on the use of an ion mirror (a reflectron) and/or a process termed 'time-lag focusing' (see *Figure 1b and c*).

As the name suggests, the ion mirror/reflectron reflects ions back from a specific location to a detector. This is achieved by the following process: a reflectron placed at the end of the flight tube has an applied voltage which is slightly higher than the accelerating voltage at the ion source. Ions entering the reflectron are decelerated by this voltage until they stop and are then reaccelerated back out of the reflectron to a second detector (see *Figure 1b*). The reflectron acts as an energy-focusing device because ions with a slightly lower kinetic energy (slower ions) do not penetrate the reflectron as far and turn around earlier, catching up with the ions of slightly greater kinetic energy which have penetrated the reflectron further. The observed effect of applying a reflectron is a decreased initial energy spread of the ions and therefore increased mass resolution.

The second approach to correct the initial spread of kinetic energies during MALDI is a technique developed by Wiley and McLaren (1953) known as 'time-lag focusing', or more recently 'delayed extraction' (Vestal *et al.*, 1995), and is implemented at the source. In *Figure 1b and c* this is represented by a second lens with applied voltage nearer the sample than the grounded continuous extraction lens shown in *Figure 1a*. In delayed extraction the MALDI ions are created in a field-free region and allowed to spread out before an extraction voltage is applied to accelerate the ions into the drift tube. This results in a significantly decreased energy spread of ions and also limits peak broadening which can occur due to metastable decomposition from ion collisions in the source during the standard continuous ion extraction. Delayed extraction increases mass resolution for peptides in a linear instrument to ~ 2000–4000 and in a reflectron instrument to ~ 3000–6000 (Brown and Lennon, 1995; Vestal *et al.*, 1995).

PSD-MALDI-MS. To obtain primary structural information in a MALDI-MS instrument individual ion packets consisting of intact and fragment ions has to be selected. The process which occurs 'post' ionization in the source is referred to as postsource decay [PSD; for a review see Chaurand *et al.* (1999)]. If peptides are subjected to PSD they will fragment predominantly along the polypeptide backbone, thus generating series of fragment ions which, in principle, contain the amino acid sequence information of the peptide (Kaufmann *et al.*, 1994). Since this metastable fragmentation occurs after accelerating the ions, all the fragment ions have the same velocity as their precursor ion, but only a fraction of its kinetic energy. All the ions, precursor and fragment, will arrive at the linear detector at essentially the same time, apparently with the same m/z ratio. Hence the fragment ions are not resolved from the precursor and are not detected in a linear TOF analyzer. The fragment ions can however be resolved by exploiting their kinetic energy differences through the use of an ion mirror (reflectron). Dual-stage or curved-field reflectron instruments have been described. The dual-stage reflectron focuses ions over a limited kinetic energy range. To bring fragment ions of a wider range into focus, the voltage of the reflectron is stepped down successively over 7–14 steps, thus focusing ions of lower and lower kinetic energy until the last spectrum is acquired (at ~ 5–10% of the initial voltage; see *Figure 1b*). The contiguous, complete fragment ion spectrum is then assembled in a data system from the calibrated fragment ion spectra generated at each reflectron voltage. The curved-field reflectron initially described by Cornish and Cotter (1994) can resolve all of the fragment ions in a single experiment. This is achieved by focusing the ions in a long reflectron consisting of a series of lenses which are kept at decreasing voltages (see *Figure 1c*). In the curved-field reflectron the voltages required to focus fragment ions of different m/z are therefore permanently present, but spatially resolved. If more than one peptide species is ionized by the same laser pulse, and in the absence of a mass filter, the fragment ions generated from different peptides cannot be distinguished and fragment ions cannot be distinguished from concurrently generated unfragmented peptide ions. Instruments equipped with either type of reflectron are therefore fitted with an ion gate which permits the selection of the precursor/fragment ion packets at low resolution ($\pm 2.5\%$; see *Figure 1b and c*). Sequence-specific fragment ion information can therefore be potentially generated from a number of peptides in an unfractionated mixture.

2.2 ESI-MS

In ESI a solution containing the analyte of interest (peptides/proteins) is passed through a narrow needle held at high potential ($\sim +1000$–$5000\,\text{V}$ for the generation of positive ions; see *Figure 2a*). The high potential causes the analyte flow to disperse as a fine spray of highly charged droplets which is directed across a small inlet orifice of a plate held at a lower potential ($\sim +100$–$1000\,\text{V}$) at the front end of the mass spectrometer. The orifice is the interface between the ion source, which is held at atmospheric pressure, and the mass spectrometer, which is evacuated. As the droplets traverse from the tip of the needle to the orifice they are desolvated by directed flows of inert gas (sometimes heated). During this process of desolvation the increasing charge density of ions in the shrinking droplet results in a Coulombic explosion releasing ions from the liquid phase of the droplet into the gas phase (Carr and Annan, 1997). After passing through the orifice, the desolvated ions are gated by a radio frequency (RF)-only quadrupole into the mass analyzer, which may be a quadrupole, hybrid quadrupole-TOF system or an ion-trap. It is characteristic of ESI to produce peptide ions bearing multiple charges. The number of charges is a function of the number of ionizable groups in the peptide. If tryptic peptides are analyzed by ESI-MS in positive ion mode, most of the peptides will carry at least two charges, one at the *N*-terminal NH_2 group and one on the side chain of the basic amino acid positioned by trypsin at the *C*-terminal of the peptide.

ESI-MS using a triple-quadrupole instrument. As the name states, a triple-quadrupole mass spectrometer consists of three sets of quadrupoles (see *Figure 2b*). Quadrupole mass filters can either transmit all ions present in the mixture if operated in RF-only mode, or can be used as mass filters, allowing ions of only a specific m/z to pass through. Quadrupoles consist of four parallel rods. Opposite pairs of rods are connected electrically carrying d.c. and a.c. voltages of opposite polarity. As ions drift through the space between the quadrupole rods the voltages are varied in a way that only allows ions of a certain m/z ratio to pass through the filter, while all the ions

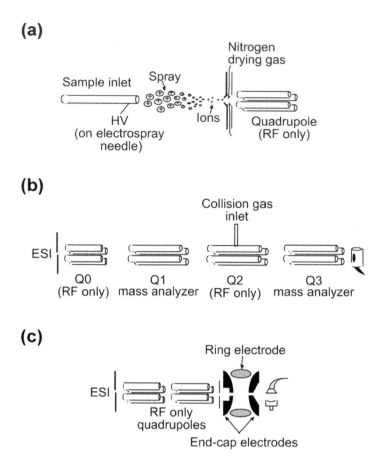

Figure 2. Electrospray ionization mass spectometers. Schematics of an: (a) ESI-MS, (b) ESI-TQ-MS and (c) ESI-IT-MS. See text for details.

with a different m/z ratio are diverted from their linear flight path and therefore eliminated from the analysis. If the applied voltages are varied over time (scanned) and the number of ions exiting the filter are recorded by use of a channel-electron multiplier as a function of the selected m/z value, a mass spectrum can be determined. Typically, quadrupole mass filters have a usable mass range of 2500–4000 u. Since the m/z ratio is recorded, rather than the mass of a peptide itself, and since ESI typically produces multiply charged peptide ions, quadrupole mass analyzers used with an ESI ion source can be used for the analysis of proteins, the mass of which significantly exceeds the nominal mass range of the quadrupoles.

In a triple-quadrupole MS the first (Q1) and third (Q3) quadrupoles are mass filters and the second quadrupole (Q2) is run in RF-only mode and serves as a collision cell for fragmentation of ions. Fragmentation is induced by colliding the peptide ions transmitted from Q1 with heavy, inert gas such as Ar or N_2, which is introduced into the collision cell. This process is referred to as collision-induced dissociation (CID).

There are five modes of operation of the triple-quadrupole instrument:

(i)   MS mode – this mode allows measurement of the masses of unfragmented ions by scanning Q1 without collision gas in Q2 and operating Q3 in RF-only mode. Alternatively, Q1 is run in RF-only mode and the mass spectrum is established by scanning Q3.

(ii)   MS/MS mode (also termed product ion scan) – at a given time point Q1 is set to transmit ions of only a selected m/z, which are subjected to CID in Q2. The masses of the resulting fragment ions are determined by scanning Q3 and recorded such that the fragment ion mass spectrum is correlated with the m/z of the peptide ion initially selected in Q1. This cycle of ion selection in Q1, CID in Q2 and analysis of the fragment ion spectrum in Q3 is then repeated by instructing Q1 to select a peptide ion of different mass. In fact, some instruments such as the TSQ 7000 (Finnigan MAT, San Jose, CA) can be programmed to automatically switch between MS and MS/MS mode and thus to select detected ions for CID and to record their fragment ion spectra independent of any user input [e.g. Ducret *et al.* (1998) and Stahl *et al.* (1996), and ICL procedure on our (DG, RA) web site (http://depts.washington.edu/ ~ruedilab/aebersold.html)]. This process is referred to as data-dependent CID. Product ion scans are the method of choice for the determination of the amino acid sequence of peptides.

(iii) Neutral loss scan mode – in neutral loss scanning mode Q1 and Q3 scan synchronously, but offset by a specific m/z value. Q2 is used as a collision cell. The offset between Q1 and Q3 corresponds to the mass of a specific neutral fragment that is collisionally eliminated in Q2. Therefore, only those ions in an ion mixture which undergo a loss of the specific group are transmitted through Q3 and to the detector. As an example, doubly charged phosphorylated peptides can be detected within a mixture of unphosphorylated peptides by scanning for a neutral loss of 50 m/z which corresponds to a loss of H_3PO (Covey *et al.*, 1991). Since the particular voltages which are applied to Q1 and Q3, respectively, are known at the time when ions are transmitted to the detector, the mass of the peptides undergoing a neutral loss is also determined.

(iv)   Precursor (or parent) ion scanning – in precursor ion mode Q1 is scanning while Q3 is set to transmit ions of a particular m/z value and Q2 acts as a collision cell. Similar to the neutral loss scan, the precursor ion scan detects those ions in a mixture of ions which lose a specific group under the collisional conditions in Q2. In contrast to the neutral loss scan, the precursor scan directly detects the group that is cleaved off (reporter ion). Since the voltages applied to Q1 are known at the time when the cleaved reporter ion is recorded at the detector, the mass of the (precursor) ion which underwent the specific fragmentation is determined. Precursor ion scanning is therefore used to detect those peptides in a peptide mixture which contain specific structural features.

(v)   In-source CID – fragmentation also occurs in the high-pressure region of an ESI source as a result of collisions with atmospheric gases. Collision conditions can be chosen such that peptides fragment along the peptide backbone (Katta *et al.*, 1991) or that specific, relatively labile groups such as phosphate are selectively cleaved off (Huddleston *et al.*, 1993). Peptide ions at charge states where $n > 1$ are preferentially fragmented because a greater amount of kinetic energy is imparted to them than is necessary to simply focus the ions into Q1. This method provides an opportunity to perform MS/MS/MS experiments on a triple quadrupole or MS/MS experiments on a single quadrupole instrument, respectively (Loo *et al.*, 1990). Any one of the above scan types, (i)–(iv), can be carried out on ions generated by in-source CID.

For the purpose of protein identification and *de novo* peptide sequencing, the instrument is most often run in MS/MS mode. The collision energy imparted to the selected ions can be varied, as can the collision gas pressure, allowing great control over the fragmentation process if time permits. In a data-dependent operation mode where the mass spectrometer has been programmed to fragment ions reaching a particular ion current exceeding a predetermined threshold value in MS-only mode, the collision energy can be ramped to try to achieve optimal fragmentation in at least one of the spectra obtained at different collision energies (Stahl *et al.*, 1996). However, when extremely low levels of analyte are present and the solution containing the sample of interest is introduced by way of a micro- or nano-spray device without sample clean-up, the chemical background and other contaminants dominate the spectrum (Wilm and Mann, 1996). Therefore to find the peptide ions present in such a mixture the instrument can be run in precursor ion scanning mode using an m/z value for Q3 corresponding to that of the immonium ion of Ile and Leu (86), which are common amino acids present in many peptides (Wilm *et al.*, 1996).

ESI-MS using an ion-trap instrument. The ion-trap mass spectrometer is a tandem mass spectrometer in which ions can be accumulated and stored prior to analysis. The ion trap, the central part of this instrument, is both a mass analyzer and collision cell. Quadrupole ion-traps use RF-only quadrupoles to guide ions from the electrospray source into the ion trap. The ion trap consists of two types of electrodes, a ring electrode and two end-cap electrodes (see *Figure 2c*). Ions enter and leave the trap through holes in the end-cap electrodes and are trapped in a field created by applying a large RF voltage to the ring electrode. The magnitude of this voltage determines frequency and motion of the trapped ions. Ions above a certain m/z remain in the trap and are collisionally cooled with residual helium gas, resulting in a collapse of the sinusoidal cycling of the ions from a larger trajectory to a smaller one over a very short period of time. A mass spectrum is generated by ramping the RF voltage linearly while applying a small voltage across the end-cap electrodes, causing 'resonance ejection' of ions of successive m/z. Resonance ejection refers to ions becoming unstable in the trap and being ejected axially through the end-cap electrodes where they are detected by an off-axis (to reduce background from neutral species) conversion dynode with an electron multiplier detector (Jonscher and Yates, 1997; Schwartz and Jardine, 1996).

Through this process of trapping and selective ejection of ions, ions of a specific m/z can be isolated in the trap. MS/MS spectra can then be generated from the isolated ions if they are fragmented in the trap and if the mass spectrum of the fragment ions is then generated by resonance ejection of all the ions in the trap. If all the fragment ions except those of a particular m/z ratio are ejected, the fragmentation/resonance ejection process can be repeated, and therefore multistage MS experiments can be performed (MS^n). An MS^{12} experiment has been documented in the literature (Louris et al., 1990) and MS^3 experiments are routinely done, particularly for the analysis of glycopeptides (Reinhold et al., 1995). The process of isolating and fragmenting an ion in the ion-trap consists of trapping the ion (m/z species) of interest by ejecting all ions having a larger m/z than the ion of interest and then ejecting all ions of a smaller m/z than the ion of interest. Thus a packet of ions of a selected m/z value (isolated ions) now resides in the trap and can be energized by altering the voltages on the electrodes so that the ions will collide with residual neutral gas (He) and fragment. The fragment ions generated can then be measured by scanning out all of the ions from the trap as described above. The amount of energy put into the trapped ion can be adjusted by varying a parameter referred to on some instruments as the relative collision energy (RCE). As described above for PSD-MALDI-MS, fragmentation of singly charged ions requires greater energy than fragmentation of doubly or multiply charged ions. In the ion trap the RCE can be manually adjusted in the ion trap to great effect in some experiments (Davis and Lee, 1997). Other parameters that can be varied on the ion-trap mass spectrometer and which affect the performance of the instrument include the number of ions allowed to enter the trap in any one event, and over how many such events the signal is averaged to increase the signal-to-noise ratio. If too many ions of very close m/z value are trapped, a phenomenon called space charging is observed. Space charging decreases the mass accuracy of the experiment because the trajectories of the closely matched ions influence each other. Thus care has to be exercised in establishing parameters that will not adversely affect the data (e.g. Courchesne et al., 1998). Commercial ion-trap mass spectrometers have instrument control software supporting data-dependent fragmentation capabilities and are therefore ideally suited to LC-MS/MS.

3. Correlative mass spectrometric-based identification strategies

3.1 Peptide-mass searching

Soon after the commercial development of MALDI-MS and the demonstration that this method was capable of measuring the masses of peptides in mixtures (notably peptides derived from the enzymatic digestion of gel-separated proteins), a number of groups developed algorithms for protein identification based upon correlating measured peptide masses with experimentally calculated peptide masses derived from proteins existing in sequence databases (see *Figure 3*; Henzel et al., 1993; James et al., 1993; Mann et al., 1993; Pappin et al., 1993; Yates et al., 1993). This approach has most often been applied to the identification of gel-separated proteins (reviewed in Aebersold and Patterson, 1998).

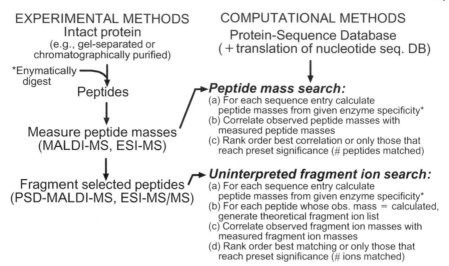

Figure 3. Protein identification by mass spectrometry.

The algorithms were all based upon the following idea for the identification of proteins whose sequences resided in sequence databases:

(i)   Peptides are generated by digestion of the protein of interest using specific cleavage reagents (usually enzymes of known cleavage specificity).

(ii)   The masses of these peptides are accurately determined experimentally using MALDI-MS (or ESI-MS).

(iii) Theoretical peptide masses are calculated for each sequence entry in the database using the same cleavage specificity as the reagent employed experimentally.

(iv)   A score (or ranking) is then calculated to provide a measure of fit between the experimentally derived and calculated peptide masses.

A number of the original groups and additional groups have made their search algorithms available on the Internet and illustrated them with examples (see *Table 1*). An obvious caveat to such a search approach is that for a protein to be identified its sequence has to exist within a sequence database (translated nucleotide or protein sequence database). Hence the approach is ideally suited to genetically well-characterized organisms, especially those whose entire genomes have been determined (e.g. see the TIGR database: http://www.tigr.org/tdb), or those for which extensive protein or cDNA sequence databases have been established. Protein identification by peptide mass searching depends on the correlation of several peptide masses derived from the same protein between the experimental dataset and the calculated datatable. For this reason the technique is neither suited to searches of translations of expressed sequence tag (EST) databases nor for the identification of proteins in mixtures of more then a couple of proteins. However, cross-species identification is possible for some proteins, and programs with this specific purpose in mind have been written (Wilkins *et al.*, 1998). ESTs represent

Table 1. Internet sites with mass spectrometry-based protein identification tools

Resource	WWW Uniform Resource Locator (http:// . . .)	Features and comments
CBRG, ETH-Zurich, Switzerland	cbrg.inf.ethz.ch/MassSearch.html	Peptide mass search program
EMBL, Heidelberg, Germany	www.mann.embl-heidelberg.de/ Services/PeptideSearch/ PeptideSearchIntro.html	Peptide-mass and fragment ion search programs
ExPasy	www.expasy.ch/tools/#proteome	Peptide-mass and fragment ion search programs
Mascot	www.matrix-science.com/cgi/ index.pl?page=/home.html	Peptide-mass and fragment ion search programs
Rockefeller University New York, USA	prowl.rockefeller.edu/	Peptide-mass and fragment ion search programs
SEQNET, Daresbury, UK	www.seqnet.dl.ac.uk/ Bioinformatics/welapp/mowse	Peptide-mass and fragment ion search programs
University of California, San Francsico, USA	prospector.ucsf.edu	Both peptide mass (MS-Fit) and fragment ion (MS-Tag) search programs
University of Washington	thompson.mbt.washington.edu/ sequest/	Instruction on how to get SEQUEST (fragment-ion search program)

only portions of coding sequence which may not be long enough to code for a significant number of peptides observed in the peptide-mass searching experiment. If mixtures of proteins are digested, the protein from which a specific subset of the peptide masses originates is not apparent. If every observed peptide could be precisely correlated with one of the proteins in a sequence database the identity of the proteins in a mixture could still be determined. This would be achieved by iterative cycles of subtracting all the peptides matching a particular, identified protein from the table of measured masses and searching the database with the remaining peptide masses. For a number of reasons listed below it is, however, rare that all of the measured peptide masses will be matched with the sequence of the protein from which they originate. This not only complicates the identification of the components in protein mixtures but also the identification of single, purified proteins. It is therefore important to understand the nature of these unmatched peptide masses and the following describes potential reasons for these apparently spurious masses with regard to whether the match was correct or not (Jensen *et al.*, 1997, 1998; Patterson and Aebersold, 1995).

(i)   The protein was identified correctly, but the additional masses are due to post-translational or artifactual modifications or post-translational processing (e.g. *N*- or *C*-terminal processing). Although it may be possible to reconcile mass differences with these particular scenarios, they should be considered tentative unless they can be confirmed experimentally.

(ii)   The protein was identified correctly, but unspecific proteolysis had occurred or a contaminating protease was present. One approach for reconciling this possibility is to determine whether the candidate protein could produce peptides of those masses excluding assumptions regarding cleavage specificity.

(iii) The protein was identified correctly, but it was part of a mixture of 'contaminating' protein(s) (e.g. even 2-DE protein spots may consist of more than one protein). If there are enough additional masses, a separate peptide-mass search can be conducted with these masses to attempt to identify the additional protein(s). Obviously control experiments would have been conducted with enzyme alone to allow identification of any enzyme-derived autolytic products.

(iv)   The protein identified may be a sequence homolog or splice variant of that reported in the database. Some search programs allow searches to be performed across species. This can be advantageous if additional confirmatory data is obtained, because matches to proteins from genetically well-characterized species may be made using proteins from less well-characterized species (Cordwell *et al.*, 1995).

(v)   The protein identified was a false positive. Without additional data and especially if the mass accuracy of the experimental data is not high, this potentially disastrous result is difficult to confirm or disprove. This can be particularly difficult when using search programs that rank their output using a score if the result is not a high score. In some cases further confidence in a match can be gained depending upon the difference in scores between the first and second highest (and subsequent) matches. The protein may also not exist in the database.

The most critical experimental parameter to control when attempting to identify proteins using peptide-mass data is the accuracy of the peptide-mass measurement (Fenyo *et al.*, 1998; Jensen *et al.*, 1996a, 1997). The greater the mass accuracy, the greater the confidence in the assignment. Another critical parameter is the specificity of the enzyme (or chemical reagent) employed. The purer and therefore more specific, the more reliable the search results will be. The most commonly employed enzyme is trypsin. However, it should be noted that even highly purified trypsin can cleave at sites other than the *C*-terminal to Lys or Arg if not followed by Pro. A problem with all proteases is that they may not cleave the substrate to completion leaving 'missed' cleavage sites and 'ragged' termini, if two or more consecutive amino acids in a protein sequence are potential cleavage sites for the enzyme. Many search programs allow the number of missed cleavage sites to be entered as a parameter. Endoproteinase LysC is another enzyme which exhibits a high degree of specificity. This enzyme produces less autodigestion products than trypsin, but it also exhibits missed cleavage sites and can cleave at sites other than after Lys (except if followed by Pro). It should also be noted that not all proteins, gel-separated or not, will be amenable to digestion with trypsin or any other specific protease. However, gel-separated proteins, due to their denatured state, are generally quite effectively digested.

While the performance of the proteases available for peptide mapping is difficult to influence, the mass accuracy of the peptide measurements is dependent on the instrument used and its operation. Optimally, peptide-mass searches should be performed with internally calibrated monoisotopic masses. In MALDI-MS the resolution required to resolve the isotopic distribution of a peptide is generally only achieved in instruments fitted with time-lag focusing/delayed extraction (as described above). The mass accuracy of 5 ppm achieved in such instruments allows assignment of monoisotopic masses and very tight tolerances to be employed in the peptide-mass search programs. Such high precision data decrease the potential for false matches (Jensen *et al.*, 1996a, 1997; Takach *et al.*, 1997). If the delayed extraction option is not available, the spectrum should be calibrated using internal standards that bracket the masses of the detected peptides. In instruments without delayed extraction, isotopic peaks are not resolved and only average masses are obtained. Monoisotopic masses can be determined with greater precision than average masses. This is due to the fact that monoisotopic masses are determined from a single isotope (C12), whereas average masses are derived from the average of the C12 and varying numbers of C13 isotopes (as well as isotopes of other elements).

To further increase the confidence of peptide-mass assignments, 'orthogonal' methods have been developed to provide additional or constraining information on the amino acid sequence or composition of individual peptides. Such analyses are usually performed on a separate aliquot of the sample in addition to that used for the original mass measurement. This orthogonal information restricts the number of peptides that match the isobaric (same mass) peptides which are identified in the original database search based on peptide mass only. The orthogonal methods can be grouped into the following categories:

(i)   Site-specific chemical modifications – methyl esterification, which adds $+14$ u for each carboxyl group [i.e. the side chains of the acidic residue, Asp, Glu or *C*-terminus, in the peptide (Pappin *et al.*, 1995)]; iodination, which adds $+126$ u for each tyrosine (Craig *et al.*, 1995); and isotopic labeling of cysteinyl residues using a 1:1 mixture of acrylamide and deuterated acrylamide (Sechi and Chait, 1998).

(ii)   Determination of partial amino acid composition of the peptide – two methods have been used. If the exchangeable hydrogens in a peptide are exchanged in solution with the heavier deuterium, each amino acid can undergo a characteristic number of exchanges (from 0 to 5 u per residue). The total increase in peptide mass therefore reflects the peptide composition (James *et al.*, 1994). Alternatively, the partial amino acid composition of a peptide can be determined through the identification of immonium ions from the MS/MS spectra.

(iii) Identification of the *N*-terminal amino acid residue – the *N*-terminal amino acids of peptides in a mixture have been determined chemically by one step of Edman (Jensen *et al.*, 1996b), or enzymatic removal of the terminal residue through the use of an aminopeptidase (Woods *et al.*, 1995).

(iv)   Identification of different cleavage sites within the peptides – to determine the presence and relative location of a specific residue in a peptide, secondary digestion of the primary digest with an enzyme of differing specificity (subdigestion) has been employed (Pappin *et al.*, 1995). Alternatively, parallel digestion of aliquots of the same sample with enzymes of different specificity has been proposed (James *et al.*, 1994).

(v)   Identification of the *C*-terminal residue(s) – similar to the identification of the *N*-terminal residue the *C*-terminal residue and in some cases one or several additional residue(s) have been removed by carboxypeptidases (Patterson *et al.*, 1995; Woods *et al.*, 1995). However, removal of one residue may not be very informative when using an enzyme with high specificity.

Fenyo *et al.* (1998) have conducted an excellent study to assess the benefits of orthogonal information for peptide-mass searches using data from the complete genome of the yeast *S. cerevisiae*, and the reader is referred to this article for a more detailed treatise of the benefits of these approaches.

3.2 Uninterpreted fragment ion searching

Just as databases can be searched with peptide masses, they can also be searched with a single peptide mass and the masses of its fragment ions. The peptide mass together with its fragment ion masses provide extremely discriminating criteria to search sequence databases including translations of EST databases (see *Figure 3*). This approach has been used extensively for the identification of gel-separated proteins following generation of peptides by enzymatic digestion (Aebersold and Patterson, 1998; Lamond and Mann, 1997; Patterson, 1997). Peptides can often be induced to fragment in the gas phase of a mass spectrometer by a number of processes described above (e.g. PSD-MALDI-MS, or CID in a triple-quadrupole or ion-trap instrument). In these instruments peptide ions fragment in a sequence-specific manner, predominantly at the peptide bond along the peptide backbone, as shown in *Figure 4*. If the positive charge of the peptide ion remains on the *N*-terminus of the fragment ion, the ion is referred to as a *b* series ion, with a subscript reflecting the number of amino acid residues present on the fragment ion, counted from the *N*-terminus. Therefore, *b* ions represent the fragments with *C*-terminal deletions and an intact *N*-terminal. If the charge remains on the *C*-terminus, the ion is referred to as a *y* series ion (with the same subscript numbering as for the *b* series ion, but counting from the *C*-terminus). Therefore, *y* ions represent the fragments with *N*-terminal deletions and an intact *C*-terminal (nomenclature as per Biemann, 1990). The CID spectrum represents the composite picture of thousands of discrete fragmentation events and so is a mix of various *b* and *y* series ions. Additional ions present in the spectrum are due to neutral losses of ammonia (-17 u from Gln, Lys and Arg) and water (-18 u from Ser, Thr, Asp and Glu) from the side chains of amino acids, and losses of carbon monoxide from *b* series ions (-28 u, resulting in *a* series ions). In addition, if a specific peptide ion undergoes multiple fragmentation events, internal fragments are generated. These are either internal acyl ions consisting of at least two amino acid residues (see *Figure 4b*) or immonium ions (see

(a)

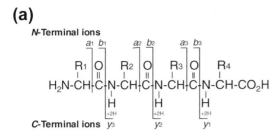

(b)

R_1R_2 (*b*) $H_2N-CH-\overset{\underset{\|}{O}}{C}-\underset{H}{N}-CH-C\equiv O\;^+$ with R_1, R_2 substituents

R_1R_2-28 (*a*) $H_2N-CH-\overset{\underset{\|}{O}}{C}-\underset{H}{N}\overset{+}{=}CH$ with R_1, R_2 substituents

(c)

$\overset{+}{H_2N}=CH$ with R substituent

Figure 4. Fragment ion nomenclature for the most common positive (a) *N*- and *C*-terminal ions, (b) internal acyl ions and (c) internal immonium ions.

Figure 4c and *Table 2*). Immonium ions represent individual amino acids and as such provide partial amino acid composition of the peptide. Not all peptide bonds have the same propensity to fragment under CID conditions and the intensity of the generated fragment ions varies significantly in the same MS/MS spectrum. The most frequent fragmentation occurs at prolyl bonds and these are generally the most intense ions in a CID spectrum. In addition, the most frequent internal acyl ions occur as a result of fragmentation at the labile prolyl bond and extend the *C*-terminal from the Pro residue.

As described above, fragmentation can be induced by PSD-MALDI-MS as well as by CID in a triple quadrupole or an ion-trap mass spectrometer. Among the three methods, fragmentation by PSD-MALDI-MS is least well controlled. The only parameters that can be varied in PSD-MALDI-MS are the laser fluence and the matrix. Furthermore, in MALDI-MS essentially only singly charged ions are generated. These require more energy to fragment than the doubly and triply charged ions generated by electrospray ionization. PSD-MALDI-MS therefore has a lower success rate and the spectra are in many cases of lower quality than the spectra of the same peptides analyzed by CID in a ESI instrument.

Fragment ion spectra contain redundant pieces of information such as over-lapping *b* and *y* series ions and multiple internal ions from the same peptide. This redundancy makes fragment ion spectra an extremely rich source of sequence-specific information, but also complicates the interpretation of the sequence because it is frequently not immediately apparent to which ion series a particular ion belongs. Apart from manually interpreting fragment ion spectra, which is described below under *de novo* sequencing, there have been two basic types of search approaches

Table 2. Residue and immonium ion masses of the 20 common amino acids[a]

Amino acid	Abbreviations – 3 letter (single letter)	Residue mass[b]	Immonium ion mass[b]
Alanine	Ala (A)	71	44
Asparagine	Asn (N)	114	87
Aspartate	Asp (D)	115	88
Arginine	Arg (R)	156	129, 112, 100, 87, 70, 43[c]
Cysteine	Cys (C)	103	76
Glutamine	Gln (Q)	128	101 (w/o 84)
Glutamate	Glu (E)	129	102
Glycine	Gly (G)	57	30
Histidine	His (H)	137	110
Isoleucine	Ile (I)	113	86
Leucine	Leu (L)	113	86
Lysine	Lys (K)	128	101, 84, 129[c]
Methionine	Met (M)	131	104
Phenylalanine	Phe (F)	147	120
Proline	Pro (P)	97	70 (w/o 112, 100 and 87)
Serine	Ser (S)	87	60
Threonine	Thr (T)	101	74
Tryptophan	Trp (W)	186	159
Tyrosine	Tyr (Y)	163	136
Valine	Val (V)	99	72

[a] The values were obtained from Jardine (1990).
[b] All masses are given as integer values and underlining indicates more abundant ions (Carr and Annan, 1997); w/o = without.
[c] Arginine and lysine both exhibit multiple immonium ions, and these are listed.

employed to match fragment ion spectra with sequences in databases. The first type relies upon a partial manual interpretation of the spectrum to identify consecutive elements of a particular (*b* or *y*) ion series (i.e. to identify fragment ions whose mass differences correspond to amino acid residue masses). This partial sequence information is used together with the mass of the first ion in the identified series and the difference in mass between the intact mass and the last ion in the identified series (these two masses define the mass of the total 'unknown' residues on either side of the sequence), the mass of the intact ion, and the cleavage specificity of the enzyme employed (both *N*- and *C*-terminal specificity or just one or neither of them). These five pieces of information constitute a 'peptide-sequence tag' (Mann and Wilm, 1994). Sequence databases can then be searched for peptides which match the components of the peptide-sequence tag. The second type of fragment ion search is a true uninterpreted fragment ion search program developed by Yates and colleagues (Eng *et al.*, 1994), called SEQUEST. This approach first searches for peptides in each sequence entry (using the sum of the masses of contiguous amino acids and, if necessary, the cleavage specificity of the protease used) whose calculated mass matches that of the intact measured peptide ion (within a given tolerance), thereby generating a list of candidate peptides. For each of these candidate peptides, the program calculates the masses of the fragment ions expected if they were subjected to the same CID conditions as the peptide in the experiment. The program then compares the experimental fragment ion spectrum with the top 500 calculated spectra using cluster-analysis algorithms. Each comparison receives a score, with the highest of these being reported. If a score of significance is achieved, the protein is identified based on the peptide fragment ion spectrum without any explicit determination of the spectrum. This automated approach can be applied to data-dependent LC-MS/MS analyses. In this approach, peptides separated by RP-HPLC are sprayed into an on-line mass spectrometer capable of data-dependent MS/MS (i.e. the mass spectrometer fragments all ions reaching a certain threshold). All of the resulting CID are recorded, batched and submitted for searching automatically.

The automated approach has several advantages, as previously reviewed (Patterson and Aebersold, 1995):

(i)   Proteins present in complex mixtures can be identified as each peptide that generates a fragment ion spectrum provides data for an independent database search (McCormack *et al.*, 1997).

(ii) In this type of analysis, the fragment ion spectrum from each peptide represents an independent database search. If a single protein has been digested, protein identification is essentially autoconfirmatory and conclusive since the same protein should be the top ranked candidate for a number of peptides (Ducret *et al.*, 1998).

(iii) As described above, the approach is easily automated (Yates *et al.*, 1995b).

(iv) The method can be adapted to find peptides carrying specified post-translational modifications by instructing the program to anticipate a modification of a specific mass on specific residues [as well as identifying the protein from which it originated; Yates *et al.* (1995a,b)].

A list of some internet sites with protein identification resources developed by these and other investigators are shown in *Table 1*.

4. *De novo* sequencing using mass spectrometric data

The previous two sections have described correlative approaches for protein identification using mass spectrometric or tandem mass spectrometric data. These methods are suitable for the identification of proteins for which the sequence is contained in a database. MS and MS/MS methods have also been developed for cases in which *de novo* sequencing is required. The MS-based approaches are based on the generation of peptide ladders in which individual peptides differ in length by one amino acid. These ladders are then analyzed by MS, typically by MALDI-TOF MS. In the MS/MS based approaches CID spectra are generated as described above for database searching, but the fragment ion spectra are manually and completely interpreted.

Peptide ladders for sequence analysis by MS are generated by chemical or enzymatic degradation of peptides, either from the *N*-terminal or *C*-terminal. The chemical methods for determining the *N*-terminal sequence employ Edman chemistry with MALDI-MS as the read-out for the generated shortened peptides. This method is generally referred to as 'ladder sequencing'. Two variations of the method have been developed. The first uses the addition of Edman reagents at each sequencing cycle to one starting aliquot of the peptide/protein substrate. The ladder is generated by partially blocking the N-terminal of a small fraction of the substrate in each cycle. The blocked peptides are then refractory to further degradation in subsequent cycles. The final mixture of products whose mass differences reflect the amino acid residue masses is analyzed by MALDI-MS (Chait *et al.*, 1993; Wang *et al.*, 1994). The second approach, also referred to as 'nested peptide sequencing', uses multiple additions of the peptide/protein at each cycle of Edman degradation. In this method the Edman chemistry is driven to completion using an excess of a volatile reagent which can be removed in each cycle by evaporation. As for ladder sequencing, a peptide ladder is generated which is analyzed by MALDI-MS (Bartlet-Jones *et al.*, 1994). Both methods require a free N-terminus, and neither can resolve the isobaric residues Ile from Leu. Gln (residue mass 128.13) and Lys (residue mass 128.17), which are difficult to distinguish in CID spectra, can be easily distinguished, however, because the amino group of the Lys side chain will become modified with the coupling reagent. In principle both methods could cope with mixtures of peptides whose intact masses are sufficiently different to avoid overlap of the degradation products (Wang *et al.*, 1994). As an alternative to the chemical step-wise degradation, peptide sequence ladders have also been generated by truncation of peptides by amino- and carboxy-peptidases. This approach has the advantage that the reactions can be conducted with very small quantities of starting material directly on the MALDI probe surface. The reactions are simply stopped by the addition of matrix (Patterson *et al.*, 1995; Woods *et al.*, 1995). By conducting a time course of reaction times, longer sequence reads can be achieved. This approach is unable to distinguish Gln and Lys as well as Ile and Leu.

If high-quality CID spectra, i.e. spectra containing complete b and y ion series can be generated, peptide sequences can be determined by *de novo* interpretation of the fragment ion masses (for a schema see Yates *et al.*, 1994), but more often than not it is a difficult process with a low confidence level in the final assignments. Major issues complicating the interpretation of CID spectra include missing fragments, i.e. incomplete ion series and difficulties in assigning the observed signals to a specific ion series. The identification of the fragments belonging to the y ion series can be facilitated by, specifically, isotopically tagging the C-terminal ions during tryptic digestion. This method takes advantage of the property of trypsin to incorporate an oxygen from a water molecule into the carboxyl group of the newly generated C-terminus. Therefore, if the tryptic digestion is conducted in a digestion buffer containing $H_2^{18}O$ all the peptides (with the exception of the peptide derived from the C-terminus) contain an isotopically tagged C-terminal carboxyl group (e.g. Schnolzer *et al.*, 1996). By conducting the trypsin digestion in a buffer composed of 50% v/v $H_2^{18}O$ / 50% v/v $H_2^{16}O$, all newly generated peptides (but not the peptide derived from the C-terminus) will be present as doublets differing in mass by 2 u (these will also be present at half the ion intensity of the same peptide generated normally in $H_2^{16}O$). Trypsin cleaves the peptide bond after Lys or Arg via the activated OH group of Ser-195 of trypsin, forming a covalent ester bond with the carboxyl group of the Lys or Arg residue, with concomitant release of the newly formed *N*-terminus. Cleavage of this intermediate through attack by an activated water molecule releases the peptide fragment whose carboxyl group contains one oxygen from solvent water (which is either $H_2^{18}O$ or $H_2^{16}O$ with equal probability; Schnolzer *et al.*, 1996). Thus when both ion species are selected in the mass spectrometer, only the ion series with an intact *C*-terminus (*y* series ions and their neutral losses) will exist as doublets separated by 2 u. Therefore, the *y* series can be distinguished and the sequence differences calculated, allowing *de novo* sequencing of the peptide. While this method simplifies the assignment of fragments to a particular ion series, it also reduces the signal intensity of each *y* ion and therefore makes sequencing at very high sensitivities more difficult. The use of high resolution MS instruments such as a quadrupole time-of-flight tandem mass spectrometer for such analyses can result in complete interpretation of peptide spectra in a single experiment (Shevchenko *et al.*, 1997), and the method has also been successfully applied using an ion-trap mass spectrometer (Qin *et al.*, 1998). An alternative chemical method to facilitate the manual interpretation of CID spectra for *de novo* sequencing employs methyl esterification of the carboxyl groups in the peptide. This reaction increases the mass of the peptide by 14 u for each carboxyl group (unmodified *C*-terminus, Asp and Glu; Hunt *et al.*, 1986). If there are no acidic residues in the peptide and the *C*-terminus has not been modified, the mass shift will be reflected in only the *y* series ions (and their neutral losses). Comparison of the fragment ion spectra obtained from derivatized and underivatized aliquots of the sample will allow the identification of the *y* series ions. The analysis becomes more complicated if there are acidic residues present, and of extremely limited value if the *N*-terminal residue is Asp or Glu (only internal fragments can then be identified). Methods to specifically tag the *N*-terminal residue with the purpose of identifying the *b* ion series have also been attempted. In these experiments peptide amino groups were derivatized with a reagent containing a permanent positive charge in the gas phase (Huang *et al.*, 1999; Roth *et al.*, 1998). While conceptually attractive, these methods have met practical difficulties, mainly due to the fact that excess reagent needs to be removed before mass spectrometric analysis of the conjugates and that the reactions cannot differentiate between the *N*-terminal amino group and the amino group of lysine side chains.

5. Separation methods for phosphorylation site analysis

5.1 Rationale for phosphorylation analysis

Among the hundreds of types of protein modifications described (Krishna and Wold, 1998), to date only a few have been shown to be reversible or to be of regulatory importance in biological processes. The best studied of these is protein phosphorylation. Protein phosphorylation has been studied extensively to understand the enzymology of the protein kinases and phosphatases which effect phosphorylation and de-phosphorylation, respectively, and the consequences of the modification on the structure and function of the target proteins (Charbonneau and Tonks, 1992; Fischer and Krebs, 1989; Hunter, 1987). The most common type of phosphorylation studied is that of the phospho-esters of serine, threonine and tyrosine. Also known to occur are phosphoramidates of arginine, histidine and lysine as well as acyl derivatives of aspartic and glutamic acid. Some of these modifications are chemically labile and they are usually not observed, unless specific precautions are taken to prevent their elimination during protein isolation (Duclos *et al.*, 1991). As an example, phosphohistidine is a relatively common modification, at least in prokaryotes, but it is completely eliminated by the acidic conditions commonly used for protein staining in polyacrylamide gels. Phosphohistidine is therefore commonly not observed in gel-separated proteins and the frequency with which the modification occurs in the cell is difficult to estimate.

The level of protein phosphorylation is modulated by two counteracting enzyme systems, phosphatases and kinases. The structures, specificities and regulation of the most common of these are well studied and reviewed (Charbonneau and Tonks, 1992; Fischer and Krebs, 1989; Hunter, 1987). There are assumed to be hundreds of protein kinases/phosphatases differing in their substrate specificities, kinetic proper-ties, tissue distribution and association with regulatory pathways. Analysis of the complete genomic DNA sequence from *S. cerevisiae* for sequence motifs that are thought to be indicative of protein kinases (and phosphatases) predicts the capacity of yeast to express 123 different protein kinases (and 40 protein phosphatases), thus suggesting that the function of 2% of expressed yeast proteins is involved in protein phosphorylation. To assess the importance of protein phosphorylation/dephos-phorylation in many regulated biological systems, the reader is pointed toward a number of reviews (Charbonneau and Tonks, 1992; Fischer and Krebs, 1989; Hunter, 1987). The functional consequences of protein phosphorylation provide insights into how biological systems are controlled at the molecular level. For instance a single phosphorylation event regulates the activity of glycogen phosphor-ylase (Johnson and Barford, 1990). Work on the protein tyrosine kinase p56lck demonstrated that phosphorylation at a single serine, possibly by mitogen-activated protein (MAP) kinase (Watts *et al.*, 1993), can alter the substrate specificity of an enzyme (Joung *et al.*, 1995). Tyrosine phosphate-induced protein interactions have been shown to provide a scaffold required for stabilizing protein complexes which perform complex biological functions, particularly in intracellular signal transduc-tion (Wange *et al.*, 1995; Watts *et al.*, 1994). It is therefore evident that reversible protein phosphorylation is an essential protein modification which controls biologi-cal activities and functions by altering the catalytic activity and substrate specificity of enzymes and by modulating the stability and subcellular location of protein complexes of regulatory function.

5.2 *Phosphorylation analysis strategies*

The principal aims of protein phosphorylation studies are three-fold: first, to determine the amino acid residues (sites) that are phosphorylated *in vivo* in a given protein in a cell that is in a specific state; second, to identify the kinase(s) responsible for the phosphorylation event; and third, to analyze the functional consequences of the observed phosphorylation events. Among these aims, the first can be directly addressed by mass spectrometry and is therefore the subject of this chapter. The second and third aims require the application of a variety of biochemical, genetic and pharmacological methods, the details of which, for reasons of space, are not further described.

Many phosphorylated proteins are present in cells in vanishingly small amounts. Even if a protein is expressed at a relatively high level, phosphoprotein analysis is complicated by the frequently low stoichiometry of phosphorylation (i.e. only a small fraction of a given protein is phosphorylated) and the presence of multiple, differentially phosphorylated forms of the same protein. For these reasons it is frequently difficult to isolate quantities of *in vivo* phosphorylated proteins which are sufficient for analysis by even the most sensitive current methods. Many protein phosphorylation studies are therefore performed with proteins modified by *in vitro* kinase reactions that can generally be scaled up to produce larger amounts of the phosphorylated protein. Before sites of phosphorylation determined by *in vitro* studies can be accepted as biologically significant, the occurrence of the same phosphorylation sites *in vivo* needs to be established. This is frequently accomplished by comparing two-dimensional phosphopeptide (2D-PP) maps of the same protein phosphorylated *in vivo* and *in vitro* (see *Figure 5*). The *in vitro* phosphorylated peptide, which can be generated in larger amounts, is analyzed chemically or mass spectrometrically and used as a reference for establishing the identity of the *in vivo* phosphorylated peptide, which may only be available in very small amounts. Comigration of the respective phosphopeptides in the map is then taken as an indication that the same site was phosphorylated *in vivo* and *in vitro* (e.g. Watts *et al.*, 1994, 1996b). This somewhat indirect method therefore establishes the identity of an *in vivo* site of phosphorylation by comigration of the sample peptide with a well-characterized reference.

Even in cases where sufficient amounts of the phosphorylated protein can be generated *in vivo* for direct (mass spectrometric) analysis, the success of the experiment critically depends on the sample preparation methods used. The following sections therefore briefly describe methods that are commonly used for the detection and isolation of phosphoproteins and phosphopeptides within mixtures consisting predominantly of non-phosphorylated polypeptides. Since these methods are not directly related to mass spectrometry they are only discussed briefly. The

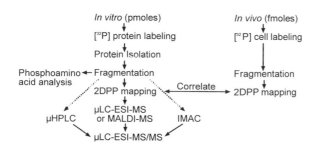

Figure 5. Phosphorylation site analysis strategies.

reader is directed to a more extensive treatment of the subject matter (Aebersold and Patterson, 1998).

5.3 Detection and isolation of phosphoproteins

Essentially any method for the analysis of the site(s) of protein phosphorylation is dependent on the purification of the phosphoprotein studied, the enzymatic or chemical fragmentation of the phosphoprotein and the isolation, separation and analysis of the generated peptides. Fortunately, the methods described in the previous sections that detail the treatment of proteins separated by polyacrylamide gel electrophoresis equally apply to the analysis of phosphoproteins. The method of choice for the generation of phosphopeptide samples is therefore separation of proteins by gel electrophoresis, fragmentation of the phosphoprotein band or spot and extraction of the generated phosphopeptide mixture for further analysis. Even though phosphorylated proteins frequently migrate in SDS-polyacrylamide gel electrophoresis (SDS-PAGE) slightly slower than their nonphosphorylated counterparts, the observation of two closely migrating bands in a one-dimensional gel or the observation of an array of spots of similar molecular mass but different pI in a 2-DE gel is insufficient to identify a particular protein as a phosphoprotein. More conclusively, phosphoproteins are positively identified, after ^{32}P radiolabeling by autoradiography or storage phosphorimaging, or by Western blotting. The incorporation of ^{32}P-radiolabel into phosphoproteins can be routinely achieved *in vivo* or *in vitro* using established protocols (e.g. Patterson and Garrels, 1994). Metabolic radiolabeling is the method of choice for *in vivo* protein phosphorylation. Cells or tissue to be labeled are incubated with ^{32}PO$_4$ for a period long enough to equilibrate the cellular ATP pool with ^{32}P. The radiolabeled ATP is then used by protein kinases to phosphorylate their respective substrates. Protein phosphorylation *in vitro* is generally performed by kinase reactions using [γ-^{32}P]ATP as the source of the radiolabel and crude fractionated cell lysates or purified kinases as the source of the kinase activity. Western blotting has been particularly successful for the detection of tyrosine phosphorylated proteins because a panel of tyrosine-phosphate specific and very sensitive antibodies have been developed [e.g. 4g10 (Upstate Biotechnology, Lake Placid, NY); py20 and RC10 (Transduction Laboratories, Lexington, KY)] and because the method is nonradioisotopic. The development of pan phospho-specific antibodies and of antibodies with specificity for phospho-serine or phospho-threonine has been less successful.

5.4 Separation of phosphopeptides

Even though the nanospray MS/MS technique has been used successfully for the analysis of phosphopeptides directly in a protein digest, some form of peptide separation prior to mass spectrometric analysis of phosphopeptides is usually advised for a number of reasons, including the following:

(i)   Phosphopeptides are frequently minor species in a protein digest and are therefore easily missed within the general background of low-intensity ions.

Peptide separation techniques also concentrate the analytes and therefore increase the signal-to-noise ratio.

(ii)   If phosphopeptides are radiolabeled to the same specific activity, the number of counts associated with each separated peptide indicates the relative amount of phosphopeptide present in the fraction. If the specific activity of the radiolabel used is known, the absolute amount of phosphopeptide present can be easily calculated.

(iii)   Reproducible patterns of separated, radiolabeled phosphopeptides can be used to quantitatively determine changes in the phosphorylation state of a protein as a function of time or cellular state.

(iv)   Peptide separation methods also effectively remove nonpeptidic contaminants, thus facilitating the detection and analysis of low-abundance phosphopeptides.

Among the separation techniques available, 2D-PP, RP-HPLC, high-resolution gel electrophoresis and immobilized metal affinity chromatography (IMAC) have been preferentially used for the separation of phosphopeptides. While it is difficult to state an explicit strategy (see *Figure 5*) for use in characterizing all phosphoproteins, some mention of a generic strategy is worth review. However, even this strategy as partially covered in *Figure 5* will not encompass all possible approaches. The details of any given characterization protocol will depend on many factors, such as personal preferences in separation and detection modes, sample availability and available equipment. For this brief discussion we assume that it was important to show similarity between *in vivo* and *in vitro* 2D-PP maps and that this was done successfully. As a general rule of thumb the first attempt to obtain sequence information can be made by direct analysis of spots from the *in vitro* 2D-PP map by either MS or MS/MS methods, as both may provide useful leads. If this is unsuccessful for some or all of the spots, then proteolysis is repeated and peptides fractionated by HPLC after which the ^{32}P-containing fractions are analyzed by MS/MS in either an automated or directed fashion. If failure to obtain sequence information from the spots of a 2D-PP map was because of poor MS signal-to-noise, then the HPLC fractions will be substantially cleaner and probably give good data. This HPLC preparative method may fail though if the stoichiometry of phosphorylation is very low. In this case a phosphopeptide may be observed, but the concentration on column will be too low to yield interpretable fragment ions. In this case the proteolysis is again repeated and phosphopeptides are enriched by IMAC. The MS/MS experiments are then carried out on the IMAC fraction(s). Regardless of the method chosen as a first approach, it is common to proceed with several iterations from the point of proteolytic fragmentation (see *Figure 5*) before all the phosphopeptides of interest are sequenced. Keep in mind that certain peptide sequences do not fragment in an easily interpretable fashion and thus, in order to establish the exact site of phosphorylation, it may be necessary to have the putative peptide synthesized. This standard is then used in comparative studies to establish the exact site of phosphorylation. When ^{32}P cannot be incorporated into a protein, the diagnostic ion scans mentioned below are used to detect phosphopeptides for sequence analysis. Each one of the separation techniques mentioned here and shown in *Figure 5* is compatible with further analysis of the separated peptides by the chemical, enzymatic and mass spectrometric methods described below.

Separation of phosphopeptides by 2D-PP. In 2D-PP mapping, peptides are separated in a first dimension by electrophoresis on a thin-layer cellulose plate and in the second dimension by thin-layer chromatography (TLC) on the same plate (Boyle *et al.*, 1991). Separated, ^{32}P radiolabeled phosphopeptides are then detected by autoradiography or storage phosphor imaging. The number of spots visualized by autoradiography provides an estimate of the maximum number of phosphorylated sites and spot intensity provides an indication of the relative stoichiometry of phosphorylation between peptides. Furthermore, every phosphopeptide present in the sample is accounted for, even if not all of the peptides are resolved. The number of observed spots does not necessarily correlate directly with the number of phosphorylation sites because incomplete proteolytic digestion is often observed with phosphoproteins. However, the 2D-PP map also provides important qualitative information about the phosphorylation state of the protein not available from any other method. This information includes:

(i)   The maximum number of phosphorylation sites as maps generally produce more spots than there are phosphorylation sites because of differential processing by proteases.
(ii)   The relative stoichiometry of phosphorylation among all phosphopeptides is provided by autoradiographic intensity.
(iii)   The relative state of hydropathy between phosphopeptides is provided by the tangential separations of electrophoresis and TLC.

Another advantage of 2D-PP mapping is that it produces purified phosphopeptides that can be analyzed, after extraction from the plate, directly by MS methods (Affolter *et al.*, 1994). If 2D-PP mapping is intended as a preparative method for MS/MS analysis, then the amount of protease added should be kept as low as is practical. Use of excess protease will be obvious when analyzing a 'purified' phosphopeptide by MS because autocatalytic products from the protease will dominate the spectra and may prevent analysis of the phosphopeptide. Furthermore, the 2D-PP method is sensitive and reproducible. Since the phosphopeptide pattern is detected by integrating the radioactive decays over potentially very long time periods, the method is very sensitive and the sensitivity achieved is only dependent on the specific activity of the radiolabel. The high degree of pattern reproducibility achieved by the method makes 2D-PP the method of choice for projects in which the state of phosphorylation of a protein under different conditions needs to be analyzed using for example time courses and induced states of activation.

Separation of phosphopeptides by RP-HPLC. Reversed-phase HPLC fractionation of phosphopeptides is reproducible, simple and does not require specialized equipment. In RP-HPLC phosphopeptides are separated based on their hydrophobicity and are detected by Cerenkov counting of collected fractions postseparation. When a graph of Cerenkov counts versus time is produced, the number of radioactive fractions is revealed. These purified phosphopeptides can then be analyzed by MS/MS using the different MS and MS/MS methods described in Section 2.2. The disadvantage of RP-HPLC over 2D-PP mapping is that very hydrophilic phosphopeptides will not stick to the column and thus will elute in the column flow-through. Conversely, very hydrophobic peptides might not elute until the end of a gradient or not at all. It is therefore possible that some of the phosphopeptides in a sample will go undetected. Generally, the resolution of RP-HPLC is also inferior to the resolution achieved by the two-dimensional peptide mapping technique. An additional note of caution is that phosphopeptides will stick to metal surfaces. Thus, significant sample losses can occur if standard metal injectors are used. A big attraction of the use of RP-HPLC for phosphopeptide analysis is the ease with which RP-HPLC systems are connected on-line to ESI mass spectrometers. The use of a mass spectrometer connected on-line to the HPLC system makes it possible to detect and characterize phosphopeptides in sample mixtures, even if the analyte is not radiolabeled. This is achieved by the implementation of specific scan functions, including the precursor ion scan, the neutral loss scan and the dissociation at the source, as described above. Phosphopeptide analysis in a LC-MS/MS system therefore is compatible with nonisotopically labeled samples, a significant advantage in cases in which phosphoproteins are isolated from human tissue.

Separation of phosphopeptides by high-resolution gel electrophoresis. A preparative approach that promises the same results as 2D-PP mapping was recently published using 2-DE to purify phosphopeptides on polyacrylamide gels (Gatti and Traugh, 1999). Nondenaturing gel isoelectric focusing was combined with alkaline 40% PAGE for phosphopeptide separation and comparative pattern analysis. Phosphopeptides were detected by autoradiography or storage phosphorimaging of ^{32}P-labeled samples. Edman sequencing was used to identify the proteins and to determine the site(s), but the method is presumably adaptable to MS-based methods. As with HPLC, but for different reasons, there is the potential for loss of specific phosphopeptides. Still, the method may appeal to those lacking the equipment for 2D-PP mapping because gel electrophoresis equipment is widespread.

Separation/enrichment of phosphopeptides by IMAC. A common difficulty with phosphopeptide analysis is caused by a low stoichiometry of phosphorylation. In such cases phosphopeptide(s) are present in the sample in very small amounts compared with the nonphosphorylated peptide with the same sequence and it is difficult to identify the phosphopeptides by MS techniques, even though their presence is confirmed by ^{32}P-label detected in a 2D-PP spot, a whole protein digest or a HPLC fraction, respectively. Data-dependent MS/MS methods fail to identify minor species in a sample because the more intense ions are selected for CID with higher priority. To alleviate this problem, selective enrichment of phosphopeptides by IMAC is frequently employed. The technique involves chelation of metals like Fe^{3+} or Ga^{3+} onto a chromatographic support consisting of iminodiacetic acid or nitrilotriacetic acid (Neville *et al.*, 1997; Nuwaysir and Stults, 1993; Tempst *et al.*, 1998). Phosphopeptides are bound selectively and with high affinity to such supports. Fractions eluted by phosphate or increased pH are therefore enriched for phosphopeptides. While the method is selective for phosphopeptides, other peptides, particularly those containing strings of acidic amino acids, are also enriched. Fractions eluted from IMAC columns are then analyzed by the various MS methods discussed in this chapter. The method has the advantage that every soluble phosphopeptide, regardless of its length, is enriched and that samples eluted from an IMAC column can be directly analyzed by RP-HPLC. This was used to establish an integrated peptide enrichment/separation system consisting of a tandem IMAC/RP column configuration (Affolter *et al.*, 1994).

5.5 *Determination of the type of phosphorylated amino acid*

Knowledge of the type of amino acid(s) phosphorylated in a peptide or protein can constrain the number of possible phosphorylated sites and therefore simplify the assignment of the phosphorylated residues within the polypeptide chain. The type of residue phosphorylated is commonly determined either by phosphoamino acid analysis or by phosphoamino acid-specific immunodetection.

Phosphoamino acid analysis employs gas- or liquid-phase hydrolysis of the peptide bonds under conditions that leave at least a fraction of the phosphate ester bonds intact. The hydrolysate of a ^{32}P-labeled phosphoprotein or phosphopeptide is then mixed with phosphoamino acid standards and separated by TLC. The unlabeled standards are detected by ninhydrin staining and the ^{32}P-amino acids by autoradiography. The type of phosphoamino acid present in the protein is then determined by correlation of the radiolabeled species with the stained standards (Hildebrandt and Fried, 1989). If larger amounts of sample are available, the type of phosphorylated amino acid can be determined directly by ninhydrin staining. Phosphoamino acid analysis, performed with phosphopeptides extracted from 2D-PP maps, is particularly useful, because such peptides frequently contain only one type of phosphoamino acid. Knowledge of the nature of the phosphorylated amino acid and additional parameters such as the peptide mass and the cleavage specificity of the enzyme used to fragment the protein can, in some cases, identify or at least suggest a phosphorylation site in a protein.

The use of antibodies specific for particular phosphoamino acids is a corroborative method to determine the type of phosphorylated amino acid that does not, however, replace phosphoamino acid analysis. Tyrosine phosphorylated proteins are easily detected by immunoblotting of gel-separated proteins with antiserum specific for tyrosine phosphate [antibodies 4g10 (Upstate Biotechnology, Lake Placid, NY) and py20; also bacterial fragment antibodies like RC10 (Transduction Laboratories, Lexington, KY)]. Such antibodies are commercially available and provide a rapid and sensitive method for monitoring tyrosine phosphorylation of proteins. Phosphoserine and phosphothreonine specific antibodies have also been described (Heffetz *et al.*, 1989) and are commercially available. Results obtained with these reagents are somewhat erratic and caution is advised when using them; however, recent studies suggest that the quality has improved (Soskic *et al.*, 1999).

5.6 Determination of the site of phosphorylation

Methods for determining sites of phosphorylation have in common that the phosphoprotein is purified, if possible to homogeneity and cleaved by specific chemical or enzymatic reactions to produce a mixture in which one or a few peptides are phosphorylated. These methods differ in the strategies used to isolate phospho-peptides, to determine their amino acid sequences and to locate the phosphorylated residues within the phosphopeptide. The separation methods described above, 2DPP mapping, HPLC or IMAC, are also the preferred micropreparative methods for the determination of phosphorylation sites.

Chemical phosphopeptide sequencing. Historically, the sites of phosphorylation were mostly determined by phosphopeptide sequencing by step-wise chemical degradation [for phosphoserine and phosphothreonine see Wettenhall *et al.* (1991); for phosphotyrosine see Aebersold *et al.* (1991)]. More recently this has largely been replaced by mass spectrometric methods. While this chapter is mainly focused on mass spectrometric methods, it is worth mentioning that both nonradioactive and radioactive phosphopeptide sequencing methods have been developed which can complement and in some cases replace mass spectrometric determination of phosphorylation sites (Watts *et al.*, 1996a). The nonradioactive methods were developed by and have been reviewed by Meyer *et al.* (1993). They involve detection of phosphoserine and phosphothreonine following β-elimination and reaction with ethanethiol to form *S*-ethylcysteine or β-methyl-*S*-ethylcysteine, which are easily extracted from automated sequencers in the form of phenylthiohydantoyl derivatives. Radioactive methods also exist that monitor, by Cerenkov counting, the release of radioactivity during the sequential cycles of Edman chemical degradation (Aebersold *et al.*, 1991; Wettenhall *et al.*, 1991). If serine or threonine phosphate residues are sequenced, the phosphate is eliminated and ^{32}P-phosphate and the respective phenylthiohydantoin derivatives are observed (Wettenhall *et al.*, 1991). In the case of a tyrosinephosphate, the phosphate ester bond is more stable and the extracted species is analyzed as phenylthiohydantoyl phosphotyrosine. If only very limited amounts of phosphopeptide are available, it may not be possible to follow the peptide sequence by chemical degradation. By following the degradation cycle, in which the radioactivity is released from the peptide, it is however frequently possible to determine which position within the peptide was phosphorylated.

Mass spectrometric analysis of phosphopeptides. A number of different strategies and methods for the mass spectrometric determination of sites of phosphorylation in phosphopeptides have been developed. They differ by the way the MS is interfaced with peptide isolation techniques and by the mass spectrometric method applied. If a phosphopeptide is present at 1 pmol or higher level, phosphopeptides can be extracted from the spots on a 2D-PP map and analyzed by automated, data-dependent, micocapillary-LC/MS/MS (Watts *et al.*, 1994; Whalen *et al.*, 1996). This method is fast, provides initial insight into the relative amount of phosphopeptide present in a spot and, through the possibility of 2D-PP pattern matching, provides the opportunity to qualitatively and quantitatively follow the course of phosphorylation at specific sites in phosphoproteins isolated from cells or tissue representing different states. For sample amounts significantly below the picomole limit, the phosphopeptide signal is usually obscured by contaminants extracted from the plate. In these cases peptide separation by capillary LC or gel electrophoresis is advised.

The mass spectrometric methods used for the determination of phosphorylation sites in phosphopeptides are variations of two basic themes. The first relies on the chemical lability of the phosphate ester bonds under the conditions found in the collision cell or at the ion source in an ESI instrument, or during PSD in MALDI-MS. Phosphopeptides are therefore identified by diagnostic ions that result from fragmentation of the phosphoester bond. The second theme relies on the detection of the mass added to a peptide by the phosphate ester group. Typically, in protein phosphorylation studies, the amino acid sequence of the protein investigated is known. Therefore phosphopeptides derived from the protein can, in principle, be detected by a net mass differential of 80 u which is caused by the addition of a phosphate group to a serine, threonine or tyrosine residue. Thus, if a peptide mass map is acquired, usually by MALDI-MS, of an enzymatic digest of the phospho-protein and if the measured masses are correlated with the masses expected from the protein, the phosphopeptides in the mixture can be identified. Both of these methods do not *a priori* identify the phosphorylated site(s) within the peptide. This can only be achieved by a product ion scan in a tandem mass spectrometer that effectively sequences the peptide. However, in cases in which the peptide sequence contains a single possible phosphorylation site, the phosphorylated residue is effectively located by the conclusive identification of a peptide as a phosphopeptide. The chances of identifying the site of phosphorylation without the need for product ion scans is therefore significantly increased if the type of amino acid phosphorylated has been determined by the methods described in Section 5.5.

The following specific MS scans, generically described above, have been success-fully used for phosphopeptide analysis. Here we describe their use for phosphopep-tide analysis specifically and provide insight into the advantages and disadvantages where appropriate. In general methods that produce some sort of phosphate specific ion (i.e. a diagnostic ion) are useful where incorporation of ^{32}P is not possible or where the label has decayed past the point of detection. If desired such scans can also be used on a radiolabeled phosphopeptide because the contribution to the mass of the phosphopeptide from the radioactive isotope of phosphate is so small that it can be ignored. It should be noted that, while some researchers are concerned about contaminating a mass spectrometer with ^{32}P samples, this possibility can be easily avoided by waiting a sufficient number of half-lives (2 weeks for ^{32}P) before conducting mass spectrometric experiments. Experiments requiring the label, such as 2D-PP mapping, can be done while there is sufficient radioactivity available.

In-source CID. Diagnostic ions generated by ESI in negative ion mode by in-source CID can be monitored to identify phosphopeptides (Huddleston *et al.*, 1993; Hunter and Games, 1994) by observation of $H_2PO_4^-$ (97 u), PO_3^- (79 u) and PO_2^- (63 u).

When in-source fragmentation (CID) is combined with on-line HPLC, a chromatographic trace is established that identifies the elution time of a phosphopeptide. If carried out as done by Carr and co-workers, then both the chromatographic marker and the phosphopeptide molecular weight are determined in the same scan (Huddleston *et al*., 1993). This is accomplished by use of a high orifice potential across the two skimmers prior to Q1, while the low m/z range is scanned for diagnostic ions. Then the orifice potential is returned to a normal voltage that does not induce fragmentation and a scan of high m/z conducted. A similar experiment can be done on instruments with a heated capillary in place of the first skimmer (Aebersold *et al*., 1998). This method uses an alternating scan approach where selected ion monitoring of appropriate diagnostic ions at a high octapole offset voltage is followed by two full scans. The first full scan is conducted at the same high offset voltage as the selected ion monitoring (SIM) experiment, providing signals for the deprotonated phosphopeptide molecular ion and the phosphopeptide molecular ion minus phosphate. Finally a second full scan is done at a normal octapole offset voltage to provide a reference to the full scan at high octapole offset. This series of three MS scans is repeated continuously throughout the LC separation. Such an experiment provides the same information as the methods of Carr and co-workers, but because SIM is used, rather than scanning, to detect diagnostic ions, one cycle of scans is faster and potentially more sensitive. The first full scan at high octapole offset is compared to the full scan at low octapole offset to provide a clue as to which peptide ion is phosphorylated when, as often occurs with peptide mixtures on microcapillary columns, peptides coelute. Such techniques are generally good down to a few femtomoles of phosphopeptide standards loaded on column, but often with real *in vitro* or *in vivo* samples sensitivities are in the picomole range.

Of course it would be advantageous in one microcapillary separation on-line with ESI-MS to detect phosphopeptides in negative ion mode and then switch to positive ion mode for a CID experiment because negative ion CID spectra generally produce insufficient fragment ions for sequence elucidation. To date this has been technically difficult to do in the same analytical run where diagnostic ion scanning is carried out, but it may be possible in the future using nonscanning mass spectrometers. The difficulty in doing such an experiment on a scanning mass spectrometer such as a quadrupole lies in the time required to switch between positive and negative ion mode in real time.

Neutral loss scan. Neutral loss scanning for phosphopeptide detection and analysis was first described by Covey *et al*. (1991) and further developed by Huddleston *et al*. (1993). It is carried out in positive ion mode with ESI in a triple quadrupole MS. Instead of using Q1 to select specific ions for fragmentation in Q2, Q1 is scanned. Q3 is also scanned but not over the same m/z range as Q1. Rather the two quadrupoles are scanned over two different m/z ranges the difference of which corresponds to the m/z value for the neutral molecule that is being lost. For neutral loss of phosphate from a $(M + 2H)^{2+}$ phosphopeptide ion that offset value is 49 m/z. From a mechanistic view only phosphoserine and phosphothreonine may undergo neutral loss of 98 by β-elimination (Gibson and Cohen, 1990). The method has not been as popular as the aforementioned in-source CID methods because of false positives and the need to know the charge state of the ion losing phosphate. An advantage of the method is that it is carried out in positive ion mode and can be used with data-dependent scanning to acquire CID in the same experiment using the detection of a neutral loss of phosphate as a trigger to initiate CID.

Precursor ion scan. This method involves negative ion ESI with continuous scanning of Q1. All ions are fragmented in Q2 and Q3 passes only one ion, which for phosphopeptides is usually 79 *m/z* corresponding to the loss of PO_3^- group. Consequently, the resultant mass spectrum shows only ions which lost 79 *m/z* (Wilm *et al.*, 1996). This greatly simplifies mixture analysis and is best done during direct infusion with a nanospray source. Again as described for the phosphate diagnostic ion scans there is a problem associated with sequencing in positive ion mode immediately after detecting the loss of phosphate in negative ion mode.

Product ion scanning. Often the information obtained by the specific scanning methods described above is not sufficient for identification of the phosphorylated residue in a phosphopeptide. In fact the above methods of in-source CID, neutral loss and precursor ion scanning are designed to identify phosphopeptides from nonphosphopeptides rather than to provide sequence information. These three methods can successfully identify a phosphorylated residue only in the case where the peptide sequence is known and it can be shown that only one of the three types of hydroxyl amino acids is present. Furthermore this reduction by default to an answer works only if there is, for instance, only one serine, threonine or tyrosine residue in the peptide sequence. Another situation that provides an answer to the location of phosphate in the peptide sequence is the case where the peptide sequence is known and phosphoamino acid analysis shows that only one of the three hydroxyl amino acids is present in the phosphoprotein. Again this reduces to the correct answer when there is only one of any of the three possible phosphorylated amino acids present in the peptide sequence. When neither of these situations exists then the generation and interpretation of CID spectra is generally required to conclusively determine the covalent structure of the phosphopeptide.

As a general trend during low-energy CID of phosphopeptides, it has been observed that phosphate tends to be lost from phosphoserine more readily than from phosphothreonine and from phosphothreonine more readily than from phosphotyrosine. Phosphate is generally lost from shorter phosphopeptides more readily than longer phosphopeptides because roughly the same amount of energy for collision is dispersed across fewer bonds. Interestingly, it is rare to observe the immonium ions for phosphoamino acids that form as a result of dehydroalanine and dehydroamino-2-butyric acid breaking down after loss of phosphate. However, using an ion trap mass spectrometer (DeGnore and Qin, 1998) and monitoring the CID of a phosphopeptide ion, dehydroamino-2-butyric acid was observed in place of threonine in the peptide fragment ion.

Post-source decay MALDI. This type of experiment is carried out on a MALDI-TOF mass spectrometer and provides a method to sequence peptides in a single-stage instrument by observing their metastable decay. It has been described above. This methodology has been successfully applied to sequence analysis of phosphopeptides (Annan and Carr, 1996; Neville *et al.*, 1997).

Enzymatic and chemical dephosphorylation. This is a method that does not receive much attention but can provide the location of phosphate in a phosphopeptide and allow a phosphopeptide to be identified in a mixture of predominantly nonphosphopeptides and involves the use of phosphatases (Zhang *et al.*, 1998). Using a simple MALDI-TOF instrument the masses of the peptides resulting from proteolytic digestion of the phosphoprotein are acquired. Then the same sample is treated with phosphatase to remove phosphate (from serine, threonine and tyrosine) and the masses acquired again. Now any mass that decreases by 80 u will be a clue as to which peptide is phosphorylated. An advantage to conducting such an experiment by MALDI is that peptide ions produced tend to be singly rather than multiply protonated and this makes interpretation easier than with ESI. In a different approach chemical treatment of the serine or threonine phosphopeptides results in removal of both the phosphate and dehydration of the serine or threonine residue carrying the modification, effectively 'tagging' the site of former modification (Patterson, 1998). Base treatment of the phosphopeptides results in β-elimination of the phosphate and concomitant dehydration of the amino acid side chain leaving amino acid residues of a unique mass (18 u less than normal). This approach has been applied to identification of sites of phosphorylation in the highly phosphorylated profilaggrin using MS/MS (Resing *et al.*, 1995).

5.7 Emerging methods and future directions in phosphoprotein analysis

The current limitations with phosphoprotein analysis are centered around the difficulties of working directly with *in vivo* ^{32}P-labeled phosphoproteins. Two issues cause these difficulties. The first is the often low stoichiometry of phosphorylation, requiring some sort of phosphopeptide enrichment prior to mass spectrometric analysis. The second is the inability to produce enough *in vivo* ^{32}P-labeled peptides for direct mass spectrometric characterization. In the second case, *in vitro* ^{32}P-labeled peptides must be used, requiring the phosphoproteins from *in vivo* and *in vitro* experiments to be shown to be identical by 2D-PP mapping. This makes the entire process from phosphoprotein purification to phosphopeptide sequencing time consuming. This also assumes that there is a kinase and conditions known which can reproduce the *in vivo* phosphorylation events *in vitro*. Therefore, methods are needed that allow direct analysis of *in vivo* phosphorylated proteins.

While it is unlikely that new methods for directly circumventing the problems associated with producing enough *in vivo* ^{32}P-labeled protein will be solved soon, more sensitive mass spectrometric methods translate into reduced sample consumption. For instance FT-ICR-MS, using either MALDI or ESI, is proving to be an order of magnitude more sensitive for peptide detection than standard triple quadrupole and ion trap technology. Such technology can routinely and accurately measure masses of peptides at the level of tens of attomoles of peptide loaded onto a MALDI probe (Solouki *et al.*, 1995). While actual MS/MS experiments in an ICR cell often return sensitivity to the femtomole range, use of broad-band dissociation of individual peptides eluting from an HPLC column can be done at the attomole level. For instance a 50 μm i.d. microcapillary HPLC column was used to introduce 9 amol of phosphorylated angiotensin II into an ESI source that was coupled to a 7 T FT-ICR-MS, allowing fragmentation and identification of the standard peptide by broad-band dissociation [Bruce, J.E., Goodlett, D.R., Smith, R.D. and Aebersold, R. (1998) Unpublished data from a collaboration at Battelle Memorial Institute's Environmental and Molecular Sciences Laboratory in Richland, WA]. This sort of sensitivity will offer advantages in terms of sample management when faced with limited amounts of *in vivo* material. Improvements in instrument control capabilities may also contribute to increased performance in phosphoprotein analysis.

It would, for example, be highly desirable if the presence of a phosphopeptide in a sample could be detected, the mass of the peptide could be determined and if the CID spectrum of the phosphopeptide could be generated all in a single automated LC-MS/MS operation. This could be achieved by implementing the capability to rapidly switch between negative ion MS and positive ion MS/MS on a tandem mass spectrometer. In such an instrument the presence of a phosphopeptide would be detected and the molecular weight measured by one of the negative ion methods mentioned, such as in-source CID. The detection of a phosphopeptide would be used as a data-dependent trigger to switch the instrument to positive ion mode and select the detected phosphopeptide ion for CID. We expect that such instrumentation will be available in the future.

A significant component of the emerging field of proteomics is the need to characterize the state of modification of all the proteins in a complex sample. An important extension and future direction in the field of phosphoprotein analysis therefore concerns the need to analyze the phosphoproteins in a sample globally and quantitatively. Currently, all the methods used for phosphoprotein analysis are focused on a single protein. It will be necessary to simplify and streamline the current methods to adapt them for high-throughput studies or to develop fundamentally different instrumentation and methods that will be suitable to perform proteome-wide phosphorylation studies.

6. Present and future challenges and opportunities

Protein identification and characterization are essential components of the discovery phase of any proteomics effort. For this to be effective it has to be performed in a high-throughput manner, efficiently and with high accuracy and sensitivity. If proteins are separated by 2-DE-separated proteins, one scenario for achieving these objectives is the implementation of a robotic system with a specialized cutting tool to excise predetermined spots from the gel and a computer-controlled protein digestion module. The spots are detected by an image analysis system which detects stained proteins in the gel or, after electrotransfer, on a membrane. The imaging system also has the capacity to do subtractive pattern analysis and therefore to highlight those proteins which are quantitatively or qualitatively different between two samples. Specific spots of interest are then marked, the excision tool is automatically directed to the coordinates of the designated spot and the protein spot is cut out. The excised spots are transferred to containers for enzymatic digestion. The whole cleavage process, which minimally consists of washing the gel slice, the addition of buffer and enzyme, extraction of peptides and sample clean-up is then usually handled by a separate liquid handling robot with the capacity to process multiple samples automatically and in parallel (Traini et al., 1998). The same robot may also spot an aliquot of each digestion onto a MALDI plate for subsequent unattended computer-controlled MALDI-MS (e.g. Onnerfjord et al., 1999; Traini et al., 1998). Masses from the MALDI-MS analysis can then be used in peptide-mass searches which in high-throughput facilities are also performed automatically. Any proteins not identified by this strategy can have the remaining enzymatic digest analyzed by a lower-throughput (but still easily automatable) technique such as LC-MS/MS for generation of fragment ion spectra in a data-dependent manner, which then are automatically searched against sequence databases (Ducret et al., 1998). The remaining enzymatic solution could also be analyzed by a still lower-throughput (but very sensitive) method of nanospray ESI-MS/MS (Shevchenko et al., 1996). Both the LC-MS/MS and the nano-ESI-MS/MS can benefit from the masses already determined to be present in the sample by MALDI-MS. The iterative strategy described here uses the highest-throughput approach first to identify as many proteins as possible before lower-throughput (and more expensive) approaches are employed. The MALDI-MS analysis can also exclude from further analysis those proteins which did yield peptides following digestion with the standardized conditions employed. Systems such as this may only reside in core facilities or specialized laboratories due to the expense and need for large numbers of samples to justify the expense. Furthermore, they are best suited for the analysis of proteins from (microbial) species for which the complete genome sequence has been determined. The difficulty of using MALDI-MS to identify proteins by correlation with EST databases currently limits the value of such systems for the identification of higher eukaryotic, particularly human and murine proteins. The development of high-throughput methods for the identification of proteins by MS/MS has progressed more slowly, mainly due to the complexities of the instrumentation and the data processing requirements involved. The development of microfabricated devices as sample delivery modules to ESI-MS/MS instruments (e.g. Figeys et al., 1998) and the possibility of identifying by MS/MS the components of relatively complex protein mixtures without the need to purify them to homogeneity (McCormack et al., 1997) are promising developments which suggest that the large-scale identification of proteins by the MS/MS route is rapidly advancing.

Higher throughput could be achieved if the proteins could be analyzed directly from the gel or following transfer to a membrane, directly on it (using MALDI-MS for analysis). This would preclude the requirement for excision of the spots from a 2-DE separation and the extensive liquid handling. However, such a method would require liquid handling for the delivery of enzymes and subsequently the matrix to the gel/membrane prior to MALDI-MS analysis in a way that does not compromise the spatial resolution achieved by gel electrophoresis. A number of groups have been advancing this idea by analyzing gel-separated proteins either directly from the gel using UV-MALDI-MS (Loo *et al.*, 1996) or following transfer to a membrane by either UV- or IR-MALDI-MS (Eckerskorn *et al.*, 1997; Patterson, 1995; Strupat *et al.*, 1994; Vestling and Fenselau, 1995). In a series of proof-of-concept experiments a 'virtual two-dimensional gel' was generated by Loo *et al.* (1996) by analyzing IEF (charge)-separated proteins directly from the gel with UV-MALDI-MS to generate the mass dimension. Matrix was spotted onto the surface of the gel to minimize diffusion and therefore retain spatial resolution. This analysis had to be performed on ultra-thin gels (0.35 mm) so that ionization could be effected. The same group demonstrated that cyanogen bromide cleavage of these proteins was also compatible with this analysis (Loo *et al.*, 1997). However, PSD-MALDI-MS analysis generally requires smaller-sized peptide fragments than are generated by cyanogen bromide, and it has yet to be demonstrated whether PSD-MALDI-MS can be performed directly from a gel.

Many ionization-suppressing contaminants from gels can be washed away following transfer of the proteins to a membrane, hence ionization of gel-separated proteins from membranes has been evaluated by a number of groups. When employing this approach, particular care has to be taken when introducing matrix as the use of organic solvents (normally used in matrices) will result in diffusion of the protein and loss of spatial resolution (as described above; Vestling and Fenselau, 1994a,b). In systematic evaluations of different membranes, PVDF membranes or derivatives of these have been found to be most suitable for this type of experiment (Vestling and Fenselau, 1995). Both UV and IR lasers have been used to ionize proteins from membranes. Both have been used successfully, but IR lasers appear to be more effective at extraction of proteins embedded in the membrane (Schreiner *et al.*, 1996; Strupat *et al.*, 1994). Eckerskorn *et al.* (1997) have demonstrated that it is possible to ionize 2-DE-separated proteins electroblotted to PVDF using IR-MALDI-MS. They were able to transform the data into a contour plot reflecting the abundance of the protein species, thereby negating the need for visualizing the proteins prior to MALDI-MS analysis. However, significant data handling issues need to be addressed before such an analysis is feasible for a full-sized 2-DE blot. In addition, the measurement of intact mass of proteins from gel-separated proteins is an important aspect of protein characterization, but does not significantly aid in protein identification. To identify the protein it must be digested (for peptide-mass searching strategies) and/or the peptides generated must be fragmented (for unin-terpreted fragment ion search strategies). Peptide digestions have been performed on proteins transferred to membranes and analyzed by either UV- or IR-MALDI-MS (Eckerskorn *et al.*, 1997; Vestling and Fenselau, 1994b). However, only PSD-(UV-) MALDI-MS data [not PSD-(IR-)MALDI-MS] have been published to date (Fabris *et al.*, 1995). Hochstrasser (1998) has introduced the concept of a 'molecular scanner' for 2-DE separated proteins electroblotted to a membrane where MALDI-MS is used to analyze all of the spots in an automated fashion. To this end Hochstrasser's group has developed a method for massively parallel digestion of all proteins separated in the 2-DE by placing a membrane with bound trypsin between the gel and the PVDF membrane. By employing a pulsed-field during the electroblotting process, proteins are slowly transferred from the gel through the trypsin-coated membrane. Remaining intact protein, together with tryptic fragments, is then captured by the PVDF membrane and can be analyzed by MALDI-MS (UV or IR has been successful; Hochstrasser, 1998). Although all of these approaches are in early stages of development they demonstrate that such analyses are possible. Once any problems with PSD-MALDI-MS analysis from peptides on membranes have been resolved, most importantly automation of the process, the power of this overall strategy can be tested fully. The single greatest problem with the approach at present is the limited number of peptides that undergo PSD yielding sufficient numbers of fragment ions for successful database searches, but this may well be addressed in the future through the use of chemical modification procedures to both enhance ionization and fragmentation of the peptides (e.g. Spengler *et al.*, 1997).

Although 2-DE is the most common form of protein separation for proteome studies due to its high resolving power, research into chromatography-based approaches, which can be quite easily integrated with on-line MS/MS, are also being developed. Opiteck *et al.* (1997b) have used 2-D chromatography (cation exchange followed by reverse phase with two separate sets of LC pumps) to separate complex protein mixtures such as an *E. coli* lysate for subsequent on-line ESI-MS analysis, all in a 2 h run time. They were able to identify an overexpressed protein in *E. coli* by performing two separate runs, using lysate from uninduced and induced *E. coli*. The same group has also demonstrated the utility of the approach for separation of peptide mixtures (Opiteck *et al.*, 1997a). Hsieh *et al.* (1996) used a computer-controlled setup with an autosampler, five columns, and three 10-port switching valves to allow a series of steps to be performed on-line, obviating the need for any manual transfers of materials. The strategy included the following series of steps for immunoaffinity capture and identification of proteins from mixtures: immunoaffinity chromatography, desalting and buffer exchange on a mixed-bed strong ion-exchange absorbent, enzymatic digestion on an immobilized trypsin column, capture of peptides on a short perfusion capillary reversed-phase column, and final separation on an analytical reversed-phase column with on-line MS/MS analysis (Hsieh *et al.*, 1996).

For the most accurate identification of protein, sequence-specific fragment ion spectra need to be generated, whether the initial protein separation was 2-DE or a targeted chromatography-based approach. In this regard, although the uninterpreted fragment ion search programs are powerful in their current format, they do not attempt to utilize the additional information content of a fragment ion spectrum – that of the relative abundances of the fragment ions. The current search programs compare mass lists of observed versus expected. When a scientist evaluates a spectrum, knowledge of the susceptibility of various peptide bonds to fragmentation is considered because a fragment ion spectrum can be thought of as a plot of the frequency of fragmentation at each peptide bond, and the peptide bonds adjacent to some residues become favored cleavage sites in the gas-phase. When search programs are able to utilize relative intensity information, the confidence level of identifications should rise to the level where identifications can truly be made automatically. This is an important goal for any large-scale proteomics effort.

Acknowledgments

D.R.G. and R.A. would like to acknowledge funding from the National Science Foundation Science and Technology Center for Molecular Biotechnology, Kinetek Pharmaceuticals and a research grant from the National Institute of Health to R.A. (no. 1RO1 A1 41109-01). D.R.G. thanks Dr J.D. Watts for helpful discussions on phosphoprotein analysis.

References

Aebersold, R. and Leavitt, J. (1990) Sequence analysis of proteins separated by polyacrylamide gel electrophoresis: towards an integrated protein database. *Electrophoresis* **11**: 517–527.

Aebersold, R. and Patterson, S.D. (1998) Current problems and technical solutions in protein biochemistry. In: *Proteins: Analysis and Design* (ed. R.H. Angeletti). Academic Press, San Diego, CA, pp. 3–120.

Aebersold, R.H., Teplow, D.B., Hood, L.E. and Kent, S.B.H. (1986) Electroblotting onto activated glass. High efficiency preparation of proteins from analytical sodium dodecyl sulfate-polyacrylamide gels for direct sequence analysis. *J. Biol. Chem.* **261**: 4229–4238.

Aebersold, R.H., Leavitt, J., Saavedra, R.A., Hood, L.E. and Kent, S.B. (1987) Internal amino acid sequence analysis of proteins separated by one- or two-dimensional gel electrophoresis after in situ protease digestion on nitrocellulose. *Proc. Natl Acad. Sci. USA* **84**: 6970–6974.

Aebersold, R., Watts, J.D., Morrison, H.D. and Bures, E.J. (1991) Determination of the site of tyrosine phosphorylation at the low picomole level by automated solid-phase sequence analysis. *Anal. Biochem.* **199**: 51–60.

Aebersold, R., Figeys, D., Gygi, S., Corthals, G., Haynes, P., Rist, B., Sherman, J., Shang, Y. and Goodlett, D. (1998) Towards an integrated technology for the generation of multidimensional protein expression maps. *J. Protein Chem.* **17**: 533–535.

Affolter, M., Watts, J.D., Krebs, D.L. and Aebersold, R. (1994) Evaluation of 2-dimensional phosphopeptide maps by electrospray-ionization mass-spectrometry of recovered peptides. *Anal. Biochem.* **223**: 74–81.

Annan, R.S. and Carr, S.A. (1996) Phosphopeptide analysis by matrix-assisted laser desorption time-of-flight mass spectrometry. *Anal. Chem.* **68**: 3413–3421.

Bartlet-Jones, M., Jeffery, W.A., Hansen, H.F. and Pappin, D.J.C. (1994) Peptide ladder sequencing by mass-spectrometry using a novel, volatile degradation reagent. *Rapid Commun. Mass Spectrom.* **8**: 737–742.

Biemann, K. (1990) Nomenclature for peptide fragment ions (positive ions). [Appendix 5]. *Methods Enzymol.* **193**: 886–887.

Boyle, W.J., van der Geer, P. and Hunter, T. (1991) Phosphopeptide mapping and phosphoamino acid analysis by two-dimensional separation on thin-layer cellulose plates. *Methods Enzymol.* **201**: 110–149.

Brown, J.L. (1979) A comparison of the turnover of amino-terminally acetylated and non-acetylated mouse L-cell proteins. *J. Biol. Chem.* **254**: 1447–1449.

Brown, J.L. and Roberts, W.K. (1976) Evidence that approximately eighty percent of the soluble proteins from Ehrlich ascites cells are amino-terminally acetylated. *J. Biol. Chem.* **251**: 1009–1014.

Brown, R.S. and Lennon, J.J. (1995) Mass resolution improvement by incorporation of pulsed ion extraction/ionization linear time-of-flight mass spectrometry. *Anal. Chem.* **67**: 1998–2003.

Carr, S.A. and Annan, R.S. (1997) Overview of peptide and protein analysis by mass spectrometry. In: *Current Protocols in Molecular Biology* (eds F.M. Ausubel, R. Brent, R.E. Kingston, D.D. Moore, J.G. Seidman, J.A. Smith and K. Struhl). Wiley, New York, pp. 10.21.1–10.21.27.

Chait, B.T., Wang, R., Beavis, R.C. and Kent, S.B.H. (1993) Protein ladder sequencing. *Science* **262:** 89–92.

Charbonneau, H. and Tonks, N.K. (1992) 1002 Protein phosphatases? *Annu. Rev. Cell Biol.* **8:** 463–493.

Chaurand, P., Luetzenkirchen, F. and Spengler, B. (1999) Peptide and protein identification by matrix-assisted laser desorption ionization (MALDI) and MALDI-post-source decay time-of-flight mass spectrometry. *J. Am. Soc. Mass Spectrom.* **10:** 91–103.

Cordwell, S.J., Wilkins, M.R., Cerpa-Poljak, A., Gooley, A.A., Duncan, M., Williams, K.L. and Humphery-Smith, I. (1995) Cross-species identification of proteins separated by two-dimensional gel electrophoresis using matrix-assisted laser desorption ionization/time-of-flight mass spectrometry and amino acid composition. *Electrophoresis* **16:** 438–443.

Cornish, T.J. and Cotter, R.J. (1994) A curved field reflectron time-of-flight mass-spectrometer for the simultaneous focusing of metastable product ions. *Rapid Commun. Mass Spectrom.* **8:** 781–785.

Courchesne, P.L. and Patterson, S.D. (1998) Identification of proteins by matrix-assisted laser desorption/ionization mass spectrometry using peptide and fragment-ion masses. *Methods Mol. Biol.* **112:** 487–511.

Courchesne, P.L., Jones, M.D., Robinson, J.R., Spahr, C.S., McCracken, S., Bentley, D.L., Luethy, R. and Patterson, S.D. (1998) Optimization of capillary chromatography ion-trap mass spectrometry for identification of gel-separated proteins. *Electrophoresis* **19:** 956–967.

Covey, T.R., Huang, E.C. and Henion, J.D. (1991) Structural characterization of protein tryptic peptides via liquid chromatography/mass spectrometry and collision induced dissociation of their doubly charged molecular ions. *Anal. Chem.* **63:** 1193–1200.

Craig, A.G., Fischer, W.H., Rivier, J.E., McIntosh, J.M. and Gray, W.R. (1995) MS based scanning methodologies applied to *Conus* venom. In: *Techniques in Protein Chemistry VI* (ed. J.W. Crabb). Academic Press, San Diego, CA, pp. 31–38.

Davis, M.T. and Lee, T.D. (1997) Variable flow liquid chromatography-tandem mass spectrometry and the comprehensive analysis of complex protein digest mixtures. *J. Am. Soc. Mass Spectrom.* **8:** 1059–1069.

DeGnore, J.P. and Qin, J. (1998) Fragmentation of phosphopeptides in an ion trap mass spectrometer. *J. Am. Soc. Mass Spectrom.* **9:** 1175–1188.

Duclos, B., Marcandier, S. and Cozzone, A.J. (1991) Chemical properties and separation of phosphoamino acids by thin-layer chromatography and/or electophoresis. *Methods Enzymol.* **201:** 10–21.

Ducret, A., van Oostveen, I., Eng, J.K., Yates, J.R. III and Aebersold, R. (1998) High throughput protein characterization by automated reverse-phase chromatography/electrospray tandem mass spectrometry. *Protein Sci.* **7:** 706–719.

Eckerskorn, C., Strupat, K., Schleuder, D., Hochstrasser, D., Sanchez, J.-C., Lottspeich, F. and Hillenkamp, F. (1997) Analysis of proteins by direct-scanning infrared-MALDI mass spectrometry after 2D-PAGE separation and electroblotting. *Anal. Chem.* **69:** 2888–2892.

Eng, J.K., McCormack, A.L. and Yates, J.R. III (1994) An approach to correlate tandem mass spectral data of peptides with amino acid sequences in a protein database. *J. Am. Soc. Mass Spectrom.* **5:** 976–989.

Fabris, D., Vestling, M.M., Cordero, M.M., Doroshenko, V.M., Cotter, R.J. and Fenselau, C. (1995) Sequencing electroblotted proteins by tandem mass spectrometry. *Rapid Commun. Mass Spectrom.* **9:** 1051–1055.

Fenn, J.B., Mann, M., Meng, C.K., Wong, S.F. and Whitehouse, C.M. (1989) Electrospray ionization for mass spectrometry of large biomolecules. *Science* **246:** 64–71.

Fenyo, D., Qin, J. and Chait, B.T. (1998) Protein identification using mass spectrometric information. *Electrophoresis* **19:** 998–1005.

Figeys, D. and Aebersold, R. (1998) High sensitivity analysis of proteins and peptides by capillary electrophoresis-tandem mass spectrometry: recent devlopments in technology and applications. *Electrophoresis* **19:** 885–892.

Figeys, D., Gygi, S.P., McKinnon, G. and Aebersold, R. (1998) An integrated microfluidics-tandem mass spectrometry system for automated protein analysis. *Anal. Chem.* **70:** 3728–3734.

Fischer, E.H. and Krebs, E.G. (1989) Commentary on "the phosphorylase b to a converting enzyme of rabbit skeletal muscle". *Biochim. Biophys. Acta* **1000:** 297–301.

Fraser, C.M. and Fleischmann, R.D. (1997) Strategies for whole microbial genome sequencing and analysis. *Electrophoresis* **18:** 1207–1216.

Gatti, A. and Traugh, J.A. (1999) A two-dimensional peptide gel electrophoresis system for phosphopeptide mapping and amino acid sequencing. *Anal. Biochem.* **266:** 198–204.

Gibson, B.W. and Cohen, P. (1990) Liquid secondary ion mass spectrometry of phosphorylated and sulfated peptides and proteins. *Methods Enzymol.* **193:** 480–501.

Heffetz, D., Fridkin, M. and Zick, Y. (1989) Antibodies directed against phosphothreonine residues as potent tools for studying protein phosphorylation. *Eur. J. Biochem.* **182:** 343–348.

Henzel, W.J., Billeci, T.M., Stults, J.T., Wong, S.C., Grimley, C. and Watanabe, C. (1993) Identifying proteins from two-dimensional gels by molecular mass searching of peptide fragments in protein sequence databases. *Proc. Natl Acad. Sci. USA* **90:** 5011–5015.

Hewick, R.M., Hunkapiller, M.W., Hood, L.E. and Dreyer, W.J. (1981) A gas–liquid solid phase peptide and protein sequenator. *J. Biol. Chem.* **256:** 7990–7997.

Hildebrandt, E. and Fried, V.A. (1989) Phosphoamino acid analysis of protein immobilized on polyvinylidene difluoride membrane. *Anal. Biochem.* **177:** 407–412.

Hochstrasser, D.F. (1998) Proteome in perspective. *Clin. Chem. Lab. Med.* **36:** 825–836.

Hsieh, Y.L., Wang, H., Elicone, C., Mark, J., Martin, S.A. and Regnier, F. (1996) Automated analytical system for the examination of protein primary structure. *Anal. Chem.* **68:** 455–462.

Huang, Z.-H., Shen, T., Wu, J., Gage, D.A. and Watson, J.T. (1999) Protein sequencing by matrix-assisted laser desorption ionization-postsource decay-mass spectrometry analysis of the *N*-Tris(2,4,6-trimethoxyphenyl)phosphine-acetylated tryptic digests. *Anal. Biochem.* **268:** 305–317.

Huddleston, M.J., Annan, R.S., Bean, M.F. and Carr, S.A. (1993) Selective detection of phosphopeptides in complex mixtures by electrospray liquid chromatography/mass spectrometry. *J. Am. Soc. Mass Spectrom.* **4:** 710–717.

Hunt, D.F., Yates, J.R. III, Shabanowitz, J., Winston, S. and Hauer, C.R. (1986) Protein sequencing by tandem mass spectrometry. *Proc. Natl Acad. Sci. USA* **83:** 6233–6237.

Hunter, A.P. and Games, D.E. (1994) Chromatographic and mass spectrometric methods for the identification of phosphorylation sites in phosphoproteins. *Rapid Commun. Mass Spectrom.* **8:** 559–570.

Hunter, T. (1987) 1001 protein kinases. *Cell* **50:** 823–829.

James, P., Quadroni, M., Carafoli, E. and Gonnet, G. (1993) Protein identification by mass profile fingerprinting. *Biochem. Biophys. Res. Commun.* **195:** 58–64.

James, P., Quadroni, M., Carafoli, E. and Gonnet, G. (1994) Protein identification in DNA databases by peptide mass fingerprinting. *Protein Sci.* **3:** 1347–1350.

Jardine, I. (1990) Molecular weight analysis of proteins. *Methods Enzymol.* **193:** 441–455.

Jensen, O.N., Podtelejnikov, A. and Mann, M. (1996a) Delayed extraction improves specificity in database searches by matrix-assisted laser desorption/ionization peptide maps. *Rapid Commun. Mass Spectrom.* **10:** 1371–1378.

Jensen, O.N., Vorm, O. and Mann, M. (1996b) Sequence patterns produced by incomplete enzymatic digestion or one-step Edman degradation of peptide mixtures as probes for protein database searches. *Electrophoresis* **17:** 938–944.

Jensen, O.N., Podtelejnikov, A.V. and Mann, M. (1997) Identification of the components of simple protein mixtures by high accuracy peptide mass mapping and database searching. *Anal. Chem.* **69:** 4741–4750.

Jensen, O.N., Larsen, M.R. and Roepstorff, P. (1998) Mass spectrometric identification and microcharacterization of proteins from electrophoretic gels: strategies and applications. *Proteins* **2** (Suppl.): 74–89.

Johnson, L.N. and Barford, D. (1990) Glycogen phosphorylase. The structural basis of the allosteric response and comparison with other allosteric proteins. *J. Biol. Chem.* **265:** 2409–2412.

Jonscher, K.R. and Yates, J.R. III (1997) The quadupole ion trap mass spectrometer – a small solution to a big challenge. *Anal. Biochem.* **244:** 1–15.

Joung, I., Kim, T., Stolz, L.A., Payne, G., Winkler, D.G., Walsh, C.T., Strominger, J.L. and Shin, J. (1995) Modification of Ser59 in the unique N-terminal region of tyrosine kinase p56lck regulates specificity of its Src homology 2 domain. *Proc. Natl Acad. Sci. USA* **92:** 5778–5782.

Karas, M. and Hillenkamp, F. (1988) Laser desorption ionization of proteins with molecular masses exceeding 10 000 daltons. *Anal. Chem.* **60:** 2299–2301.

Katta, V., Chowdhury, S.K. and Chait, B.T. (1991) Use of a single-quadrupole mass spectrometer for collision-induced dissociation studies of multiply charged peptide ions produced by electrospray ionization. *Anal. Chem.* **63:** 174–178.

Kaufmann, R., Kirsch, D. and Spengler, B. (1994) Sequencing of peptides in a time-of-flight mass spectrometer: evaluation of postsource decay following matrix-assisted laser desorption ionization (MALDI). *Int. J. Mass Spectrom. Ion Processes* **131:** 355–385.

Krishna, R.G. and Wold, F. (1998) Posttranslational modifications. In: *Proteins: Analysis and Design* (ed. R.H. Angeletti). Academic Press, San Diego, CA, pp. 121–206.

Lamond, A.I. and Mann, M. (1997) Cell biology and the genome projects – a concerted strategy for characterizing multiprotein complexes by using mass spectrometry. *Trends Cell Biol.* **7:** 139–142.

Loo, J.A., Edmonds, C.G. and Smith, R.D. (1990) Primary sequence information from intact proteins by electrospray ionization tandem mass spectrometry. *Science* **248:** 201–204.

Loo, R.R.O., Stevenson, T.I., Mitchell, C., Loo, J.A. and Andrews, P.C. (1996) Mass-spectrometry of proteins directly from polyacrylamide gels. *Anal. Chem.* **68:** 1910–1917.

Loo, R.R.O., Mitchell, C., Stevenson, T.I., Martin, S.A., Hines, W., Juhasz, P., Patterson, D., Peltier, J., Loo, J.A. and Andrews, P.C. (1997) Sensitivity and mass accuracy for proteins analyzed directly from polyacrylamide gels: implications for proteome mapping. *Electrophoresis* **18:** 382–390.

Louris, J.N., Brodbelt-Lustig, J.S., Cooks, R.G., Glish, G.L., van Berkel, G.J. and McLuckey, S.A. (1990) Ion isolation and sequential stages of mass spectrometry in a quadrupole ion trap mass spectrometer. *Int. J. Mass Spectrom. Ion Processes* **96:** 117–137.

Mann, M. and Wilm, M. (1994) Error-tolerant identification of peptides in sequence databases by peptide sequence tags. *Anal. Chem.* **66:** 4390–4399.

Mann, M., Hojrup, P. and Roepstorff, P. (1993) Use of mass spectrometric molecular weight information to identify proteins in sequence databases. *Biol. Mass Spectrom.* **22:** 338–345.

Matsudaira, P. (1987) Sequence analysis from picomole quantities of proteins electroblotted onto polyvinylidene difluoride membranes. *J. Biol. Chem.* **262:** 10 035–10 038.

McCormack, A.L., Schieltz, D.M., Goode, B., Yang, S., Barnes, G., Drubin, D. and Yates, J.R. III (1997) Direct analysis and identification of proteins in mixtures by LC/MS/MS and database searching at the low-femtomole level. *Anal. Chem.* **69:** 767–776.

Meyer, H.E., Eisermann, B., Heber, M., Hoffmann-Posorske, E., Korte, H., Weigt, C., Wegner, A., Hutton, T., Donella-Deana, A. and Perich, J.W. (1993) Strategies for nonradioactive methods in the localization of phosphorylated amino acids in proteins. *FASEB J.* **7:** 776–782.

Moritz, R.L., Eddes, J., Ji, H., Reid, G.E. and Simpson, R.J. (1995) Rapid separation of proteins and peptides using conventional silica-based supports: identification of 2-D gel proteins following in-gel proteolysis. In: *Techniques in Protein Chemistry VI* (ed. J.W. Crabb). Academic Press, San Diego, CA, pp. 311–319.

Neville, D.C., Rozanas, C.R., Price, E.M., Gruis, D.B., Verkman, A.S. and Townsend, R.R. (1997) Evidence for phosphorylation of serine 753 in CFTR using a novel metal-ion affinity resin and matrix-assisted laser desorption mass spectrometry. *Protein Sci.* **6:** 2436–2445.

Nuwaysir, L.M. and Stults, J.T. (1993) Electrospray ionization mass sepctrometry of phospho-peptides isolated by on-line immobilized metal-ion affinity chromatography. *J. Am. Soc. Mass Spectrom.* **4:** 662–669.

Onnerfjord, P., Ekstrom, S., Bergquist, J., Nilson, J., Laurell, T. and Marko-Varga, G. (1999) Homogenous sample preparation for automated high throughput analysis with matrix-assisted laser desorption/ionization time-of-flight mass spectrometry. *Rapid Commun. Mass Spectrom.* **13:** 315–322.

Opiteck, G.J., Jorgenson, J.W. and Anderegg, R.J. (1997a) Two-dimensional SEC/RPLC coupled to mass spectrometry for the analysis of peptides. *Anal. Chem.* **69:** 2283–2291.

Opiteck, G.J., Lewis, K.C., Jorgenson, J.W. and Anderegg, R.J. (1997b) Comprehensive on-line LC/LC/MS of proteins. *Anal. Chem.* **69**: 1518–1524.

Pappin, D.J.C., Hojrup, P. and Bleasby, A.J. (1993) Rapid identification of proteins by peptide-mass fingerprinting. *Curr. Biol.* **3**: 327–332.

Pappin, D.J.C., Rahman, D., Hansen, H.F., Bartlet-Jones, M., Jeffery, W. and Bleasby, A.J. (1995) Chemistry, mass spectrometry and peptide-mass databases: Evolution of methods for the rapid identification and mapping of cellular proteins. In: *Mass Spectrometry in the Biological Sciences* (eds A.L. Burlingame and S.A. Carr). Humana Press, Totowa, NJ, pp. 135–150.

Patterson, D.H., Tarr, G.E., Regnier, F.E. and Martin, S.A. (1995) C-Terminal ladder sequencing via matrix-assisted laser-desorption mass-spectrometry coupled with carboxypeptidase-Y time-dependent and concentration-dependent digestions. *Anal. Chem.* **67**: 3971–3978.

Patterson, S.D. (1994) From electrophoretically separated protein to identification: strategies for sequence and mass analysis. *Anal. Biochem.* **221**: 1–15.

Patterson, S.D. (1995) Matrix-assisted laser-desorption/ionization mass spectrometric approaches for the identification of gel-separated proteins in the 5–50 pmol range. *Electrophoresis* **16**: 1104–1114.

Patterson, S.D. (1997) Identification of low to subpicomolar quantities of electrophoretically separated proteins: towards protein chemistry in the post-genome era. *Biochem. Soc. Trans.* **25**: 255–262.

Patterson, S.D. (1998) Protein identification and characterization by mass spectrometry. In: *Current Protocols in Molecular Biology* (eds F.M. Ausubel, R. Brent, R.E. Kingston, D.D. Moore, J.G. Seidman, J.A. Smith and K. Struhl). Wiley, New York, pp. 10.22.1–10.22.24.

Patterson, S.D. and Aebersold, R. (1995) Mass spectrometric approaches for the identification of gel-separated proteins. *Electrophoresis* **16**: 1791–1814.

Patterson, S.D. and Garrels, J.I. (1994) Two-dimensional gel electrophoresis of post-translational modifications. In: *Cell Biology: a Laboratory Handbook* (ed. J.E. Celis). Academic Press, New York, pp. 249–257.

Pluskal, M.G., Przekop, M.B., Kavonian, M.R., Vecoli, C. and Hicks, D.A. (1986) Immobilon™ PVDF transfer membrane: a new membrane substrate for western blotting of proteins. *BioTechniques* **4**: 272–283.

Qin, J., Herring, C.J. and Zhang, X. (1998) *De novo* peptide sequencing in an ion-trap mass spectrometer with ^{18}O labeling. *Rapid Commun. Mass Spectrom.* **12**: 209–216.

Reinhold, V.N., Reinhold, B.B. and Costello, C.E. (1995) Carbohydrate molecular weight profiling, sequence, linkage, and branching data: ES-MS and CID. *Anal. Chem.* **67**: 1772–1784.

Resing, K.A., Johnson, R.S. and Walsh, K.A. (1995) Mass spectrometric analysis of 21 phosphorylation sites in the internal repeat of rat profilaggrin, precursor of an intermediate filament associated protein. *Biochemistry* **34**: 9477–9487.

Roth, K.D.W., Huang, Z.H., Sadagopan, N. and Watson, J.T. (1998) Charge derivatization of peptides for analysis by mass spectrometry. *Mass Spectrom. Rev.* **17**: 255–274.

Schnolzer, M., Jedrzejewski, P. and Lehmann, W.D. (1996) Protease-catalyzed incorporation of O-18 into peptide-fragments and its application for protein sequencing by electrospray and matrix-assisted laser desorption/ionization mass-spectrometry. *Electrophoresis* **17**: 945–953.

Schreiner, M., Strupat, K., Lottspeich, F. and Eckerskorn, C. (1996) Ultraviolet matrix-assisted laser-desorption ionization-mass spectrometry of electroblotted proteins. *Electrophoresis* **17**: 954–961.

Schwartz, J.C. and Jardine, I. (1996) Quadrupole ion trap mass spectrometry. *Methods Enzymol.* **270**: 552–586.

Sechi, S. and Chait, B.T. (1998) Modification of cysteine residues by alkylation. A tool in peptide mapping and protein identification. *Anal. Chem.* **70**: 5150–5158.

Shevchenko, A., Jensen, O.N., Podtelejnikov, A.V., Sagliocco, F., Wilm, M., Vorm, O., Mortensen, P., Shevchenko, A., Boucherie, H. and Mann, M. (1996) Linking genome and proteome by mass spectrometry: large-scale identification of yeast proteins from two dimensional gels. *Proc. Natl Acad. Sci USA* **93**: 14 440–14 445.

Shevchenko, A., Chernushevich, I., Ens, W., Standin, K.G., Thomson, B., Wilm, M. and Mann, M. (1997) Rapid 'de novo' peptide sequencing by a combination of nanoelectrospray, isotopic labeling and a quadrupole/time-of-flight mass spectrometer. *Rapid Commun. Mass Spectrom.* **11**: 1015–1024.

Solouki, T., Marto, J.A., White, F.M., Guan, S. and Marshall, A.G. (1995) Attomole biomolecule mass analysis by matrix-assisted laser desorption/ionization Fourier transform ion cyclotron resonance. *Anal. Chem.* **67**: 4139–4144.

Soskic, V., Gorlach, M., Poznanovic, S., Boehmer, F.D. and Godovac-Zimmerman, J. (1999) Functional proteomics analysis of signal transduction pathways of the platelet-derived growth factor β receptor. *Biochemistry* **38**: 1757–1764.

Spengler, B., Luetzenkirchen, F., Metzger, S., Chaurand, P., Kaufmann, R., Jeffery, W., Bartlet-Jones, M. and Pappin, D.J.C. (1997) Peptide sequencing of charged derivatives by postsource decay MALDI mass spectrometry. *Int. J. Mass Spectrom. Ion Processes* **169/170**: 127–140.

Stahl, D.C., Swiderek, K.M., Davis, M.T. and Lee, T.D. (1996) Data-controlled automation of liquid-chromatography tandem mass-spectrometry analysis of peptide mixtures. *J. Am. Soc. Mass Spectrom.* **7**: 532–540.

Strupat, K., Karas, M., Hillenkamp, F., Eckerskorn, C. and Lottspeich, F. (1994) Matrix-assisted laser desorption ionization mass spectrometry of proteins electroblotted after polyacrylamide gel electrophoresis. *Anal. Chem.* **66**: 464–470.

Takach, E.J., Hines, W.M., Patterson, D.H., Juhasz, P., Falick, A.M., Vestal, M.L. and Martin, S.A. (1997) Accurate mass measurements using MALDI-TOF with delayed extraction. *J. Protein Chem.* **16**: 363–369.

Tempst, P., Link, A.J., Riviere, L.R., Fleming, M. and Elicone, C. (1990) Internal sequence analysis of proteins separated on polyacrylamide gels at the submicrogram level: improved methods, applications and gene cloning strategies. *Electrophoresis* **11**: 537–553.

Tempst, P., Posewitz, M.C. and Erdjument-Bromage, H. (1998) Micro-sample handling as a front-end to mass spectrometry. In: *Proceedings of the 46th American Society for Mass Spectrometry Conference on Mass Spectrometry and Allied Topics*, Orlando, FL, 31 May–4 June, 1998, p. 437.

Traini, M., Gooley, A.A., Ou, K., Wilkins, M.R., Tonella, L., Sanchez, J.-C., Hochstrasser, D.F. and Williams, K.L. (1998) Towards an automated approach for protein identification in proteome projects. *Electrophoresis* **19**: 1941–1949.

Vandekerckhove, J., Bauw, G., Puype, M., Van Damme, J. and Van Montagu, M. (1985) Protein-blotting on polybrene-coated glass-fiber sheets. *Eur. J. Biochem.* **152**: 9–19.

Vestal, M.L., Juhasz, P. and Martin, S.A. (1995) Delayed extraction matrix-assisted laser desorption time-of-flight mass spectrometry. *Rapid Commun. Mass Spectrom.* **9**: 1044–1050.

Vestling, M.M. and Fenselau, C. (1994a) Poly(vinylidene difluoride) membranes as the interface between laser desorption mass spectrometry, gel electrophoresis, and in situ proteolysis. *Anal. Chem.* **66**: 471–477.

Vestling, M.M. and Fenselau, C.C. (1994b) Protease digestions on PVDF membranes for matrix-assisted laser desorption mass spectrometry. In: *Techniques in Protein Chemistry V* (ed. J.W. Crabb). Academic Press, San Diego, CA, pp. 59–67.

Vestling, M.M. and Fenselau, C. (1995) Surfaces for interfacing protein gel electrophoresis directly with mass spectrometry. *Mass Spectrom. Rev.* **14**: 169–178.

Wang, R., Chait, B.T. and Kent, S.B.H. (1994) Protein ladder sequencing: towards automation. In: *Techniques in Protein Chemistry V* (ed. J.W. Crabb). Academic Press, San Diego, CA, pp. 19–26.

Wange, R.L., Isakov, N., Burke, T.R., Otaka, A., Roller, P.P., Watts, J.D., Aebersold, R. and Samelson, L.E. (1995) F-2(Pmp)(2)-Tam-Zeta(3), a novel competitive inhibitor of the binding of Zap-70 to the T-cell antigen receptor, blocks early T-cell signaling. *J. Biol. Chem.* **270**: 944–948.

Watts, J.D., Welham, M.J., Kalt, L., Schrader, J.W. and Aebersold, R. (1993) IL-2 stimulation of T lymphocytes induces sequential activation of mitogen-activated protein kinases and phosphorylation of p56lck at serine-59. *J. Immunol.* **151**: 6862–6871.

Watts, J.D., Affolter, M., Krebs, D.L., Wange, R.L., Samelson, L.E. and Aebersold, R. (1994) Identification by electrospray ionization mass spectrometry of the sites of tyrosine phosphorylation induced in activated Jurkat T cells on the protein tyrosine kinase ZAP-70. *J. Biol. Chem.* **269**: 29 520–29 529.

Watts, J.D., Brabb, T., Bures, E.J., Wange, R.L., Samelson, L.E. and Aebersold, R. (1996a)

Identification and characterization of a substrate specific for the T cell protein tyrosine kinase Zap-70. *FEBS Lett.* **398**: 217–222.

Watts, J.D., Affolter, M., Krebs, D.L., Wange, R.L., Samelson, L.E. and Aebersold, R. (1996b) Electrospray ionization mass spectrometric investigation of signal transduction pathways: determination of sites of inducible protein phosphorylation in activated T-cells. In: *Biochemical and Biotechnological Applications of Electrospray Ionization Mass Spectrometry* (ed. A.P. Snyder), ACS Symposium Series. American Chemical Society, Washington, DC, pp. 381–407.

Wettenhall, R.E., Aebersold, R.H. and Hood, L.E. (1991) Solid-phase sequencing of 32P-labeled phosphopeptides at picomole and subpicomole levels. *Methods Enzymol.* **201**: 186–199.

Whalen, S.G., Gingras, A.C., Amankwa, L., Mader, S., Branton, P.E., Aebersold, R. and Sonenberg, N. (1996) Phosphorylation of eIF-4E on serine 209 by protein kinase C is inhibited by the translational repressors, 4E-binding proteins. *J. Biol. Chem.* **271**: 11 831–11 837.

Wiley, W.C. and McLaren, I.H. (1953) Time-of-flight mass spectrometer with improved resolution. *Rev. Sci. Instrum.* **26**: 1150–1157.

Wilkins, M.R., Gasteiger, E., Wheeler, C.H., Lindskog, I., Sanchez, J.-C., Bairoch, A., Appel, R.D., Dunn, M.J. and Hochstrasser, D.F. (1998) Multiple parameter cross-species protein identification using MultiIdent – a World Wide Web accessible tool. *Electrophoresis* **19**: 3199–3206.

Wilm, M.S. and Mann, M. (1994) Electrospray and Taylor–Cone theory, Dole's beam of macromolecules at last? *Int. J. Mass Spectrom. Ion Processes* **136**: 167–180.

Wilm, M. and Mann, M. (1996) Analytical properties of the nanoelectrospray ion-source. *Anal. Chem.* **68**: 1–8.

Wilm, M., Neubauer, G. and Mann, M. (1996) Parent ion scans of unseparated peptide mixtures. *Anal. Chem.* **68**: 527–533.

Woods, A.S., Huang, A.Y.C., Cotter, R.J., Pasternack, G.R., Pardoll, D.M. and Jaffee, E.M. (1995) Simplified high-sensitivity sequencing of a major histocompatibility complex class I-associated immunoreactive peptide using matrix-assisted laser-desorption ionization mass-spectrometry. *Anal. Biochem.* **226**: 15–25.

Yates, J.R. III, Speicher, S., Griffin, P.R. and Hunkapiller, T. (1993) Peptide mass maps: a highly informative approach to protein identification. *Anal. Biochem.* **214**: 397–408.

Yates, J.R. III, McCormack, A.L., Hayden, J.B. and Davey, M.P. (1994) Sequencing peptides derived from the Class II major histocompatibility complex by tandem mass spectrometry. In: *Cell Biology: a Laboratory Handbook* (ed. J.E. Celis). Academic Press, New York, pp. 380–388.

Yates, J.R. III, Eng, J.K. and McCormack, A.L. (1995a) Mining genomes – correlating tandem mass-spectra of modified and unmodified peptides to sequences in nucleotide databases. *Anal. Chem.* **67**: 3202–3210.

Yates, J.R. III, Eng, J.K., McCormack, A.L. and Schieltz, D. (1995b) Method to correlate tandem mass-spectra of modified peptides to amino-acid-sequences in the protein database. *Anal. Chem.* **67**: 1426–1436.

Zhang, X., Herring, C.J., Romano, P.R., Szczepanowska, J., Brzeska, H., Hinnebusch, A.G. and Qin, J. (1998) Identification of phosphorylation sites in proteins separated by polyacrylamide gel electrophoresis. *Anal. Chem.* **70**: 2050–2059.

Chapter 6

Image analysis of two-dimensional gels

Klaus-Peter Pleissner, Helmut Oswald and Susan Wegner

1. Introduction

Two-dimensional gel electrophoresis (2-DE) is capable of separating thousands of polypeptides from a complex protein solution according to their electrical charge and molecular weight. Using staining procedures, proteins can be visualized as spots with varying properties or features, such as size, darkness/brightness and location on the gel (see Chapter 4). Several gel types and staining methods are available and differ from laboratory to laboratory. However, the common outcome is a two-dimensional image of a gel and a start for further processing to reveal the information on the thousands of proteins that the images contain. The large amount of data and the need for an objective, reproducible and quantitative analysis necessitate computer-assisted methods to process these images. Such gel images with their particular features, that is the spots, can be processed with a combination of general and special image-processing techniques. A typical work flow of 2-DE gel image analysis comprises the following steps:

(i)   scanning of gel image (data acquisition);
(ii)   image processing;
(iii) spot detection and quantitation;
(iv)   gel matching;
(v)   data analysis;
(vi)   data presentation and interpretation;
(vii) creation of 2-DE databases.

The scanning process converts the 'analog' gel image into a digital representation for further computer-based processing. It is important to define the parameters employed to extract the intrinsic information from protein spots. The overall goal of the subsequent image processing is the detection of the exact spot positions on the gel and the determination of their shape and intensity as a measure for the protein abundance. Computer imaging and the analysis of the 2-DE gel images are used as tools to provide such information. The computer imaging encompasses:

(i)   the pre-processing of the image along with noise and artifact removal; and
(ii)   detection and quantitation of the individual spots.

Proteomics, edited by S.R. Pennington and M.J. Dunn
© 2001 BIOS Scientific Publishers, Oxford.

Beyond the information obtained for the protein spots in a single image, the possible changes in the same protein spot across a series of different gels are of great interest for biological analysis. For such a comparison, registration of images by adequate landmarks is required to achieve comparable values. Landmarks are used to determine polynomes for geometric transformation and spots can be matched to each other hereafter. In this way corresponding protein spots can be pursued through a series of gels. Density profiles show the alterations of the spot intensity within gels. Following the analysis of profiles it is possible to characterize alteration trends. Profiles can be presented as graphs (bar charts), tables, scattergrams or maps. The results of the analysis may then be maintained in databases that link this information to the original gel images.

For cooperation between proteome research groups to promote synergistic data sharing and exchange, it is necessary to utilize software and database systems according to a paradigm of openness and federativeness. Gels, their images and databases of information extracted from them must be in a format that allows access and integration. Here, access to Internet technology can help to overcome the boundaries of individual laboratories.

All existing computerized gel analysis systems, the development of most of which was started in the late 1970s, follow the above-mentioned procedure for extraction of data from 2-DE gel images. Today, a number of noncommercial and commercial systems, such as QUEST/QUEST II/PDQUEST, TYCHO/KEPLER, GELLAB II, BioImage, HERMeS, PHORETIX and ELSIE/MELANIE II, are available and have found much acceptance in the 2-DE scientific community. The related hardware for these systems has undergone dramatic changes in the last 20 years, and is still evolving rapidly, but the basic algorithms used remain the same. Many aspects of computerized gel analysis have been described before (for instance, Anderson and Anderson, 1996; Anderson *et al.*, 1981; Appel *et al.*, 1991, 1997a,b; Bossinger *et al.*, 1979; Collins and Blose, 1992; Garrels, 1979, 1983, 1989; Lemkin and Lipkin, 1981a–c; Lemkin *et al.*, 1979, 1982; Miller, *et al.*, 1982; Olson and Miller, 1988; Prehm *et al.*, 1987; Serpico and Vernazza, 1987; Skolnick, 1982; Vincens and Tarroux, 1987a,b; Vo *et al.*, 1981). The present chapter gives an overview of 2-DE image analysis and the main features that are important for a better understanding of it.

2. Data acquisition

2.1 Image acquisition devices

Data are acquired from 2-DE gels by scanning and converted into a digital presentation. Obviously, data acquisition significantly predetermines the final quality of the analysis. Two technical parameters gear the satisfactory performance of conversion: (i) the spatial resolution for digitization; and (ii) the densitometric resolution for quantitation of the gray values. Some image acquisition devices serve for the conversion of gels into data and contribute to the quality of the results; such devices include cameras, densitometers (scanners), storage phosphor imagers and fluorescent imagers (see *Table 1*).

Table 1. Image acquisition devices mainly used for 2-DE.

	Scanning device			
Information carrier	Camera TV/CCD	Densitometer White-light laser	Storage phosphor imager	Fluorescent imager
X-ray film (radiolabeled sample)	X	X X		
Phosphor image plate			X	
Gel silver-, color-stained	X	X X		
Gel (fluorescent-labeled)	X			X

Various types of cameras are used, mostly for the documentation of gels but also for scanning. The quality of cameras depends on their input/output characteristic. Densitometers or scanners that can be distinguished in white light and laser scanners are most appropriate for the task of acquiring digital data from the 2-DE gel image. They are characterized by high densitometric and geometric (spatial) resolution, as well as almost linear characteristics, and may be used for wet or dried gels that have been stained with silver or Coomassie blue. Two-dimensional gel electrophoresis patterns of radioactively labeled samples (^{35}S, ^3H, ^{14}C or ^{32}P) are captured with X-ray films or storage phosphor imagers. When using X-ray films it is necessary to extend the dynamic range in densitometric or gray value resolution by using a number of X-ray film exposures, with increasing exposure time of the same gel, to visualize as many protein spots as possible. The differentially exposed films are merged to a composite digital 2-DE gel image. The individual gray values are linearly combined to increase the dynamic density range. To correlate the optical density with the radioactivity (disintegration per minute, d.p.m. value), a calibration strip, exposed together with the film, is measured and standardization curves are produced. Storage phosphor imagers provide the means to replace X-ray films and the merging of various exposed films, as they have a high dynamic range over five orders of magnitude for autoradiographic imaging (Patterson and Latter, 1993). An application of a storage phosphor imaging system combined with the QUEST II system has been described previously (Amess and Tolkovsky, 1995). Nevertheless, the usage of phosphor imagers is limited to radiolabeled protein mixtures.

3. Digital image processing

Image enhancement is widely used in the field of applied computer graphics with smoothing, contrast enhancement, edge detection and background subtraction being the most familiar graphical tools for the processing of images. The importance of these operators depends on the field of application. Whereas in some fields they are mainly used as image enhancers to improve the optical quality of images, analytical sciences may rely on these operators to provide exact and unequivocal data. One example of such a use is the analysis of protein spots in digitized images of 2-DE gels. Here, smoothing, contrast enhancement, edge detection and background subtraction allow an accurate approach to determining the shape and size of protein spots.

The different operators are normally defined within a 'neighborhood' of an image point or pixel. The commonly used neighborhoods are 3×3, 5×5 or 7×7 pixels. Based on a regular grid that represents the digital image, a 3×3 neighborhood, for example, is composed of those pixels that border the central image point.

Smoothing operators are used to eliminate statistical noise in the image. The main approach is the determination of the Gaussian, the average or the median value in a defined neighborhood of an image point. Contrast enhancement redefines gray values of an image to obtain higher contrasts between spots and image background. The purpose of edge detection is the localization of gray value changes in the image.

In general, the first or second derivative of the image function is computed. Various operators can be used in small neighborhoods around image points to determine the derivatives. Typical techniques are the gradient operator for the first, and the Laplace operator for the second derivative. Derivatives are apparently very noise sensitive, therefore they are often used in combination with Gaussian smoothing.

Background subtraction is usually applied to eliminate meaningless changes of the background, for example uneven or various background levels in different areas of the image. One approach is averaging of the darkest and of the brightest regions of the background and applying the average across the entire image. A more sophisticated method is known as 'rolling disk' (Sternberg, 1986) which removes background as well as streaks from a gel image (see *Figure 1*). It is possible to influence the efficacy of the removal, since a small radius produces a very clear image but also wipes out the tiniest spots.

4. Protein spot detection and quantitation

Remembering that the main target of this exercise is the analysis of 2-DE protein gels, spot detection and modeling of their shape is an important task. Spot detection results in an *n*-tuple of x/y-positions, shape parameters and integrated spot intensity. Protein spots have several optical features. They may appear separately, touch each other or overlap within the original gels and their subsequent computer images. They also may appear within streaks and blurs which occur due to technical problems of running and staining the gels and which differ from the background. They may also have no clear-cut contours and hence appear as a continuous distribution with circular or elliptical shape. The morphology of spots depends on the gel preparation, staining procedures and the digitization error of the gel analysis system. The variability and lack of homogeneity of gray values within individual spots and background require competent algorithms for spot detection and quantitation. A broad range of such algorithms does exist and these can be subdivided into Gaussian fitting and Laplacian of Gaussian (LOG) spot detection (Anderson and Anderson, 1981; Appel *et al.*, 1997b; Lemkin and Lipkin, 1981a,b; Wu *et al.*, 1994).

Another approach to spot detection is accomplished by the line analysis and chain

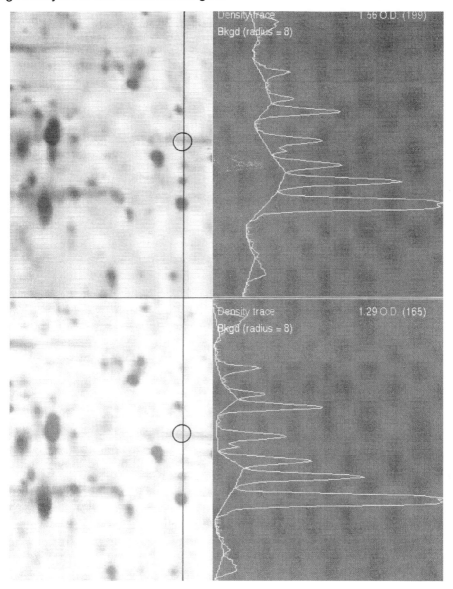

Figure 1. Background subtraction and streak removal using the geometrical model of the 'rolling disk'. The radius of the ball is adapted to the size of the spots. The upper part of the figure shows the background trace and the density trace of the spots along a selected line. The lower part illustrates the same features after background subtraction.

analysis algorithms first described by Garrels (1979). In this method, each vertical scan line is scrutinized for density peaks. The peaks on adjacent scan lines are assembled into 'chains' and the position of their centers (spot centers) is determined. After having located spots, their volumes (integrated spot intensities) can be determined by fitting a mathematical model to the shape of the spot. Two-dimensional Gaussian modeling is often used. Spots that deviate significantly from a single Gaussian shape can be modeled as a group of overlapping Gaussian shapes.

An alternative spot detection procedure that is mainly applied to camera-digitized gel imaging has been described (Prehm *et al.*, 1987). The preprocessing consists of shading correction and nonrecursive separable low-pass smoothing filter. A segmentation step assumes that spots belong to areas with a convex* curvature of the gray level surface. Applying a segmentation template (9×9 convex area detector), the pixel describe a spot kernel if the degree of curvature in the central pixel of the 9×9 neighborhood is greater than zero. Otherwise, the spot on the image belongs to the background.

Spots located extremely close to each other, or those that are 'shouldered' or 'back-packing' (a small spot appended to a larger one), are often not separated, since the area of the gray levels between the spots may not indicate a negative (concave) curvature. However, the magnitude of curvature in these areas is lower than in spot areas. Hence, in order to separate these spots, a template with 7×7 pixels is used and the differences in the magnitude of curvature are calculated between the central pixel and the pixel on the rim of the 7×7 template in eight directions of the 3×3 neighborhood. The central pixel is assigned to the background if the two differences between the magnitude of curvature are smaller than a fixed threshold in at least one direction. The spot kernels found by this segmentation process are enlarged by two pixels. The spot volume is determined as the sum of the optical density of all pixels within the enlarged spot kernel.

Other approaches for spot detection are based on the watershed transformation (WST) (Bettens *et al.*, 1997; Meyer and Beucher, 1990; Pleissner *et al.*, 1999; Vincent, 1993; Wegner *et al.*, 1998). Adapting the geographical principle of watersheds to images, grayscale pictures are considered as topographic reliefs, representing the elevation at a point by the numerical value. For the algorithmic computation we use the interpretation as an immersion process. This may be envisioned as drilling holes in each regional minimum of the relief, and immersing the surface into a lake, ensuring a constant water level in all the valleys of the gray-level mountains. Starting from the minima of the lowest altitude, the water will fill up the different valleys. At the end of this immersion procedure each minimum is completely surrounded by watersheds, which delimit its associated valley. The valleys correspond to the regions whereas the watersheds define the optimal contours of an image. Watersheds employed in image segmentation of the gradient of the image have to be interpreted as reliefs. The assumption is that homogeneous regions of the image are characterized by low contrast and hence a low gradient, whereas the high values of the gradient image are created by the high contrast contours of the original image. The regions are obtained using WST on the gradient image. Each region is then described by a feature achieving a mosaic image. The selected features should be similar to the

---

* Note that the term convex is used in a mathematical sense. Here, convex is related to the viewpoint of the observer. When considered from the rear, the shape of the spot pulls out towards the viewing person. Thus, the spot area is convex.

gray levels of the original image, for instance the average of a region's gray values in the original image.

The WST produces closed contours. Each minimum (valley) describes a small region which is surrounded by local maxima (ridges). The consequence of such a segmentation may be a strong oversegmentation. That means that, as well as the relevant region contours, contours which are mainly due to image noise are also found.

Either of the following two types of spot regions have to be found to identify the required protein spots within the oversegmented mosaic image: regions that correspond to a complete spot or those that only cover part of a spot. To reduce the number of candidate spot regions within the mosaic image, two threshold steps are applied. These are based on the assumption that spot regions have significantly higher gray values than the background and that they border on a background region. The required threshold values are adapted automatically for each new gel image.

After using these criteria there are still some regions remaining that are neither spot regions nor partial spot regions. Regarding a gel image as a surface, the shape of a spot is obviously convex, and so are partial spot regions. Since the second derivative describes the curvature in a topographic relief of a watershed, spot and partial spot regions can be determined mathematically. Spot regions and partial spot regions are identical to convex curvature. Finally, all partial spot regions of neighboring spot groups are examined and merged if they descend from the same spot. A spot should have an approximately elliptical shape, and partial spot regions of one spot should have a local convex curvature in a small neighborhood along their watershed lines. The combination of these spot characteristics is used as the merging criterion. The main steps of this spot detection algorithm are shown in *Figure 2*.

5. Gel matching

Apart from the detection and quantification of spots in individual gel images, the comparison of identical spots in serial gels is important. Therefore, the comparative analysis of alterations in protein spot expression, under various experimental conditions, becomes an important question. Inhomogeneities within gels may occur as a consequence of differences in the sample preparation, variations of electrophoretic conditions or temperature, and unequal mobility of proteins in different gel regions. These do not allow a faultless one-to-one mapping of gels. Therefore, in order to analyze the changes of proteins, spot patterns must be matched.

A prerequisite for matching images or spot patterns is their registration. The registration is a procedure that renders data comparable by transformation. Normally, the position of a protein spot on a gel depends to a large extent on varying parameters, such as texture and density of the gel. Despite many research ambitions to obtain and run a 'perfect' gel, they still vary significantly in chemical and physical features. Therefore, protein positions are not exactly the same when gels are compared and their information value is not therefore directly comparable.

(match sets). The matching process leads to the tracing of individual spots. Spots and their changes in intensity can then be pursued through the series of gels. Gel matching is one of the most critical steps in the analysis of gel images. Most of the algorithms are only successful as long as gels are very similar, that is, when they are produced in the same laboratory under the same conditions. The comparison of gel images from different sources often leads to incompatibility.

In some systems, data analysis is supported by match sets that are composed of any group of gels (control group, experimental group) that has already been processed by matching. Sometimes an additional feature is provided by a 'standard image' that is derived from a template image. The standard image that comprises of Gaussian modeled spots contains all the spots and data of the serial gels included in the match set.

6. Data analysis

Data analysis is mainly driven by three questions: (i) the evaluation of quality assurance; (ii) the determination of qualitative and (iii) the determination of quantitative changes in protein expression. Knowledge of the number of valid protein spots resolved in each gel, the number of intensity-saturated spots in each gel, and the number of spots matched to every gel is fundamental in evaluation of the quality of the results. The relation between the total numbers of spots and the matched spots, which should preferably be found in all gels, defines the quality measure; the higher the total number of spots, the less it matters when a particular protein spot cannot be found in all the gels of a set.

For the determination of quantitative and qualitative changes in the protein expressions, the intensity values of all individual spots in a set of gels are represented in a table, an $N \times M$ matrix (*Figure 4*). N gives the number of spots (rows) and M the number of gels matched (columns). The raw data on the original $N \times M$ matrix are represented by the intensities of spots. Since the mean intensities of the gel images can differ, for example different exposure time, a normalization has to be performed for quantitative analysis (Burstin *et al.*, 1993). The normalization of the individual spot intensities is based on the average intensity of the related gel image.

For the determination of quantitative alterations in protein spots one is searching, for example, for those spots that are increased or decreased in their intensities or are not changed. The quantitative analysis can be performed with various statistical methods, such as the alterations between group parameters calculated as mean value and deviation of spot intensity. Furthermore, the alterations between groups can be performed by statistical tests in a univariate manner (for instance, Student's t-test, log t-test and Mann–Whitney test) or a multivariate manner (correspondence analysis).

Whereas the univariate tests consider each row independently in the $N \times M$ data matrix, the multivariate approach considers the correlation of spots (the rows). The correspondence analysis can detect the similarities between gels and identify typical spot features. The mathematical background of the correspondence analysis for the

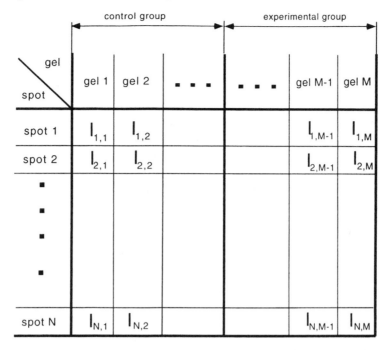

Figure 4. Table (matrix) for data analysis after matching. The raw data on spots (integrated optical density) are stored in an $N \times M$ matrix, where N represents the number of spots (number of rows), and M represents the number of gels (number of columns).

2-DE was first described by Pun *et al.* (1988). The procedure aims at the determination of the spots which characterize the gel best. Schmid *et al.* (1995) described correspondence analysis in detail and applied it for cell classification using 2-DE protein patterns. We have made the additional approach of correspondence analysis for data generated by the PDQUEST system (Pleissner *et al.*, 1998). Overall, the univariate and multivariate statistical approaches are important tools for the analysis of the alterations of protein expression in 2-DE gels.

Qualitative differences in protein expression between gel groups can be found by looking for spots that are present in one group only. These spots are often called 'on/off' spots, since proteins are switched 'on/off' under particular experimental conditions and they are therefore important, as they are evident candidates for a protein alteration.

An important feature of the analyzing power of a gel analysis system is the data handling of user sets, which are user-defined combinations of spots. Additionally, the results of different analyzing procedures can be combined by, for example, Boolean operators. In this way it is possible to determine nonsaturated spots that could be found and matched in all gels. Moreover, matched spots in user sets may increase or decrease due to protein changes, that may, in addition, be found to be statistically significant. The application of such data analysis tools has been described for the analysis of human heart disease (Pleissner *et al.*, 1997b).

7. Data presentation

For a precise presentation of results, some prerequisites have to be considered. First, each detected protein spot should have an assigned molecular mass, M_r, and isoelectric point, pI. Therefore, an M_r/pI grid is overlaid on the gels using some characteristic protein spots (reference spots), for which the M_r/pI values are known. Such reference spots are proteins that are well known and easy to detect on gels. Usually, M_r/pI values are listed in supplier catalogs, and they can also be obtained from SWISS-PROT (ExPASy molecular biology server), as long as the protein name or protein accession code is known. On gels of human heart samples, for instance, proteins such as serum albumin, actin, tropomyosin, myosin light chain 1 and myoglobin are clearly recognizable and may serve as markers for the generation of M_r/pI grids (Knecht *et al.*, 1994).

Using these marker proteins, all other protein M_r/pI values are evaluated by assuming a linear interpolation along the pI-axis (horizontal axis) and an exponential interpolation along the M_r-axis (vertical axis). Clearly, the precision of M_r/pI values depends on the algorithm used for the calculation of the M_r/pI grid.

Graphs or bar charts can be used to compare spot quantitation and reflect alteration trends of spots. Ideally such alterations should result from changes in the underlying protein expression. Moreover, graphs and bar charts are also useful in detecting unmatched spots and spots with a quantitation that is suspiciously above average. Bar charts present the gel spot number on the x-axis and the intensity (spot volume) on the y-axis. Additionally, if gels are forming replicate groups the average intensity values and the standard deviations of groups can be illustrated as bars. Using this presentation the trend of alteration between the groups is clearly recognizable. Thus, such charts can provide a powerful graphical tool to represent the data stored in the matching table. Some types of these graphs, showing the spot variations between a control and experimental group of gels, are illustrated in *Figure 5*. Additionally, the output of intensities as tables is usually applied for further evaluation of data by EXCEL, StatView, SSP, SAS, among others.

For the assessment of the similarity between spot intensities, scattergrams are often used to show the intensity relatedness of two gels or two gel groups. Using such plots, the correlation coefficient can be calculated as a measure of correlation of the intensities. Ten control gels versus 10 experimental gels were analyzed to construct a scattergram in a study on the effects of hypertension (Pleissner *et al.*, 1998). A remarkable similarity of both groups can be demonstrated when comparing the average intensities for all matched spots in both groups (*Figure 6*).

8. Databases

8.1 Construction of 2-DE databases

The previously described analysis gives plenty of descriptive information about the protein *res.* protein spots in the analyzed gel images, that is, protein name, M_r/pI values and also bibliographic and other references. Generally, the information is

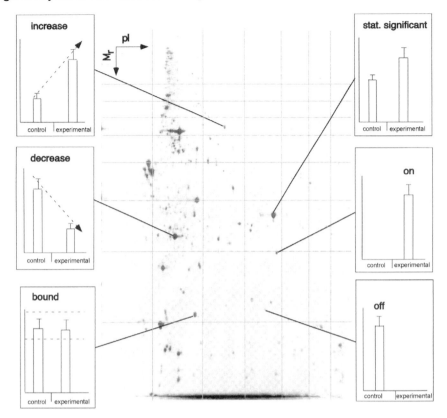

Figure 5. Some types of graphs used to express protein alterations. Each group (control, experimental) consists of several individual gels. The bars represent, for instance, the mean values and standard deviations of spot intensities within each group.

saved in 2-DE databases. Databases normally give unequivocal reference to a representative gel image of the analyzed set of images. The representative image can be one of the analyzed gel images or a synthetically created standard image. This image-to-database and database-to-image links imply that:

(i) if a protein spot is marked on the image, the descriptive information should appear; and

(ii) if the protein name or other descriptive information is known, the location of this protein on the image has to be shown and annotated.

Some gel analysis systems include their own internal database. In the PDQUEST system, for instance, a database spot number (DSN), in addition to the standard spot number (SSP), indicates the general location of the spot, since the numbering is based on the M_r/pI grid. To each spot various annotation categories can be attributed. The annotations are directly associated with the spot position on the standard gel and can be retrieved by simple queries. Although such internal databases (Lipkin, 1980; Jungblut *et al.*, 1994) are helpful, they are very system-

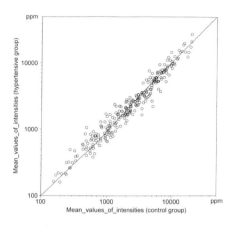

Figure 6. Scattergram showing the dependency of the mean values for the spot intensities of an experimental (hypertensive) group vs a control group. Number of spots matched to every gel equals 269; number of gels equals 20.

specific and do not allow the gel images and descriptive information to be shared worldwide.

Access by Internet technology helps to overcome the laboratory borders. The creation of 2-DE databases on the World Wide Web (WWW) has been described previously (Appel *et al.*, 1993; Boutell *et al.*, 1994; Evans *et al.*, 1997; Latter *et al.*, 1995; Lemkin, 1997a,b; Monardo *et al.*, 1994; Müller *et al.*, 1996; Pleissner *et al.*, 1996). Currently, about 30 WWW-accessible 2-DE databases are indexed in the WORLD-2DPAGE index (http://www.expasy.ch/ch2d/2d-index.html) on the ExPASy server. An overview of construction, interfacing and integrating 2-DE WWW databases is given in Appel (1997c).

Currently, some software packages for the construction of 2-DE databases are available free, such as the Make2dbb package (Hoogland *et al.*, 1997). This software package can be used independently of the MELANIE II system together with a new tool that converts a TIFF image of a gel, and a text file that contains the x/y-coordinates of protein spots and the accession number to SWISS-PROT, into the specific image format (http://www.expasy.ch/ch2d/tiff2mel.html).

8.2 Comparing Internet-distributed 2-DE databases

The comparison of gel images via the Internet along with knowledge transfer is an important task in proteome research and supports or replaces expensive identification of proteins. Creating database master gels and using the SWISS-PROT accession number of a protein, the spots can be localized and identified in federated 2-DE databases (Appel *et al.*, 1996). The visual interlaboratory comparison of two gel images featured in a 2-DE metadatabase using warping transformation (matching), contrast enhancement, smoothing filters and flickering technique is realized by a Flicker-server (http://www-lmmb.ncifcrf.gov/flicker; Lemkin, 1997a). Flickering means that two images (source and target image) can be visualized alternatively.

Problems with the automated comparison/matching of 2-DE gel images that are available in WWW databases are described in detail in Pleissner *et al.* (1999). In this case, spot parameters (spot position and intensity) are employed for point pattern matching using an approach which is known from computational geometry. More-

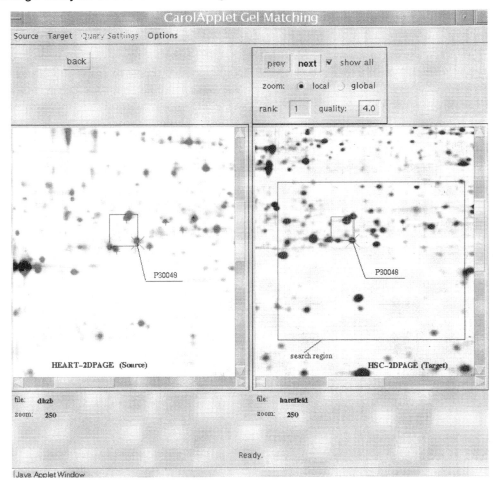

Figure 7. The CAROL user interface applied to the comparison of images from the HEART-2DPAGE (source) and HSC-2DPAGE (target) databases. The search pattern (+) given in the source image and the corresponding pattern after the matching. For one match (X) the SWISS-PROT AC found in both databases is shown to illustrate the correctness of the matching result.

over, spot detection and matching procedures are directly performed on the Internet by means of a user interface, invoked in any Java-capable Internet browser. This matching approach compares point patterns of a source image with those in a target image by checking the slopes and lengths of edges between points of both images, and also takes the relation of spot intensities into account. A central idea for the matching algorithm stems from the use of incremental Delaunay triangulation, known from computational geometry (Alt and Guibas, 2000). The transformation of the absolute intensity values to a discrete intensity integer, and the usage of a discrete threshold, enable a matching approach that has only a weak dependence on gray values (absolute densitometric resolution).

This matching approach has been implemented in the CAROL software system (http://gelmatching.inf.fu-berlin.de). The program will be able to match gel images from 2-DE databases over the entire Internet. The user can open GIF-images from any 2-DE database, carry out the spot detection and perform a local matching between the source and target image. In *Figure 7* the CAROL user interface applied to the comparison of images from the World-2DPAGE indexed 2-DE databases HEART-2DPAGE (left, source) and HSC-2DPAGE (right, target) is demonstrated.

9. Concluding remarks

This chapter has outlined the main principles of computerized image analysis of 2-DE gels and 2-DE database establishment and their comparison in the WWW. There is no doubt that an exact and reproducible evaluation of protein spot data cannot be performed without computerized image analysis. Various commercial gel analysis systems are currently available and used for the determination of protein spot alterations under defined conditions. Specific protein markers can be detected and used for further investigation once enough experimental data on protein alterations is available. Also, concepts of database construction and comparison were described to enable protein data and gel images available via the Internet to be stored, shared and compared. In this context, digital image processing and analysis improve the comparability of different 2-DE databases.

Acknowledgments

The authors would like to thank Vera Regitz-Zagrosek, Bernhard Krüdewagen and Eckart Fleck from the German Heart Institute Berlin and Frank Hoffmann, Klaus Kriegel, Carola Wenk and Helmut Alt from the Free University Berlin for scientific collaboration and support.

References

Alt, H. and Guibas, L. (2000) Discrete geometric shapes: matching, interpolation, and approxima-tion, a survey. In: *Handbook for Computational Geometry* (ed. J. Urrutia and J.-R. Sack). North Holland, Amsterdam, pp. 121–154.
Amess, B. and Tolkovsky, A.M. (1995) Programmed cell death in sympathetic neurons: a study by two-dimensional polyacrylamide gel electrophoresis using computer image analysis. *Electro-phoresis* **16:** 1255–1267.
Anderson, N.G. and Anderson, N.L. (1996) Twenty years of two-dimensional electrophoresis: past, present and future. *Electrophoresis* **17:** 443–453.
Anderson, N.L., Taylor, J., Scandora, A.E., Coulter, B.P. and Anderson, N.G. (1981) The TYCHO system for computer analysis of two-dimensional gel electrophoresis patterns. *Clin. Chem.* **27:** 1807–1820.
Appel, R.D. (1997) Interfacing and integrating databases. In: *Proteome Research: New Frontiers in Functional Genomics* (eds M.R. Wilkins, K.L. Williams, R.D. Appel and D.F. Hochstrasser). Springer, Berlin, pp. 149–175.
Appel, R.D., Hochstrasser, D.F., Funk, M., Vargas, J.R., Pellegrini, C., Muller, A.F. and Scherer,

J.-R. (1991) The MELANIE project: from a biopsy to automatic protein map interpretation by computer. *Electrophoresis* **12**: 722–735.

Appel, R.D., Sanchez, J.-C., Bairoch, A., Golaz, O., Miu, M., Vargas, J.R. and Hochstrasser D.F. (1993) SWISS-2DPAGE: a database of two-dimensional electrophoresis images. *Electrophoresis* **12**: 1232–1238.

Appel, R.D., Bairoch, A., Sanchez, J.-C., Vargas J.R., Golaz, O., Pasquali, C. and Hochstrasser, D.F. (1996) Federated two-dimensional electrophoresis database: a simple means of publishing two-dimensional electrophoresis data. *Electrophoresis* **17**: 540–546.

Appel R.D., Palagi, P.M., Walther, D., Vargas, J.R, Sanchez, J.-C., Ravier, F., Pasquali, C. and Hochstrasser, D.F. (1997a) Melanie II – a third-generation software package for analysis of two-dimensional electrophoresis images: I. Features and user interface. *Electrophoresis* **18**: 2724–2734.

Appel, R.D., Vargas, J.R., Palagi, P.M., Walther, D. and Hochstrasser, D.F. (1997b) Melanie II – a third-generation software package for analysis of two-dimensional electrophoresis images: II. Algorithms. *Electrophoresis* **18**: 2735–2748.

Bettens, E., Scheunders, P., VanDyck, D., Moens, L. and Van Osta, P. (1997) Automatic segmentation and modelling of two-dimensional electrophoresis gels. *Electrophoresis* **18**: 792–798.

Bossinger, J., Miller, M.J., Vo, K.-P., Geiduschek, P. and Xuong, N. (1979) Quantitative analysis of two-dimensional electrophoretograms. *J. Biol. Chem.* **254**: 7986–7998.

Boutell, T., Garrels, J.I., Franza, B.R., Monardo, P.J. and Latter, G.I. (1994) REF52 on the global gel navigator: an internet accessible two-dimensional gel electrophoresis database. *Electrophoresis* **15**: 1487–1490.

Burstin, J., Zivy, M., de Vienne, D. and Damerval C. (1993) Analysis of scaling methods to minimize experimental variations in two-dimensional electrophoresis quantitative data: application to the comparison of maize inbred lines. *Electrophoresis* **14**: 1067–1073.

Collins, P.J. and Blose, S.H. (1992) Todays densitometer – multi-application densitometry for biotechnology. *Int. J. Biotechnol.* **9**: 202–205.

Evans, G., Wheeler, C.H., Corbett, J.M. and Dunn, M.J. (1997) Construction of HSC-2DPAGE: a two-dimensional gel electrophoresis database of heart proteins. *Electrophoresis* **18**: 471–479.

Garrels, J.I. (1979) Two-dimensional gel electrophoresis and computer analysis of proteins synthesized by clonal cell lines. *J. Biol. Chem.* **254**: 7961–7977.

Garrels, J.I. (1983) Quantitative two-dimensional gel electrophoresis of proteins. *Methods Enzymol.* **100**: 411–423.

Garrels, J.I. (1989) The QUEST system for quantitative analysis of two-dimensional gels. *J. Biol. Chem.* **264**: 5259–5282.

Hoogland, C., Baujard, V., Sanchez, J.-C., Hochstrasser, D.F. and Appel, R.D. (1997) Make2ddb: a simple package to set up a two-dimensional electrophoresis database for the World Wide Web. *Electrophoresis* **18**: 2755–2758.

Jungblut, P., Otto, A., Zeindl-Eberhart, E., Pleissner, K.-P., Knecht, M., Regitz-Zagrosek, V., Fleck, E. and Wittmann-Liebold, B. (1994) Protein composition of human heart: the construction of a two-dimensional electrophoresis database. *Electrophoresis* **15**: 685–707.

Knecht, M., Regitz-Zagrosek, V., Pleissner, K.-P., Emig, S., Jungblut, P., Hildebrandt, A. and Fleck, E. (1994) Computer-assisted analysis of endomyocardial biopsy protein patterns by two-dimensional gel electrophoresis. *Eur. J. Clin. Chem. Clin. Biochem.* **32**: 615–624.

Kuick, R.D., Skolnick, M.M., Hanash, S. and Neel, J.V. (1991) A two-dimensional electrophoresis-related laboratory information processing system: spot matching. *Electrophoresis* **12**: 736–746.

Latter, G. I., Boutell, T., Monardo, P.J., Kobayashi, R., Futcher, B., Mclaughlin, C.S. and Garrels J.I. (1995) A *Saccharomyces cerevisiae* Internet protein resource now available. *Electrophoresis* **16**: 1170–1174.

Lemkin, P.F. (1997a) Comparing two-dimensional electrophoretic gel images across the Internet. *Electrophoresis* **18**: 461–470.

Lemkin, P.F. (1997b) The 2DWG meta-database of two-dimensional electrophoretic gel images on the Internet. *Electrophoresis* **18**: 2759–2773.

Lemkin, P.F. and Lipkin, L.E. (1981a) GELLAB: a computer system for 2D gel electrophoresis analysis I. Segmentation of spots and system preliminaries. *Comput. Biomed. Res.* **14**: 272–297.

Lemkin, P.F. and Lipkin, L.E. (1981b) GELLAB: a computer system for 2D gel electrophoresis analysis. II. Pairing spots. *Comput. Biomed. Res.* **14**: 355–380.

Lemkin, P.F. and Lipkin, L.E. (1981c) GELLAB: a computer system for two-dimensional gel electrophoresis analysis. III. Multiple two-dimensional gel analysis. *Comput. Biomed. Res.* **14**: 407–446.

Lemkin, P., Merril, C., Lipkin, L., Van Keuren, M., Oertel, W., Shapiro, B., Wade, M., Schultz, M. and Smith, E. (1979) Software aids for the analysis of 2D gel electrophoresis images. *Comput. Biomed. Res.* **12**: 517–544.

Lemkin, P.F., Lipkin, L.E. and Lester E.P. (1982) Some extensions to the GELLAB two-dimensional electrophoretic gel analysis system. *Clin. Chem.* **28**: 840–849.

Lipkin, L.E. (1980) Data-base techniques for multiple two-dimensional polyacrylamide gel electrophoresis analyses. *Clin. Chem.* **26**: 1403–1412.

Meyer, F. and Beucher, S. (1990) Morphological segmentation. *J. Visual Commun. Image Repres.* **1**: 21–46.

Miller, M.J., Vo, K.-P., Nielsen, C., Geiduschek, E.P., Xuong, N.H. (1982) Computer analysis of two-dimensional gels: semi-automatic matching. *Clin. Chem.* **28**: 867–875.

Monardo, P.J., Boutell, T., Garrels, J.I. and Latter, G.I. (1994) A distributed system for two-dimensional gel analysis. *Comput. Appl. Biosci.* **10**: 137–143.

Müller, E.-C., Thiede, B., Zimny-Arndt, U., Scheler, C., Prehm, J., Müller-Werdan, U., Wittmann-Liebold, B., Otto, A. and Jungblut, P. (1996) High-performance human myocardial two-dimensional electrophoresis database: edition 1996. *Electrophoresis* **17**: 1700–1712.

Olson, A.D. and Miller, M.J. (1988) Elsie4: quantitative computer analysis of sets of two-dimensional gel electrophoretograms. *Anal. Biochem.* **169**: 49–70.

Patterson, S.D. and Latter, G.I. (1993) Evaluation of storage phosphor imaging for quantitative analysis of 2-D gels using the Quest II system. *BioTechniques* **15**: 1076–1083.

Pleissner, K.-P., Sander, S., Oswald, H., Regitz-Zagrosek, V. and Fleck, E. (1996) The construction of the World Wide Web-accessible myocardial two-dimensional gel electrophoresis protein database 'HEART-2DPAGE': a practical approach. *Electrophoresis* **17**: 1386–1392.

Pleissner, K.-P., Sander, S., Oswald, H., Regitz-Zagrosek, V. and Fleck, E. (1997a) Towards design and comparison of World Wide Web-accessible myocardial two-dimensional gel electrophoresis protein databases. *Electrophoresis* **18**: 480–483.

Pleissner, K.-P., Söding, P., Sander, S., Oswald, H., Neuss, M., Regitz-Zagrosek, V. and Fleck, E. (1997b) Dilated cardiomyopathy-associated proteins and their presentation in a WWW-accessible two-dimensional gel protein database. *Electrophoresis* **18**: 802–808.

Pleissner, K.-P., Regitz-Zagrosek, V., Krüdewagen, B., Trenkner, J., Hocher, B. and Fleck, E. (1998) Effects of renovascular hypertension on myocardial protein patterns – analysis by computer-assisted two-dimensional gel electrophoresis. *Electrophoresis* **19**: 2043–2050.

Pleissner, K.-P., Hoffmann, F., Kriegel, K., Wenk, C., Wegner, S., Sahlström, A., Oswald, H., Alt, H. and Fleck, E. (1999) New algorithmic approaches to protein spot detection and pattern matching in 2-DE gel databases. *Electrophoresis* **20**: 255–265.

Prehm, J., Jungblut, P. and Klose, J. (1987) Analysis of two-dimensional electrophoretic protein patterns using a video camera and a computer II. Adaptation of automatic spot detection to visual evaluation. *Electrophoresis* **8**: 562–572.

Pun, T., Hochstrasser, D., Appel, R., Funk, M., Villars-Augsburger, V. and Pellegrini C. (1988) Computerized classification of two-dimensional gel electrophoretograms by correspondence analysis and ascendant hierarchical clustering. *Appl. Theor. Electrophoresis* **1**: 3–9.

Schmid, H.-R., Schmitter, D., Blum, P., Miller, M. and Vonderschmitt, D. (1995) Lung tumor cells: a multivariate approach to cell classification using two-dimensional protein pattern. *Electrophoresis* **16**: 1961–1968.

Serpico, S.B. and Vernazza, G. (1987) Problems and prospects in image processing of two-dimensional gel electrophoresis. *Opt. Engng* **26**: 661–668.

Skolnick, M.M. (1982) An approach to completely automatic comparison of two-dimensional electrophoresis gels. *Clin. Chem.* **28**: 979–986.

Skolnick, M.M. and Neel, J.V. (1986) An algorithm for comparing two-dimensional electrophoretic gels with particular reference to the study of mutation. *Adv. Hum. Genet.* **15**: 55–160.

Sternberg, S.R. (1986) Gray scale morphology. *Comput. Vision Graph. Image Processing* **35**: 333–355.

Tarroux, P., Vincens, P. and Rabilloud, T. (1987) HERMeS: a second generation approach to the automatic analysis of two-dimensional electrophoresis gels. Part V: data analysis. *Electrophoresis* **8**: 187–199.

Vincens, P. and Tarroux, P. (1987a) HERMeS: a second generation approach to the automatic analysis of two-dimensional electrophoresis gels. Part III: spot list matching. *Electrophoresis* **8**: 100–107.

Vincens, P. and Tarroux, P. (1987b) HERMeS: a second generation approach to the automatic analysis of two-dimensional electrophoresis gels. Part IV: data base organization and management. *Electrophoresis* **8**: 173–186.

Vincent, L. (1993) Morphological grayscale reconstruction in image analysis: applications and effective algorithms. *IEEE Trans. Image Processing* **22**: 176–201.

Vo, K.-P., Miller, M.J., Geiduschek, P., Nielsen, C., Olson, A. and Xuong, N. (1981) Computer analysis of two-dimensional gel. *Anal. Biochem.* **112**: 258–271.

Wegner, S., Sahlström, A., Pleissner, K.-P., Oswald, H. and Fleck, E. (1998) Eine hierarchische Wasserscheidentransformation für die Spotdetektion in 2D Gel-elektrophorese-Bildern. In: *Informatik Fachberichte Heft 10, Bildverarbeitung in der Medizin.* Springer, Berlin, pp. 134–138.

Wu, Y., Lemkin, P.F. and Upton, K. (1994) A fast spot detection algorithm for 2D electrophoresis analysis. *Electrophoresis* **14**: 1341–1356.

Chapter 7

Enhancing high-throughput proteome analysis: the impact of stable isotope labeling

Manfredo Quadroni and Peter James

1. Introduction

Proteome analysis is concerned with the global changes in protein expression as visualized most commonly by two-dimensional gel electrophoresis (2-DE) and analyzed by mass spectrometry. Recent commercial interest has resulted in the development of several integrated systems that attempt to provide a streamlined work flow from sample separation by 2-DE to protein identification by mass spectrometry. Simplification and standardization of the procedures in a manner that is both user-friendly and robust may now open the way for many more laboratories to successfully use this technique. At the same time, novel tagging methodologies have been introduced which can greatly simplify quantification and data interpretation and may allow the gel electrophoresis step to be eliminated and replaced by simple chromatography steps. In this chapter we outline the various stages in a typical approach to proteome analysis and summarize the new developments that have occurred since we last reviewed the field in late 1998 (Quadroni and James, 1999).

The availability of complete genome sequences (Devine and Wolfe, 1995) and extensive expressed sequence tag (EST; Adams *et al.*, 1991) libraries potentially allows the entire potential protein complement of organisms to be defined (the proteome). Interest is now focused on trying to interpret the massive influx of new sequence data and understand how the vast array of chemical species in the cell interact with one another to create the molecular machinery of the cell. The focus of biological problem solving must now move from a reductionist to a global approach. Instead of dissecting a process out to identify a putative single effector, more subtle analyses based on monitoring the entire genome expression can be carried out. Such a description of a biological system by a qualitative and quantitative analysis of mRNA expression has been made possible by the development of a variety of DNA expression analysis methods and genome-wide studies (Lashkari *et al.*, 1997; Velculescu *et al.*, 1997). A prerequisite to these large-scale mRNA expression studies and to the sequencing of genomic DNA is a high degree of automation. Reproduc-

Proteomics, edited by S.R. Pennington and M.J. Dunn

ibility is essential for a statistical analysis of these complex datasets and automation and high-throughput data accumulation are essential for such large undertakings. DNA and mRNA are physico-chemically very homogenous and 'easy' to handle, can be amplified by polymerase chain reaction methods and are hence amenable to automation. Proteins, in contrast to nucleic acids, are vastly more diverse and a universal handling method is unlikely to be found. Currently the only systematic methods for analyzing the state of expression of the majority of proteins in a cell are those based on 2-DE (Klose, 1975; O'Farrell, 1975; O'Farrell *et al.*, 1977). Recently, isotopic labeling methods (Gygi *et al.*, 1999; Oda *et al.*, 1999) have been developed which make quantification of proteins separated by 2-DE much more reliable. These methods open the way for the development of alternatives to 2-DE and greatly extend the dynamic range of protein expression measurement. In this chapter we will outline the approaches and pitfalls in trying to automate protein identification and quantification methods for comprehensive proteome analysis.

2. Sample preparation

Perhaps the single most critical point in obtaining useful proteome analyses is reproducible sample preparation. The cells or tissue must be efficiently disrupted and the contents of the cells solubilized reproducibly in order to obtain a representative protein population sample. Physical disruption methods such as sonication, mechanically driven rapid pressure changes, homogenization and shearing-based techniques are often used to disrupt the cells or tissues prior to protein extraction with a urea-based solution containing non-ionic detergents, reducing agents and a protease inhibitor cocktail. The extent of recovery of membrane and cytoskeletal proteins is very variable, with some proteins being completely solubilized yet up to 10% of the cell protein remaining in the pellet after extraction. The development of small 2-DE gels which allow samples to be run in one day (from sample loading to gel staining and scanning) is an important step. This allows a rapid evaluation of a large number of preparation methods in parallel. The extremely high degree of complexity of eukaryotic tissues often requires that a prefractionation step be carried out in order to reduce the complexity of the protein mixture being analyzed and allow the resolution and analysis of minor components. When dealing with a mixed cell population such as a tissue, prefractionation of cells using a fluorescence-activated cell sorter (FACS, Madsen *et al.*, 1988) can allow small subpopulations of cells to be specifically isolated, thus greatly increasing the sensitivity of the analysis. A further important advance has been the development of free-flow electrophoresis methods to separate cell organelles/compartments (Anderson, 1967; Burggraf *et al.*, 1995). This is an important first step in the development of protein expression maps that will show where and under what conditions a protein is to be found in a cell (Blackstock and Weir, 1999). This will allow the development of theoretical models in which the status of the cell is defined in terms of a set of interacting molecular machines.

3. Two-dimensional gel separation and analysis

3.1 Automating 2-D PAGE

The standard denaturing 2-DE gel system has remained an essentially manual operation since it was established over 20 years ago and as yet there is no valid alternative protein-separation technique with an equivalent dynamic range and resolving power. A detailed study using a prototype semiautomated instrument clearly showed the advantages of mechanical reproducibility (Harrington *et al.*, 1993), but until recently the only developments in large-scale automated production of 2-DE gels had been carried out in commercial enterprises such as Large Scale Biology and Oxford Glycosciences. Three companies have now announced the development of integrated systems for proteome analysis: Amersham-Pharmacia, BioRad and Genomic Solutions. We will outline these at the end of this section, however we must emphasize that these are very new and sometimes only partially complete and so experience with them is very limited.

3.2 Protein detection and quantification by scanning methods

Usually proteins are visualized by their UV absorption; however, since polyacrylamide absorbs in the same UV range, proteins must be stained in order to be visualized. The staining method should be effective for all proteins in order to allow quantification and should be compatible with further analysis steps. By far the most commonly used protocols are Coomassie blue and silver staining of proteins. Although both methods are protein-sequence sensitive (especially silver), they are both linear over a wide dynamic range for most proteins. An alternative is the use of fluorescence labeling, either covalently introduced into the proteins during equilibration between the first and second dimensions or by using fluorescently labeled SDS analogs or fluorescent dyes (Steinberg *et al.*, 1996). Again, linearity of response and general applicability are the limiting factors for most fluorescent dyes, although the extended sensitivity range down to the mid-attomole level is exceptional. None of the methods suffer from side reactions that disturb further analysis; even silver staining (Shevchenko *et al.*, 1996) shows little chemical modification (see Chapter 4). Most commonly gels are scanned by laser densitometers in the absorbance mode, which gives a linear response from 0 to 3.5 optical density (OD) units. Detection methods based on labeling with radioactive isotopes are exquisitely sensitive (subattomole), but are not well suited for automation due to the difficulty of correlating the positions of the spots on a film (or the computer printout of an image from a $BaFBr:Eu^{2+}$ storage screen) with the corresponding gel matrix.

Regardless of the visualization method used, 2-DE gel separations suffer from several limitations. The advent of ultrasensitive MS techniques for protein identification has highlighted one of the shortcomings of wide-pH-range 2-DE gel separations: whilst this single gel format offers an immediate overview of the state of the proteome, many of the spots are inadequately separated. Partially overlapping spots can be resolved by computational approaches but now it is becoming clear that many symmetrical spots contain two or more proteins. For prokaryotes, about 20%

of all spots contain one or more proteins (Link *et al.*, 1997); for eukaryotes, the number approaches 40%. A collaborative study by Proteome Inc. and Perseptive Biosystems (Parker *et al.*, 1998) has shown in detail the extent and nature of spot cross-contamination in 2-DE gels of yeast extracts. Many of the spots contained three or more proteins. Although MS techniques are powerful enough to resolve the protein mixtures (Arnott *et al.*, 1998), multiple proteins per spot make interpretation of many experiments extremely difficult.

3.3 Protein quantification from 2-DE gels by stable isotope labeling

In vivo labeling. A commonly used technique in protein analysis is the use of radiolabeled amino acids (such as ^{35}S methionine or, less commonly, cysteine) to increase the detection sensitivity of 2-DE and to allow pulse-chase experiments to determine rates of protein synthesis and degradation. Recently the use of stable isotopes for MS analysis of proteins has been an area of intense interest. One approach is to use isotopically depleted media (i.e. media low in heavy isotopes such as ^{15}N, $^{2 \text{ and } 3}$H and ^{13}C). This allows the isotopic cluster of an ion to be collapsed into a single peak, thereby increasing sensitivity and accuracy (Marshall *et al.*, 1997). Whole-cell isotopic labeling has been used as a method to quantificate relative changes in protein expression and modification (Oda *et al.*, 1999; *Figure 1*). Cells are grown in either a 'light' medium, which consists of compounds with a normal isotope distribution, or in 'heavy' medium, which is highly enriched in ^{15}N (95%). The protein extracts from the cells grown under different conditions are pooled and partially separated, usually by 2-DE. The spot of interest is excised, digested and the extracted peptides analyzed by mass spectroscopy. This allows one to differentiate between the peptides originating from the two cell pools since the ^{15}N incorporation shifts one of the pools upwards by one mass unit. The peptides thus appear as doublets and can be quantitated by the relative height of the peaks. This ratio was found to be linear over an abundance ratio of two orders of magnitude. This procedure also lends itself well to defining changes in the level of posttranslational modifications (*Figure 2*). This method is essentially limited by material costs since it is impractical to label entire animals and is often not applicable to eukaryotic cell cultures due to the need for serum-derived factors usually obtained from fetal calf serum.

Postgel labeling. An alternative approach is to use postseparation isotopic labeling for relative quantification (*Figure 3*). A very promising alternative to 2-DE gel analysis as a comprehensive method for comparative proteomics was recently described by the group of Aebersold and is called isotope-coded affinity tagging (ICAT; Gygi *et al.*, 1999). An alternative approach is to label proteins separated from extracts of cells obtained under different conditions separated by 2-DE. Individual spots are selected and excised and then the lysine residues are succinylated and the proteins digested with Asp/GluC protease. Proteins obtained from state 1 are specifically labeled at the *N*-terminal with the light reagent H4NicNHS [1-(H4-nicotinoyloxy) succinimide] and those from state 2 are marked

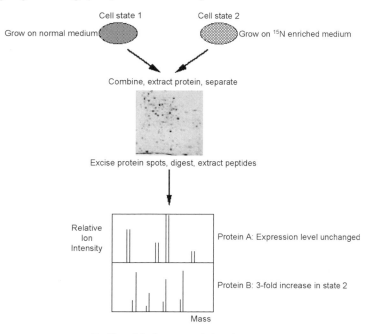

Figure 1. *In vivo* stable isotope labeling for 2-DE gel analysis. Cells were grown under standard and modified conditions in normal and 15N enriched medium, respectively. The proteins were extracted from the cells and combined prior to separation by 2-DE. Spots were excised, digested and analyzed by MALDI-MS. The ratio of 14N to 15N isotopes in an ion cluster allowed quantification of the change in expression level between the two states.

with the deuterated heavy reagent D4NicNHS. The digests are combined and a fraction is analyzed by MALDI-MS in order to quantificate the relative amount of the proteins from the D4/H4 ratios of the individual peptides (*Figure 4*). The advantage of this method is that it can be applied to all protein extract types and is inexpensive in comparison to cell culture in isotopically depleted or enriched media. One important application of the technique is that it allows the relative quantification of proteins which are only partially separated as multiple proteins in either a 1-D band or 2-D spot. This is especially useful when studying membrane proteins which can be partially separated easily by 1-DE but which cannot be run on 2-DE. The labeling also has the effect of allowing a greatly simplified interpretation of the MS/MS spectra resulting from the labeled peptides.

3.4 Isolation of protein spots from 2-DE gels

Once a series of gels has been analyzed, the spots selected for further analysis may be cut out by a robotic spot picking system for subsequent processing of the individual protein spots (Walsh *et al.*, 1998). The Australian Proteome Analysis Facility (APAF) has developed such a system in collaboration with a robotics company,

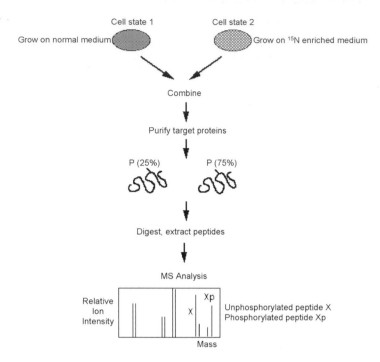

Peak ratios allow relative quantitation of site specific phosphorylation levels

Figure 2. Isotope labeling and identification of post-translational modifications. The basic strategy for isotope labeling and expression quantification can be applied to determine the level of changes in post-translational modifications such as phosphorylation.

ARRM (Advanced Rapid Robotics Manufacturing, Kent Town, South Australia), which is now commercially available. The robot excises spots from either gels or PVDF membranes and places them in a 96-well plate for automated proteolysis by a second robotics system (marketed by Canberra Packard, Downer's Grove, Il, USA), which subsequently loads the digests together with matrix onto a multisample MALDI target plate. The analysis of 288 protein spots from a single 2-DE membrane blot was achieved within 10 working days (Walsh *et al.*, 1998). Ideally the 2-DE analysis software should select the spots for analysis and download the spot coordinates to the robot for subsequent excision. This is simple to implement for PVDF membranes but gels are more difficult since they deform easily, especially during spot excision, and they are prone to drying out. A prerequisite therefore is a mechanically stable gel. Other automatic digestion devices have been described (Davis *et al.*, 1995a; Houthaeve *et al.*, 1997). Recently three commercial 'proteome systems' have been announced by Amersham-Pharmacia, BioRad and Genomic Solutions. The basic components of all three are very similar, involving a gel cutter to excise spots of interest from a stained gel according to the results of computer-aided 2-DE gel analysis based on criteria defined by the user (i.e. spots appearing or disappearing) The basic structure of such a 'complete' system is shown diagramma-tically in *Figure 5*.

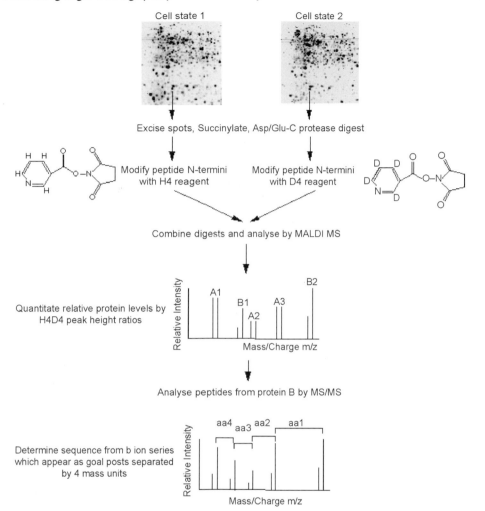

Figure 3. Postextraction protein labeling for 2-DE gel analysis. In order to circumvent problems of protein labeling *in vivo*, proteins can be isotopically labeled after 2-DE separation. This method allows for facilitated MS/MS spectra interpretation since only the *N*-termini of the peptides are labeled.

3.5 Protein digestion/fragmentation

Currently the major limiting factor in biological protein and peptide analysis is sample handling. Enzymatic digestion of proteins at concentrations lower that 500 fmol/μl is very inefficient. This is simply due to the fact that proteases show 50% maximal activity in the 5–50 pmol/μl range. A digest of 50 pmol/20 μl BSA produces around 60 main peptides and the percentage mole recovery is around 95%. In contrast for 100 fmol/20 μl, only six peptides are seen, with a recovery of around 5%. The alternative is either to develop chemical methods (but these still suffer from sample handling losses) or to carry out the fragmentation directly in the mass

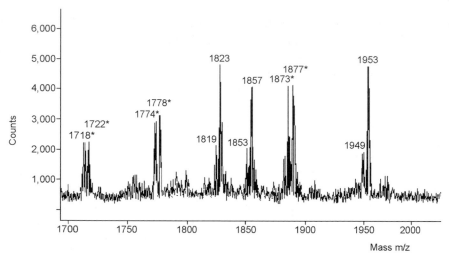

Figure 4. MALDI-MS spectrum of two proteins found in a single spot. After digestion and labeling, the relative expression level of a protein under different conditions can be determined. In this spectrum, the peaks marked with an asterisk belong to a protein whose expression level remains constant. In the same spot, however, there is a second protein (peptide peaks unmarked) whose expression level is being increased by a factor of three.

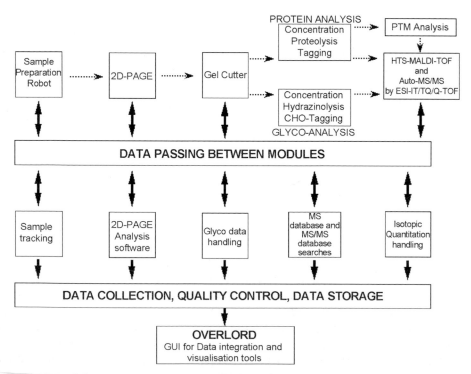

Figure 5. A modular proteome analysis system. A two-tier proteome analysis workstation software that controls the mechanical devices such as 2-DE running, spot excision, digestion and MS analysis as well as carrying out the data analysis and sample tracking.

spectrometer. MS/MS of intact proteins electrosprayed into various MS instruments has been achieved with a wide range of techniques such as collision-induced dissociation (Senko *et al.*, 1994), infrared multiphoton dissociation (IRMPD; Little *et al.*, 1994), ultraviolet laser-induced fragmentation and surface-induced fragmentation (Williams *et al.*, 1990). All of these techniques can be carried out by Fourier transform ion cyclotron resonance (FT-ICR) where the high mass accuracy, resolution and ability to carry out MS^n experiments allow one to make sense of the complex fragmentation patterns. It has already been demonstrated that intact proteins can be identified from their fragmentation in an FT-ICR by using a combination of the exact intact mass with a series of sequence tags extrapolated from the MS^n experiments (Mortz *et al.*, 1996). Recently Li and Marshall have demonstrated on-line identification of proteins by LC/ESI FT-ICR MS. A normal scan is first used to extract the exact mass of the intact protein and on alternate scans IRMPD is used to fragment one selected m/z ion from the protein. The intact mass is used with a wide mass window to select a subset of the database entries. The list of mostly *b*- and *y*-fragment ions, as well as any small sequence tags obtained from IRMPD, is then matched against this set to identify the protein. At the moment though, FT-ICR-MS instruments are extremely expensive and automation has not yet been achieved.

4. Mass spectrometry: protein identification using MS data

In order to achieve high-throughput protein identification, a hierarchical approach must be taken. The first level is very rapid, low-information-content data accumulation to identify the bulk of the proteins, followed by a second level of lower-throughput, high-information-content data generation to define the rest. Finally a third level of labor-intensive data accumulation can be carried out to define post-translational modifications. In practical terms, MALDI-based peptide mass fingerprinting is the ideal first step, followed by MS/MS peptide fragment fingerprinting/partial sequencing as the second.

4.1 Peptide mass fingerprinting

In 1993, five groups independently proposed the idea of peptide mass fingerprinting (Henzel *et al.*, 1993; James *et al.*, 1993; Mann *et al.*, 1993; Pappin *et al.*, 1993; Yates *et al.*, 1993). The concept is that the set of peptide masses obtained by mass spectrometric analysis of a digestion of a protein with a specific protease can act as a fingerprint, and this property is unique to that protein. Therefore the set of masses can be used to search a protein database in which the sequences have been replaced by the calculated fingerprints to find a similar pattern. The most rapid method of obtaining peptide masses from digests is by MALDI-MS, usually in combination with a time-of-flight analyzer (TOF). Most commerical MALDI-TOF instruments are equipped with sample loading targets that accommodate up to 1024 loading positions and software to allow unattended data accumulation. Sample preparation is critical to the success of this technique. If the sample is homogenous a few laser shots are sufficient to obtain a good spectrum without having to search the sample position for a spot with an optimal peptide:substrate ratio. Recently a novel approach to sample preparation for MALDI-MS has been put forward using a predeposited seed layer of matrix crystals (Onnerfjord *et al.*, 1999). The sample homogeneity allows mass spectra to be obtained on each individual sample within 15 s using around 16 laser shots. High-density target plates using the seed-layer method were prepared by spotting approximately 100 pl droplets onto the target, resulting in sample spots 500 μ in diameter using a flow-through piezo-electric microdispenser. Since virtually all of the peaks above 1000 are singly charged peptides, data extraction is simple and high signal-to-noise ratios can be achieved. Mass errors can now be kept at 5 ppm or lower using 'natural' internal standards (such as known tryptic autolytic fragments) over the range 800–3000 m/z. Peptide fingerprinting allows a rapid identification of the protein if it is represented in the database; however the technique is not very successful when searching genomic or EST databases. The random noise element in the database search can be minimized by using two or more orthogonal datasets (James *et al.*, 1994). Data from two protease digests with differing specificities (e.g. AspN and LysC) can be combined. Alternatively a single digest can be measured in the native state and again after carrying out a chemical modification on the sample plate (such as methylation, acetylation or deuterium exchange). This greatly increases the confidence level but doubles the analysis time.

4.2 Peptide fingerprinting with small sequence tags

An alternative to using dual digestions or native and chemically modified digests is to exploit the sequence information that is already present in a simple peptide fingerprint. Trypsin cuts at Lys and Arg, two of the most abundant amino acids in proteins. Consequently there are a comparatively large number of repetitive sequences, such as KK, RR, KKR, in proteins. Since trypsin cleaves randomly at these sites and has an intrinsically low exoprotease activity, ragged *C*-termini are produced. On average 2.5 such sequence tags are found per peptide mass spectrum (Jensen *et al.*, 1996). Ragged *N*- or *C*-termini can be efficiently generated by sequential endo- and exopeptidase digestions and the extracted data used to search sequence databases (Korostensky *et al.*, 1998). The confidence level of protein identification is much higher than simple fingerprinting alone. However, the method suffers the same drawback as enzymatic digestion, namely the effectiveness of the exoprotease activity drops rapidly with decreasing peptide concentrations and increasing numbers of peptides in a digest. Also, it is not easy to extract the sequence tags since exopeptidase digests are very unpredictable and the tag may follow the parent ion or start five amino acids later, making it difficult to assign a tag to a particular parent in a complex digest.

An alternative to exo- or endopeptidase ragged termini generation is the use of sequential chemical degradation steps. The use of a single-step manual Edman degradation of peptide mixtures to generate a single *N*-terminal amino acid tag has been shown to be effective (Jensen *et al.*, 1996), but the derivatization causes a loss in sensitivity by a factor of about 10 and the washing step required can cause extensive sample loss. Essentially, the method is an abbreviated form of 'ladder sequencing', in which automated Edman degradation using phenylisothiocyanate was carried out in the presence of a low amount of sequence terminator (phenylthiocyanate) and the resulting ladder of peptide fragments was analyzed by MALDI-MS (Chait *et al.*, 1993). A modified form of this procedure without a chain terminator generates the peptide ladder by adding equal aliquots of starting peptide each cycle and driving both the coupling and cleavage reactions to completion (Bartlet-Jones *et al.*, 1994). The main drawback is the extra number of handling steps, although the retention of the peptide terminal amine allows for subsequent modification with quaternary ammonium alkyl NHS esters to improve sensitivity.

An alternative degradation chemistry based on thioester degradation that can be carried out using aqueous reagents on peptides immobilized on C-18 reverse-phase beads immobilized in a Teflon membrane has been developed. The procedure can be carried out on 100 protein digests simultaneously on a single membrane fixed in a MALDI target which can be inserted directly into a mass spectrometer for analysis. Essentially the proteins are first chemically or proteolytically digested and immobilized on C-18 particles embedded in a Teflon membrane. The digest is divided into two aliquots, one for MALDI-TOF-MS analysis as the intact peptide mixture (*Figure 6a*), the other for measurement after chemical degradation (*Figure 6b*). The

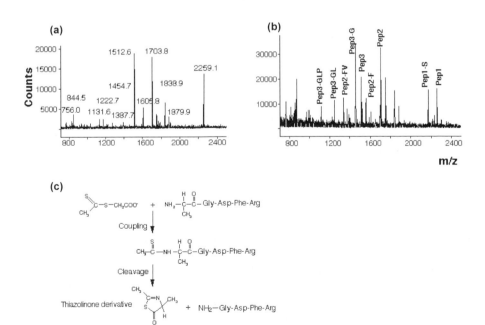

Figure 6. Chemtag chemistry and MALDI-MS. (a) The MALDI-MS spectrum of a protein spot isolated from a 2-DE gel and digested with trypsin. Since only a few peptides are present and two proteins are present in the spot, peptide mass fingerprinting fails to identify any protein present. (b) The digest after three rounds of degradation with thioacetylthioglycollic acid. The sequence tags associated with their parent ions are shown and could be used to identify the two proteins present in the spot.

degradation is based on the use of thioacylthioesters such as thioacetylglycollic acid (*Figure 6c*). Trimethylamine vapor is first passed through the membrane to establish alkaline conditions, at which point thioacetylglycollic acid in an *N*-ethylmorpholine buffer is passed through. The peptides react to form the *N*-thioacetyl derivative. The membrane is then washed with water then 0.1% trifluoroacetic acid (TFA) in water before passing TFA vapor through in order to cleave off the first amino acid. By carefully choosing the amount of water vapor present, only 25% amino acid cleavage takes place; the other 75% undergoes hydrolysis of the thioacetyl group. This is repeated several times and a ladder of peptide products are built up from which a sequence tag and parent mass can be read by MALDI analysis and used for database searching. The advantage of this method is the speed of data accumulation, since the reactions are carried out in parallel and the minimization of sample loss since the peptide digests are immobilized immediately after digestion up to analysis. More-over nonsequence specific enzymatic and chemical digests can be used to generate data for database searching. Thus MALDI-MS in combination with automated digestion and other peptide modification chemistries provides a high-throughput approach to protein characterization. Many of the steps involved, such as target loading, sample location and spectral acquisition as well as extraction of peak data for database searching are now standard for most commercial mass spectrometers.

5. Mass spectrometry: protein identification using MS/MS data

5.1 MS/MS data acquisition

Conceptually similar approaches to protein identification by peptide mass finger-printing using the MS/MS fragmentation pattern from a single peptide have been proposed (Eng *et al.*, 1994; Mann and Wilm, 1994). Tandem mass spectrometry of peptides was pioneered in the 1980s by the groups of Don Hunt for low-energy, triple-quadrupole instrumentation (Hunt *et al.*, 1986) and by Klaus Biemann (1990) for high-energy, four-sector magnetic instruments. The first mass scanning stage is used to isolate a single peptide before acceleration of the ions through a region of higher pressure containing a collision gas such as argon. The second mass scanning stage is used to measure the masses of the fragments arising from the peptide as a result of collision-induced/activated dissociation (CID). Postsource decay (PSD) spectra can be obtained using TOF instruments equipped with ion gates and mirrors by increasing the laser power by a factor of two over that needed to obtain a normal spectrum. A large fraction of the desorbed ions obtained by MALDI undergo 'delayed' fragmentation before reaching the detector as a result of multiple collisions of the peptides with the matrix during plume expansion and ion acceleration. The set of ions produced by either technique can be used to determine the sequence of the peptide. The set of fragment masses acts as a fingerprint for the peptide and can be used to search sequence databases for similar peptides. PSD analysis should be amenable to automation, and curved field reflectrons, which allow an entire fragmentation spectrum to be accumulated in one scan, have made it simpler to carry out. However it suffers from the disadvantage that PSD spectra are very difficult to obtain from peptide mixtures and some form of fractionation should be carried out prior to analysis (Gevaert *et al.*, 1996).

The accumulation of CID-MS/MS fragmentation data is somewhat more time consuming than simple peptide fingerprinting, although the data obtained has a much higher information content. Essentially two approaches to obtaining high-sensitivity MS/MS data can be taken using an electrospray interface to a triple quadrupole, ion trap or quadrupole-TOF mass spectrometer: either the digest is sprayed as a mixture directly from a capillary with a 1–10 μM exit tip at very low flow rates (< 50 nl/min; Andren *et al.*, 1994; Gale and Smith, 1993; Valaskovic *et al.*, 1995; Wahl *et al.*, 1993; Wilm and Mann, 1994), or via a nanospray online separation method such as capillary zone electrophoresis (CZE) or HPLC. The peptide mixture can then be analyzed automatically using a program to pick out the ions for MS/MS, which has the advantage that certain impurities such as autolytic fragments and gel-derived contaminants can be filtered out using a look-up table (Davis *et al.*, 1995b). This approach has been successfully used to sequence entire proteins (Piccinni *et al.*, 1994). The dynamic CZE/HPLC approach is much more sensitive than the static nanospray method, since the peptides are more concentrated, by a factor of 50 or more (Davis and Lee, 1997). Both methods have the advantage that they can be combined with parent ion scanning. Scans for the parents of common fragmentation products of peptides such as immonium or phosphate ions allow determination of peptide ion masses even when these ions have signal-to-noise ratios of 1 in the mass spectrum (Wilm *et al.*, 1996). However the greatest advantage of the dynamic methods is that they can easily be interfaced with an autosampler, allowing automated sample clean-up.

5.2 MS/MS database searching

Algorithms for automated searching sequence of databases using uninterpreted MS/MS spectra have been developed over the past 4 years. Probably the most widely used of these is SEQUEST, developed by John Yates and Jimmy Eng at the University of Washington, Seattle (Eng *et al.*, 1994). The program was originally intended for searching protein databases using MS/MS fragmentation spectra of unmodified peptides, but has been subsequently extended to allow searching with post-translationally modified peptides (Yates *et al.*, 1995b), DNA database searching (Yates *et al.*, 1995a), and searching with high-energy CID or PSD data (Griffin *et al.*, 1995; Yates *et al.*, 1996). Essentially the program searches for all peptides in the database (a protease can, but does not have to be defined) which have the same mass as the parent ion (within a defined mass window) and then matches the predicted MS/MS spectrum with the experimentally determined one, and finally carries out a cross-correlation analysis of the best scoring peptides in order to determine the best match. Several other programs have now been described which are available on the World Wide Web, which can use uninterpreted spectra, such as Fragfit (part of the Prowl software suite; http://prowl.rockefeller.edu/) at the Rockefeller Insititute, New York, and MS-Tag (part of the ProteinProspector suite; http://prospector.ucsf.edu/) at the University of California, San Francisco.

Another algorithm for protein identification in sequence databases, Peptide-Search, was developed by Matthias Mann (Mann and Wilm, 1994). The MS/MS spectrum must be manually inspected to find a group of ions which form a series from which a small sequence (the tag) can be read and used with the intact peptide mass, the mass from the N-terminal to the start of the tag sequence, and the mass from the end to the tag to the C-terminal, to search the databases. Since the search can be carried out using only the tag and the N-terminal mass, for example, the algorithm can identify peptides carrying undefined post-translational modifications. The program can also carry out mixed searches, for example a tag search with a peptide mass search, and then combine the results. Other programs have been developed which combine various elements of SEQUEST and PeptideSearch, such as MassFrag (Gevaert *et al.*, 1996), No-name (Patterson *et al.*, 1996), Fragfit (Arnott *et al.*, 1998) and MASCOT from Matrix Science (http://www.matrixscience.com/).

5.3 *Automated MS/MS interpretation and database searching*

The original approach to protein identification, the manual interpretation of a MS/MS spectrum to obtain a peptide sequence for use in a homology search such as BLASTA or TFASTA (Altschul *et al.*, 1990, 1994), is receiving renewed attention. Early attempts at automated MS/MS spectra interpretation used a partial correlation method to fit increasingly longer sequences to a spectrum (such as SEQPEP, Johnson and Biemann, 1989) or used a pattern-matching approach such as that described by Hines *et al.* (1992). Recently a program has been developed, Lutkefisk (Taylor and Johnson, 1997), which uses the set of possible sequence interpretations of an MS/MS spectrum generated by SEQPEP to search sequence databases using a modified FASTA approach. An alternative to this method is to generate sequence ions in the MS/MS spectrum that contain an isotopic signature that allows them to be immediately identified as N-terminally derived b- or C-terminal y-series ions. Digestion of proteins in 50:50 $^{18}O/^{16}O$-labeled water produces peptides which appear as doublets in the mass spectrum with a spacing of two mass units between the peaks. This is due to the introduction of a water molecule from the solvent during the hydrolysis of the peptide bond (Takao *et al.*, 1991). The b-ions in the MS/MS spectrum thus appear as a singlet, whilst the y-ions appear as doublets. This has been used as the base for automated 'de novo' sequence tag extraction for database searching using a high-resolution quadrupole/time-of-flight mass spectrometer (Shevchenko *et al.*, 1997). Several alternative chemical approaches have been used to specifically label the N-terminus of a peptide with an isotopic label, although this has been less successful because of a certain lack of discrimination between alpha (N-terminal) and epsilon (lysine) amino groups and the drop in detection sensitivity incurred by removing a basic site. We have recently introduced a method which greatly facilitates the deduction of sequences from MS/MS spectra. The method for labeling was discussed in the section 'Postgel labeling', above. Briefly, proteins are labeled with a 50:50 mixture of H4 or D4 nicotinic acid. The reaction is absolutely specific for the N-terminal since the amino group of lysine side chains are succinylted prior to derivatization. The MALDI spectra show that all the peptides occur as

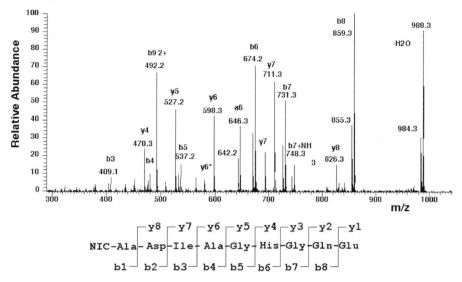

Figure 7. MS/MS of D4H4-labeled peptides. The MS/MS spectrum of the peptide ADIAGHGQE is shown. The *b*-ions appear as doublets separated by four mass units, whilst the *y*-ions appear as singlets. The presence of the nicotinic acid moiety on the *N*-terminal favors the formation of *b*-ions and allows the sequence to be easily deduced from the spectrum.

doublets separated by four mass units. The parent ion is selected with a wide enough mass window to accept the entire isotope envelope and the resulting MS/MS spectrum is dominated by ions originating from the *N*-terminal (*b*- and some *a*-ions) which all appear as doublets (*Figure 7*). All the ions (*y* or internal) arising from the *C*-terminal appear as singlets. It is thus fairly easy to extract the sequence. Even if an unknown post-translational modification is present, this does not affect the ability to use the sequence (containing an X at the appropriate point) in a (T)FASTA sequence homology search. The sequence inference is fairly easy to automate and will allow very fast and high-confidence-level protein identification.

5.4 Data interpretation tools

The automation process does not stop once the database searches have been carried out. High-throughput protein analysis requires the development of software tools to aid data interpretation to present meaningful trends and conclusions to the experimentalist. In the case of HPLC or CZE-autoMS/MS runs, the resulting data files are enormous and not much can be obtained from manual viewing. In order to compare various files, powerful data analysis software packages are required. One of the earliest programs, which was widely used and very user-friendly, was MacPro-Mass (Lee and Vemuri, 1990) developed in Terry Lee's laboratory for interpreting mass spectral data obtained from MS protein analysis. A second-generation program, Sherpa, was developed at the University of Washington, Seattle by Alex Taylor (Taylor *et al.*, 1996). The program can search concurrently against multiple protein sequences, compare two LC-MS files concurrently, search for glycosylated and phosphorylated peptides, and give a simple evaluation of the quality of a match between the data and a prediction. Such tools are invaluable aids to analysis. Recently John Yates has described a variant of the SEQUEST program to compare collision-induced dissociation spectra of peptides for library searching and subtractive analysis of tandem mass spectra obtained during LC-MS/MS experiments (Yates *et al.*, 1998).

Thus most areas in proteome analysis have been partially automated. The main areas which still need a great deal of attention are: digestion of very small amounts (> 10 fmol) of material; a universal approach to identifying post-translation modifications, especially the complex area of glycosylation; and an MS method whereby samples can be rapidly and highly efficiently concentrated after digestion for peptide fingerprinting and, if necessary, tandem mass spectrometry from the same sample (a MALDI ion trap would be ideal). Sadly lacking is a comprehensive laboratory information management system to link all the various forms of data together in an easily accessible manner as well as one that allows high-order data interpretation. Stable isotope labeling goes a long way to simplifying many of these problems, such as multiple proteins in a single spot and their quantification, as well as an easy method for compacting the information content of MS/MS datafiles. Much needs to be done, but most of the problems are trivial – the impact of nanotechnology and better data mining tools will soon transform this field.

References

Adams, M.D., Kelley, J.M., Gocayne, J.D., Dubnick, M., Polymeropoulos, M.H., Xiao, H., Merril, C.R., Wu, A., Olde, B., Moreno, R.F. *et al.* (1991) Complementary DNA sequencing: expressed sequence tags and human genome project. *Science* **252:** 1651–1656.

Altschul, S.F., Gish, W., Miller, W., Myers, E.W. and Lipman, D.J. (1990) Basic local alignment search tool. *J. Mol. Biol.* **215:** 403–410.

Altschul, S.F., Boguski, M.S., Gish, W. and Wootton, J.C. (1994) Issues in searching molecular sequence databases. *Nature Genet.* **6:** 119–129.

Anderson, N.G. (1967) Preparative zonal centrifugation. *Methods Biochem. Anal.* **15:** 271–310.

Andren, P.E., Emmet, M.R. and Caprioli, R.M. (1994) Micro-electrospray: zeptomole/attomole per microliter sensitivity for peptides. *J. Am. Mass Spectrom.* **5:** 867–869.

Arnott, D., Henzel, W.J. and Stults, J.T. (1998) Rapid identification of comigrating gel-isolated proteins by ion trap-mass spectrometry. *Electrophoresis* **19:** 968–980.

Bartlet-Jones, M., Jeffery, W.A., Hansen, H.F. and Pappin, D.J. (1994) Peptide ladder sequencing by mass spectrometry using a novel, volatile degradation reagent. *Rapid Commun. Mass Spectrom.* **8:** 737–742.

Biemann, K. (1990) Sequencing of peptides by tandem mass spectrometry and high-energy collision-induced dissociation. *Methods Enzymol.* **193:** 455–479.

Blackstock, W.P. and Weir, M.P. (1999) Proteomics: quantitative and physical mapping of cellular proteins. *Trends Biotechnol.* **17:** 121–127.

Burggraf, D., Weber, G. and Lottspeich, F. (1995) Free flow-isoelectric focusing of human cellular lysates as sample preparation for protein analysis. *Electrophoresis* **16:** 1010–1015.

Chait, B.T., Wang, R., Beavis, R.C. and Kent, S.B. (1993) Protein ladder sequencing. *Science* **262:** 89–92.

Davis, M.T. and Lee, T.D. (1997) Variable flow liquid chromatography tandem mass spectrometry and the comprehensive analysis of complex protein digest mixtures. *J. Am. Soc. Mass Spectrom.* **8:** 1059–1069.

Davis, M.T., Lee, T.D., Ronk, M. and Hefta, S.A. (1995a) Microscale immobilized protease reactor columns for peptide mapping by liquid chromatography/mass spectral analyses. *Anal. Biochem.* **224**: 235–244.

Davis, M.T., Stahl, D.C., Hefta, S.A. and Lee, T.D. (1995b) A microscale electrospray interface for on-line, capillary liquid chromatography/tandem mass spectrometry of complex peptide mixtures. *Anal. Chem.* **67**: 4549–4556.

Devine, K.M. and Wolfe, K. (1995) Bacterial genomes: a TIGR in the tank. *Trends Genet.* **11**: 429–431.

Eng, J.K., McCormack, A.L. and Yates, J.R. (1994) Correlating tandem mass spectral data of peptides to sequences in a protein database. *J. Am. Soc. Mass Spectrom.* **5**: 976–989.

Gale, D.C. and Smith, R.D. (1993) Small volume and low flow rate electrospray ionisation mass spectrometry of aqueous solutions. *Rapid Commun. Mass Spectrom.* **7**: 1017–1021.

Gevaert, K., Verschelde, J.L., Puype, M., Van Damme, J., Goethals, M., De Boeck, S. and Vandekerckhove, J. (1996) Structural analysis and identification of gel-purified proteins, available in the femtomole range, using a novel computer program for peptide sequence assignment, by matrix-assisted laser desorption ionization-reflectron time-of-flight-mass spectrometry. *Electrophoresis* **17**: 918–924.

Griffin, P.R., MacCoss, M.J., Eng, J.K., Blevins, R.A., Aaronson, J.S. and Yates, J.R. III (1995) Direct database searching with MALDI-PSD spectra of peptides. *Rapid Commun. Mass Spectrom.* **9**: 1546–1551.

Gygi, S.P., Rist, B., Gerber, S.A., Turecek, F., Gelb, M.H. and Aebersold, R. (1999) Quantitative analysis of complex protein mixtures using isotope-coded affinity tags. *Nature Biotechnol.* **17**: 994–999.

Harrington, M.G., Lee, K.H., Yun, M., Zewert, T., Bailey, J.E. and Hood, L. (1993) Mechanical precision in two-dimensional electrophoresis can improve protein spot positional reproducibility. *Appl. Theor. Electrophoresis* **3**: 347–353.

Henzel, W.J., Billeci, T.M., Stults, J.T., Wong, S.C., Grimley, C. and Watanabe, C. (1993) Identifying proteins from two-dimensional gels by molecular mass searching of peptide fragments in protein sequence databases. *Proc. Natl Acad. Sci. USA* **90**: 5011–5015.

Hines, W.M., Falick, A.M., Burlingame, A.L. and Gibson, B.W. (1992) Pattern based algorithm for peptide sequencing from tandem high energy collision induced dissociation mass spectra. *J. Am. Soc. Mass Spectrom.* **3**: 326–336.

Houthaeve, T., Gausepohl, H., Ashman, K., Nillson, T. and Mann, M. (1997) Automated protein preparation techniques using a digest robot. *J. Protein Chem.* **16**: 343–348.

Hunt, D.F., Yates, J.R. III., Shabanowitz, J., Winston, S. and Hauer, C.R. (1986) Protein sequencing by tandem mass spectrometry. *Proc. Natl Acad. Sci. USA* **83**: 6233–6237.

James, P., Quadroni, M., Carafoli, E. and Gonnet, G. (1993) Protein identification by mass profile fingerprinting. *Biochem. Biophys. Res. Commun.* **195**: 58–64.

James, P., Quadroni, M., Carafoli, E. and Gonnet, G. (1994) Protein identification in DNA databases by peptide mass fingerprinting. *Protein Sci.* **3**: 1347–1350.

Jensen, O.N., Vorm, O. and Mann, M. (1996) Sequence patterns produced by incomplete enzymatic digestion or one-step Edman degradation of peptide mixtures as probes for protein database searches. *Electrophoresis* **17**: 938–944.

Johnson, R.S. and Biemann, K. (1989) Computer program (SEQPEP) to aid in the interpretation of high-energy collision tandem mass spectra of peptides. *Biomed. Environ. Mass Spectrom.* **18**: 945–957.

Klose, J. (1975) Protein mapping by combined isoelectric focusing and electrophoresis of mouse tissues. A novel approach to testing for induced point mutations in mammals. *Humangenetik* **26**: 231–243.

Korostensky, C., Staudenmann, W., Dainese, P., Hoving, S., Gonnet, G. and James, P. (1998) An algorithm for the identification of proteins using peptides with ragged *N*- or *C*-termini generated by sequential endo- and exopeptidase digestions. *Electrophoresis* **19**: 1933–1940.

Lashkari, D.A., DeRisi, J.L., McCusker, J.H., Namath, A.F., Gentile, C., Hwang, S.Y., Brown, P.O. and Davis, R.W. (1997) Yeast microarrays for genome wide parallel genetic and gene expression analysis. *Proc. Natl Acad. Sci. USA* **94**: 13 057–13 062.

Lee, T.D. and Vemuri, S. (1990) MacProMass: a computer program to correlate mass spectral data to peptide and protein structures. *Biomed. Environ. Mass Spectrom.* **19**: 639–645.

Link, A.J., Hays, L.G., Carmack, E.B. and Yates, J.R. III (1997) Identifying the major proteome components of *Haemophilus influenzae* type-strain NCTC 8143. *Electrophoresis* **18**: 1314–1334.

Little, D.P., Speir, J.P., Senko, M.W., O'Connor, P.B. and McLafferty, F.W. (1994) Infrared multiphoton dissociation of large multiply charged ions for biomolecule sequencing. *Anal. Chem.* **66**: 2809–2815.

Madsen, P.S., Hokland, M., Ellegaard, J., Hokland, P., Ratz, G.P., Celis, A. and Celis, J.E. (1988) Major proteins in normal human lymphocyte subpopulations separated by fluorescence-activated cell sorting and analyzed by two-dimensional gel electrophoresis. *Leukemia* **2**: 602–615.

Mann, M., Hojrup, P. and Roepstorff, P. (1993) Use of mass spectrometric molecular weight information to identify proteins in sequence databases. *Biol. Mass Spectrom.* **22**: 338–345.

Mann, M. and Wilm, M. (1994) Error-tolerant identification of peptides in sequence databases by peptide sequence tags. *Anal. Chem.* **66**: 4390–4399.

Marshall, A.G., Senko, M.W., Li, W., Li, M., Dillon, S., Guan, S. and Logan, T.M. (1997) Protein molecular mass to 1 Da by 13C 15N double depletion and FT-ICR mass spectrometry. *J. Am. Chem. Soc.* **119**: 433–434.

Mortz, E., O'Connor, P.B., Roepstorff, P., Kelleher, N.L., Wood, T.D., McLafferty, F.W. and Mann, M. (1996) Sequence tag identification of intact proteins by matching tanden mass spectral data against sequence data bases. *Proc. Natl Acad. Sci. USA* **93**: 8264–8267.

Oda, Y., Huang, K., Cross, F.R., Cowburn, D. and Chait, B.T. (1999) Accurate quantification of protein expression and site-specific phosphorylation. *Proc. Natl Acad. Sci. USA* **96**: 6591–6596.

O'Farrell, P.H. (1975) High resolution two-dimensional electrophoresis of proteins. *J. Biol. Chem.* **250**: 4007–4021.

O'Farrell, P.Z., Goodman, H.M. and O'Farrell, P.H. (1977) High resolution two-dimensional electrophoresis of basic as well as acidic proteins. *Cell* **12**: 1133–1141.

Onnerfjord, P., Ekstrom, S., Bergquist, J., Nilsson, J., Laurell, T. and Marko-Varga, G. (1999) Homogeneous sample preparation for automated high throughput analysis with matrix-assisted laser desorption/ionisation time-of-flight mass spectrometry. *Rapid Commun. Mass Spectrom.* **13**: 315–322.

Pappin, D.J.C., Hojrup, P. and Bleasby, A.J. (1993) Rapid identification of proteins by peptide-mass fingerprinting. *Curr. Biol.* **3**: 327–332.

Parker, K.C., Garrels, J.I., Hines, W., Butler, E.M., McKee, A.H., Patterson, D. and Martin, S. (1998) Identification of yeast proteins from two-dimensional gels: working out spot cross-contamination. *Electrophoresis* **19**: 1920–1932.

Patterson, S.D., Thomas, D. and Bradshaw, R.A. (1996) Application of combined mass spectrometry and partial amino acid sequence to the identification of gel-separated proteins. *Electrophoresis* **17**: 877–891.

Piccinni, E., Staudenmann, W., Albergoni, V., De Gabrieli, R. and James, P. (1994) Purification and primary structure of metallothioneins induced by cadmium in the protists *Tetrahymena pigmentosa* and *Tetrahymena pyriformis*. *Eur. J. Biochem.* **226**: 853–859.

Quadroni, M. and James, P. (1999) Proteomics and automation. *Electrophoresis* **20**: 664–677.

Senko, M.W., Speir, J.P. and McLafferty, F.W. (1994) Collisional activation of large multiply charged ions using Fourier transform mass spectrometry. *Anal. Chem.* **66**: 2801–2808.

Shevchenko, A., Wilm, M., Vorm, O. and Mann, M. (1996) Mass spectrometric sequencing of proteins silver-stained polyacrylamide gels. *Anal. Chem.* **68**: 850–858.

Shevchenko, A., Chernushevich, I., Ens, W., Standing, K.G., Thomson, B., Wilm, M. and Mann, M. (1997) Rapid 'de novo' peptide sequencing by a combination of nanoelectrospray, isotopic labeling and a quadrupole/time-of-flight mass spectrometer. *Rapid Commun. Mass Spectrom.* **11**: 1015–1024.

Steinberg, T.H., Haugland, R.P. and Singer, V.L. (1996) Applications of SYPRO orange and SYPRO red protein gel stains. *Anal. Biochem.* **239**: 238–245.

Takao, T., Hori, H., Okamoto, K., Harada, A., Kamachi, M. and Shimonishi, Y. (1991) Facile assignment of sequence ions of a peptide labelled with ^{18}O at the carboxyl terminus. *Rapid Commun. Mass Spectrom.* **5**: 312–315.

Taylor, J.A. and Johnson, R.S. (1997) Sequence database searches via de novo peptide sequencing by tandem mass spectrometry. *Rapid Commun. Mass Spectrom.* **11:** 1067–1075.

Taylor, J.A., Walsh, K.A. and Johnson, R.S. (1996) Sherpa: a Macintosh-based expert system for the interpretation of electrospray ionization LC/MS and MS/MS data from protein digests. *Rapid Commun. Mass Spectrom.* **10:** 679–687.

Valaskovic, G.A., Kelleher, N.L., Little, D.P., Aaserud, D.J. and McLafferty, F.W. (1995) Attomole-sensitivity electrospray source for large-molecule mass spectrometry. *Anal. Chem.* **67:** 3802–3805.

Velculescu, V.E., Zhang, L., Zhou, W., Vogelstein, J., Basrai, M.A., Bassett, D.E. Jr, Hieter, P., Vogelstein, B. and Kinzler, K.W. (1997) Characterization of the yeast transcriptome. *Cell* **88:** 243–251.

Wahl, J.H., Goodlett, D.R., Udseth, H.R. and Smith, R.D. (1993) Use of small-diameter capillaries for increasing peptide and protein detection sensitivity in capillary electrophoresis-mass spectrometry. *Electrophoresis* **14:** 448–457.

Walsh, B.J., Molloy, M.P. and Williams, K.L. (1998) The Australian Proteome Analysis Facility (APAF): assembling large scale proteomics through integration and automation. *Electrophoresis* **19:** 1883–1890.

Williams, E.R., Henry, K.D., McLafferty, F., Shabanowitz, J. and Hunt, D.F. (1990) Surface-induced dissociation of peptide ions in Fourier-transform mass spectrometry. *J. Am. Soc. Mass Spectrom.* **1:** 413–416.

Wilm, M. and Mann, M. (1994) Electrospray and Taylor–Cone theory, Dole's beam of macromolecules at last. *Int. J. Mass Spectrom. Ion Processes* **136:** 167–180.

Wilm, M., Neubauer, G. and Mann, M. (1996) Parent ion scans of unseparated peptide mixtures. *Anal. Chem.* **68:** 527–533.

Yates, J.R. III., Speicher, S., Griffin, P.R. and Hunkapiller, T. (1993) Peptide mass maps: a highly informative approach to protein identification. *Anal. Biochem.* **214:** 397–408.

Yates, J.R. III., Eng, J.K. and McCormack, A.L. (1995a) Mining genomes: correlating tandem mass spectra of modified and unmodified peptides to sequences in nucleotide databases. *Anal. Chem.* **67:** 3202–3210.

Yates, J.R. III., Eng, J.K., McCormack, A.L. and Schieltz, D. (1995b) Method to correlate tandem mass spectra of modified peptides to amino acid sequences in the protein database. *Anal. Chem.* **67:** 1426–1436.

Yates, J.R. III., Eng, J., Clauser, K.R. and Burlingame, A.L. (1996) Searching sequence databases with uninterpreted high-energy CID spectra of peptides. *J. Am. Soc. Mass Spectrom.* **7:** 1089–1098.

Yates, J.R. III., Morgan, S.F., Gatlin, C.L., Griffin, P.R. and Eng, J.K. (1998) Method to compare collision-induced dissociation spectra of peptides: potential for library searching and subtractive analysis. *Anal. Chem.* **70:** 3557–3565.

Chapter 8

The automation of proteomics: technical and informatic solutions for high-throughput protein analysis

Marc R. Wilkins

1. A historical introduction

The mid 1970s saw the introduction of two-dimensional (2-DE) gel electrophoresis (Kenrick and Margolis, 1970; Klose 1975; O'Farrell, 1975; Scheele, 1975). These authors described the combination of a first dimension of isoelectric focusing with a second dimension of separation by mass. The resolving power of these separations was astonishing, with over 1000 *Escherichia coli* proteins being visualized on a single gel (O'Farrell, 1975). This gave researchers a new qualitative and quantitative glimpse of the protein complexity that exists in biological systems. The potential of the technique was immediately obvious to early adopters, who saw that the technique would allow the visualization of many of the proteins from an organism. In addition to this, differential display of 2-DE separations was seen and illustrated as a means of finding proteins with amino acid changes or potential post-translational modifications (O'Farrell and Goodman, 1976; Steinberg *et al.*, 1977).

Subsequent to the description of the 2-DE protein separation technique, a flurry of applications were described for a multiplicity of sample types and for many different purposes. Some samples were grown or prepared under different conditions in order to reveal changes in gene expression (e.g. O'Farrell and Ivarie, 1979). Other samples compared gene expression in mutant and normal strains of organisms (e.g. Reeh *et al.*, 1976). Yet other researchers applied 2-DE techniques to the understanding of protein–protein interactions in protein complexes (Briggs and Capaldi, 1977). The first 2-DE gel image analysis packages were also constructed some time ago (Anderson *et al.*, 1981). So in many ways, the potential of 2-DE electrophoresis as a means to investigate complex samples was available almost 20 years ago. The synthesis of these ideas is exemplified by the proposal that a human protein index – what we would now term a human proteome based on 2-DE electrophoretic separation of tissues – be generated (Anderson and Anderson, 1982).

1.1 Even protein chemists get the blues

The potential offered in the 1970s for the understanding of tissues or organisms by

Proteomics, edited by S.R. Pennington and M.J. Dunn
© 2001 BIOS Scientific Publishers, Oxford.

2-DE separation was not realized in the 1980s. There are a number of reasons for this. Firstly, there was a lack of nucleic acid and protein sequence information available in databases. Thus, whilst there was a good understanding of many physiological and biological processes in a variety of organisms, this understanding was not reflected in sequence information in databases. This was because molecular biology techniques were in their infancy, with dideoxy nucleic acid sequencing (Sanger *et al.*, 1977) being a new and manual technique, as was the amplification of DNA by the polymerase chain reaction (Saiki *et al.*, 1985). The sequencing of the first eukaryotic chromosome (Oliver *et al.*, 1992), which would rely on automated DNA sequencers using fluorescent dye chemistry (Prober *et al.*, 1987), was still many years away.

A second, fundamental reason why the potential offered by 2-DE was not realized is that the techniques available for protein identification were limiting. Methods for blotting proteins to nitrocellulose had been described (Towbin *et al.*, 1979) and detection of blotted proteins with antibodies showed promise as a sensitive identification method (Burnette, 1981). However, these approaches were limited as relatively few proteins had antibodies raised against them, and the production of monoclonal or polyclonal antibodies was a slow and labor intensive task. The sequencing of proteins by automated Edman degradation (Edman and Begg, 1967) potentially offered a means of generating new sequences for proteins or a means of protein identification by linking with genes via sequence databases. However, early liquid-phase sequenators required several hundred-fold more protein than was purified on a single 2-DE gel. It was not until 1987 that Matsudaira (1987) described a combination of blotting to polyvinylidene difluoride membranes and gas phase sequencing, to allow picomole quantities of proteins to be analyzed. One of the most effective ways for identifying proteins, at least in model organisms, remained the comigrating of pure known proteins with unknown proteins on a 2-DE gel (e.g. Phillips *et al.*, 1980).

As protein identification was such a difficult task, some researchers in the 1980s shifted their emphasis to the quantitative examination of changes that occur in protein expression under different conditions. Large databases of costimulated and coregulated proteins were described for *E. coli* (VanBogelen *et al.*, 1990), for the REF52 rat cell line (Garrels and Franza, 1989) and for mouse embryos (Latham *et al.*, 1991). Similarly, toxicological studies monitored changes in large numbers of mouse liver proteins after treatment with numerous toxic agents (Anderson *et al.*, 1987). In these cases, between 300 and 1600 proteins from different treatments were subjected to cross-matching and cross-correlation. These studies were impressive in that they showed large groups of proteins behaving in a coordinated manner. As such, they were many years ahead of similar studies based on microarray nucleic-acid hybridization techniques (Schena *et al.*, 1995; Wodicka *et al.*, 1997). However, the scientific utility of coregulated and corepressed proteins can only be realized when proteins involved in these groups are identified and characterized. Without this, the proteins were little more than spots on an image.

1.2 Evolution to the new science of proteomics

The 1990s saw a number of advances in different fields, all of which coalesced to form a new science. Two-dimensional gels became truly micropreparative with the widespread adoption of immobilized pH gradient technology (Bjellqvist *et al.*, 1993). Genome sequences of model organisms and other organisms were completed (e.g. those for *Mycoplasma genitalium*, *E. coli*, *Saccharomyces cerevisiae*, *Caenorhabditis elegans*), resulting in an explosion in the number of gene and protein sequences in databases. Mass spectrometry changed from being a specialist's tool to one available in almost every protein chemistry facility. This, in conjunction with a technique called peptide mass fingerprinting (Henzel *et al.*, 1993) which simplified the linking of a protein to a gene sequence in a database, revolutionized the speed and sensitivity of protein identification (for reviews see Kuster and Mann, 1998; Yates, 1998a,b). Amongst the above technical advances, the concept of the proteome was coined (Wilkins *et al.*, 1995). This term has served to focus and legitimize a science that, as compared to advances made in molecular biology, had previously not been given due recognition.

At the turn of a new millennium, proteomics is a field on the brink of a massive expansion. The *Drosophila melanogaster* genome has been completed, and the human genome sequence will be completed in the year 2000. Genome sequences for further model organisms such as the mouse and *Arabidopsis thaliana* are expected to soon follow. Making sense of these organisms with the benefit of genomic sequences – an endeavor otherwise termed functional genomics or functional proteomics – is biology's next big step. Functional genomics will rely heavily on use of high-density microarray technology (Schena *et al.*, 1995; Wodicka *et al.*, 1997). This approach is appealing in its capacity for high throughput, and simultaneous assaying of mRNA expression states. Proteomics, however, can offer far more in its ability to investigate protein expression, protein processing and modifications, protein–protein interactions, and protein function (Wilkins *et al.*, 1997a). The challenge for proteomics is thus, firstly, in the implementation of automated, high-throughput techniques, and, secondly, in the construction of databases and tools to allow the massive amounts of proteomic data to be efficiently queried and understood. This review will briefly address these two areas.

2. The automation of proteomics – technical solutions

Until recently, the study of proteins was very much a one-protein-at-a-time science. Researchers typically had a single protein, which was the focus of their research effort. Proteins were purified one by one, and studied one by one. Proteomics has seen a change in this approach, as researchers have broadened their foci to the simultaneous investigation of many proteins – whether it is for the characterization of a protein complex (e.g. Rigaut *et al.*, 1999), or the identification of all proteins expressed in an organism (e.g. Fountoulakis *et al.*, 1997, 1998a,b; Langen *et al.*, 1997). This broadening of focus has been possible only with changes in speed, sensitivity and precision of protein identification and characterization technology.

This has given protein researchers the opportunity to capitalize on the ability of 2-DE gel electrophoresis to purify proteins in a parallel, many-at-a-time fashion.

The recent technology that has impacted most on proteomics has been the mass spectrometer. The use of either electrospray ionization (ESI) or matrix-assisted laser desorption ionization (MALDI) methods, in conjunction with either quadrupole, ion-trap or time-of-flight mass detectors, has revolutionized the ease of protein analysis (for a review see Yates, 1998b). Importantly, this has allowed single proteins from single 2-DE gels to be identified and characterized, even if the proteins are only visible with silver staining (Wilm et al., 1996). However, what frustrated researchers until recently was that, whilst hundreds to thousands of proteins could be purified in a parallel on a 2-DE gel, there was no facile route for the preparation of large numbers of samples prior to their analysis by mass spectrometry. Furthermore, the introduction of large sample numbers into a mass spectrometer was difficult, especially with ESI mass spectrometers. Faced with these challenges, a variety of means of automation of protein analysis have been developed. These can be grouped into technologies teamed with 2-DE gels, technologies teamed with 1-DE gels, and alternatives to gel-based separations. These technologies, as well as efforts to automate the 2-DE separation procedure itself, will be discussed in turn.

2.1 Automating 2-DE gels

The automation of 2-DE gels will be necessary for high-throughput proteomics, providing that the 2-DE gel remains the method of choice for protein separation. Over the years, there have been numerous automation initiatives. Small-format gel separations (50 × 43 mm) have been completely automated (Brewer et al., 1986); however these separations are not thought to provide the necessary resolution or quantities of proteins needed for proteomics. Efforts have been made to automate large-format gels (Harrington et al., 1993; Nokihara et al., 1992), and a platform termed BioMetre[TM] has been under construction at Large Scale Biology Corporation for many years to automate the entire 2-DE process. Furthermore, patents have been lodged protecting means of linking first- and second-dimensional separations (Hochstrasser 1989, 1998) – a method which would facilitate the production of an all-in-one, and potentially automatable 2-DE gel. However, despite these efforts, an automated process for the separation of 2-DE gels is not available. Instead, manufacturers and researchers have concentrated on automating single steps in the procedure (e.g. gel staining), or increasing throughput with parallel, rather than serial approaches. This has resulted in the production of gel separation tanks for first and second dimensions that can carry, and separate, many first- or second-dimensional gels at once. A benefit arising from this is increases in reproducibility of separations.

2.2 Automated technologies teamed with 2-DE gels

After separation of a proteome on a 2-DE gel, there may be hundreds to thousands of proteins of interest to be analyzed by mass spectrometric techniques. A number of different approaches have been proposed, and in some cases commercialized, to automate the mass spectrometric analysis of 2-DE separated proteins or their peptides.

Spot picking robots for MALDI-MS protein characterization. One approach has been to automate the steps previously undertaken by human operators, to streamline the process of protein identification (*Figure 1*). To excise protein spots from gels, a number of robotic devices have been described in the literature (e.g. Traini *et al.*, 1998) or released commercially (e.g. the Flexys Robotic Workstation from Genomic Solutions). Generally, these robots image the gel with a digital camera, excise protein spots with a cutting head, and deliver gel pieces to 96-well plates. Subsequently, liquid-handling devices are applied to wash gel pieces and add endoproteinases. This has been achieved either through the modification of existing liquid-handling stations (e.g. Packard Multiprobe 104 in Traini *et al.*, 1998), in which case the same robot was used to ultimately deliver peptides to a MALDI sample plate for analysis, or the development of new robot technology (e.g. the ProGest robot from Genomic Solutions, which is a commercialization of work by Houthaeve *et al.*, 1997). The ProGest robot is teamed with a second robot, the ProMS, which undertakes parallel desalting of peptides and delivery to MALDI sample plates. The subsequent mass spectrometric analysis of many samples on recent MALDI mass spectrometers is relatively straightforward, as most manufacturers have implemented automatic acquisition methods; and as MALDI sample plates hold 100–384 samples, depending on the instrument used, large numbers of samples can be analyzed with little or no operator intervention. An early example of automated MALDI-MS analysis can be seen in Shevchenko *et al.* (1996).

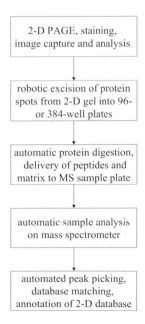

Figure 1. Automation of proteomics can be done by streamlining current processes. This approach is an automation of methods and techniques that have been effective when undertaken manually.

One advantage of the above approach is that the tracking of data for the mass spectrometry procedure is facile. The x–y positions of spots on gels are first recorded in association with a TIF or other image format. These are in turn associated with a certain position in a 96-well plate, and the 96-well plate number is carried through to be used as the identifier on the sample plate for MALDI analysis. In our laboratories, we use techniques similar to those described here, and have found them to be robust and very high throughput. Over a period of 2 days, we have prepared and analyzed 300 proteins on a single MALDI-TOF mass spectrometer for the purpose of protein identification. This level of throughput is currently difficult to achieve in other automated systems.

Gridding of 2-DE gels for MALDI-MS protein characterization. A variation on the approach described above has been developed by scientists at Hoffmann–La Roche (Langen *et al.*, 1998). Instead of excising individual spots from a 2-DE gel, entire gels are cut with a gridding tool into small pieces (e.g. 1 × 1 mm). Gel pieces are then automatically delivered into 96- or 384-well plates for digestion prior to analysis on mass spectrometers. The philosophy of this approach is very much holistic, as every piece of gel is subjected to digestion and analysis, whether protein is present or not. In this manner, faint spots that might not have been detected or cut by an excision robot will be included for analysis. The disadvantage of the gridding approach, compared to the spot excision approach, is that more gel pieces would be expected to contain greater than one protein, and that many single-protein spots will be arbitrarily divided into many pieces. This presents challenges for data analysis, where data for single-protein spots need to be correctly compiled. An interesting feature that does arise, however, is that the total ion count (TIC) from spectra for each gridded piece of acrylamide can be used to generate a virtual image of the 2-DE gel that was analyzed. The resolution of the image will, however, be determined by the size of grid used to generate the acrylamide gel pieces.

Automated characterization of 2-DE separated proteins via membranes and MALDI-MS. The two above automation approaches involve quite extensive manipulation of gel pieces in order to achieve protein characterization. Whilst this has been shown to be effective, these approaches are somewhat inelegant in the number of different steps required. Because of this, some groups have been investigating the use of membranes as the medium on which to carry and analyze proteins (*Figure 2*). Early work showed that masses of whole proteins could be determined by MALDI-MS when desorbed directly from polyvinylidene difluoride (PVDF) membranes. This involved attaching membranes to a sample plate, applying matrix, and desorbing protein from the membrane surface by ultra-violet (UV) or infra-red (IR) lasers (Eckerskorn *et al.*, 1992; Vestling and Fenselau, 1994). This idea has since been extrapolated such that portions of membranes are scanned at high resolution in a mass spectrometer, such as pixel sizes of 30 × 30 microns, to generate a virtual 2-DE image of material blotted or otherwise applied to a membrane (Eckerskorn *et al.*, 1997; Stoeckli *et al.*, 1999). With the development of protein digestion methods that maintain spatial integrity of resulting peptides on membranes (Bienvenut *et al.*,

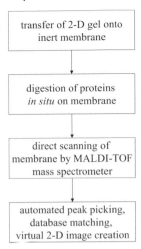

Figure 2. Automation of proteomics can be done by transferring 2-D gels onto membranes, digesting to peptides, and then scanning membranes in a MALDI-TOF mass spectrometer. This is a different approach to the simple automation of manual steps as in Fig. 1.

1999), the scanning of membranes has resulted in the production of virtual images of protein spots from 2-DE gels, as well as peptide masses which are useful for protein identification (Binz *et al.*, 1999). The scanning of membranes in MALDI mass spectrometers is automatable, and therefore potentially of high throughput for protein identification.

The advantages of the membrane scanning approaches are that proteins can be digested *in situ* on membranes (Binz *et al.*, 1999; Schleuder *et al.*, 1999), potentially in parallel, and that the only sample handling involved concerns the placing of a membrane onto a mass spectrometer plate. As proteins attached to membranes are stable over long periods of time, the membranes also present a convenient means of archiving. Yet a current disadvantage of the scanning technique is that it remains slow. For example, the complete analysis of a 16 × 16 cm 2-DE gel with current methods would take 36 days (Binz *et al.*, 1999). A further drawback is that the numbers of peptides from a protein obtained by on-membrane digestion and scanning MALDI-TOF-MS appear to be lower than if the protein had been analyzed by standard digestion and MALDI-TOF-MS techniques. This may limit identification efficacy and the ability to characterize proteins for post-translational modifications. Nevertheless, variations on this method may prove useful with developments in computer and mass spectrometer hardware and software.

Automated characterization of 2-DE separated proteins via LC-MS. The analysis of protein digests from 2-DE gels by liquid chromatography mass spectrometry (LC-MS) presents an alternative to analysis by MALDI-TOF-MS. As well as the ability to measure peptide masses, LC-MS also presents the possibility for peptides to be fragmented by collision-induced dissociation (CID) techniques. This can produce peptide sequence tags (Mann and Wilm, 1994) that are useful for protein identification, and in cases of efficient peptide fragmentation allow *de novo* peptide sequences to be generated (e.g. Qin *et al.*, 1998).

The generation and analysis of data via LC-MS is very efficient once a sample is introduced for analysis, and many LC-MS analyses of proteins from 2-DE gels have been reported (e.g. Gygi *et al.*, 1999). However, throughput on these systems remains at least an order of magnitude lower than analysis via MALDI-TOF-MS, for example, about 20 samples per day (Link *et al.*, 1997). This is because it is technically desirable to work with capillary electrospray or so-called nanospray sources (Wilm and Mann, 1996) that can increase analytical sensitivity and the amount of data generated from a sample, but capillary and nanospray systems are prone to blockage and are inherently difficult to automate. Nevertheless, commercial autosampler interfaces between 96-well plates and LC systems are appearing, and these are likely to be robust and allow increases in sample throughput. More excitingly, some researchers have described microfabricated devices with electroosmotic pumps, which have been interfaced to ion trap or quadrupole-based mass spectrometers (Figeys *et al.*, 1997, 1998). These also have the potential to aid the automation of analysis from gel to LC-MS, but it is not yet clear how robust these systems will ultimately be.

2.3 Automated technologies as alternatives to 2-DE

The increased interest in proteomics has led to discussion about the best means of visualizing and separating a proteome of a cell or tissue. Two-dimensional gel electrophoresis currently has the highest resolving power of any protein separation technique. When undertaken in a large-size format (whether achieved on a physically large gel or by undertaking multiple separations on narrow pI-range immobilized gradient strips), close to 10 000 proteins from a sample can be resolved (Klose, 1999). Use of sequential sample extraction techniques adds hydrophobicity as a third dimension to the analysis (Herbert, 1999; Klose, 1999), and can visualize at least 20% further proteins. Yet despite these features, and the fact that a 2-DE gel is a parallel purification technique, 2-DE does have some limitations. Most notable of these are the difficulty in visualizing very low abundance proteins, very hydrophobic proteins, and proteins that have a very basic pI (Wilkins *et al.*, 1996, 1998a). Many users are also concerned about the inter- and intralaboratory reproducibility of separations, even in the light of data which shows that very high reproducibility is possible (Corbett *et al.*, 1994; Lopez and Patton, 1997). These issues have fueled research into separation technologies that present an alternative to 2-DE gels. These novel technologies may be classified into those that team with 1-DE gels, and those which are nongel-based separations.

Automated technology teamed with 1-D gels. The concept of direct scanning of surfaces by MALDI-TOF-MS has been applied to isoelectric focusing gels to create virtual 2-DE gel images (Ogorzalek Loo *et al.*, 1997; Loo *et al.*, 1999). In this case, the first-dimensional separation is carried out on-gel, which allows a protein's pI to be calculated from its position along an immobilized pH gradient (IPG) strip. The second-dimensional mass measurement is achieved by the mass spectrometer, thus replacing the separation by SDS–PAGE. Scientifically this approach has merit as the

IPG strip is an equilibrium-based means of separation, and the mass spectrometer should be considerably more accurate than a non-equilibrium SDS–PAGE second-dimensional separation. If data acquisition and analysis can be automated, the process presents a novel means of generating high-accuracy mass data and virtual 2-DE maps for large numbers of proteins.

Technically, the approach first involves the separation of proteins using standard isoelectric focusing techniques on IPG strips. After the process of washing, soaking in sinapinic acid matrix and air-drying, the IPG strips are then introduced directly into a MALDI-TOF mass spectrometer. Scans are taken at close intervals along the length of the IPG strip (e.g. every 30 μm), whereupon numerous proteins are ionized and their masses detected at each position. Scanning across the strip produces a series of spectra, which are finally deconvoluted to yield an image similar to a 2-DE gel (Ogorzalek Loo et al., 1999).

It is currently difficult to compare the results of the direct scanning of IPG strips with that obtained by normal 2-DE gel separations, as virtual 2-DE gel images are yet to be published for an entire IPG strip. However, there are likely to be some limitations encountered in the procedure. Firstly, it is known that certain proteins ionize well in MALDI-TOF, but others, such as those of high mass, do not. This will almost certainly affect efforts to quantitate proteins in a meaningful way. Secondly, it is not apparent if the direct MALDI-TOF technique will have the power to resolve as many different proteins as an SDS–PAGE gel. Finally, whilst the scanning technique may be of use for generating high-accuracy virtual 2-DE gel images, it currently offers no facile means of identifying proteins of interest by, for example, peptide mass fingerprinting. It is possible to cut a protein slice from an IPG gel and digest proteins to peptides with cyanogen bromide (Loo et al., 1999); however, a mixture of proteins will almost certainly be present in any one part of an IPG strip. However, it is interesting to note that accurate measurements of protein pI and mass alone may be sufficiently stringent in small organisms with sequenced genomes to allow proteins to be identified (Cavalcoli et al., 1997; Wilkins et al., 1998b), and it is these parameters that will be readily determined by the IPG scanning approach.

Automated, nongel-based separations of proteomes. Some of the more novel alternatives to gel-based separations have involved one or two-dimensional chromatography, coupled with either ultra-violet protein detection and/or electrospray mass spectrometers. Some researchers have taken whole proteomes, and separated proteins by a combination of automated size exclusion chromatography and reversed-phase chromatography (e.g. Opiteck et al., 1998; or see Isobe et al., 1991 for early work). These approaches produced two-dimensional UV chromatographs, with a first dimension of apparent protein size and the second dimension of protein hydrophobicity (Opiteck et al., 1998). The separation is complete in 6–17 h, depending on run conditions. Whilst the intention of the method has been to display the contents of a proteome, the number of protein species discernible for the separation of E. coli cell lysate was about 250. This is approximately an order of magnitude lower than that seen for E. coli on a single silver-stained 2-DE gel (Tonella *et al.*, 1998). Clearly, a further dimension of separation would be required to achieve the resolution necessary for proteome analysis.

A different approach to 2-D chromatography has been developed by other researchers, which aims to analyse peptides generated from the proteins present in a sample, rather than the proteins themselves (*Figure 3*). The technique uses the ability of triple quadrupole, ion-trap or quadrupole time-of-flight mass spectrometers to automatically fragment peptides and identify the proteins from which they were derived by comparison with sequence databases (for a review see Yates, 1998b). In this manner, the proteins present in a relatively simple sample such as a protein complex (e.g. Link *et al.*, 1999) or potentially an entire proteome (S.D. Patterson, unpublished) can be investigated. The method first takes a sample, and digests all proteins to peptides using an enzyme such as trypsin. Then, peptides are loaded onto reversed-phase columns in order to achieve a degree of separation, and eluted directly into an electrospray mass spectrometer capable of peptide fragmentation. These mass spectrometers have the ability to select an ion of interest from a mixture of peptides, automatically fragment this ion to generate a sequence tag or complete sequence, and continue this process iteratively with other ions present. Many hundreds of peptides can be fragmented in a single liquid chromatography run, leading to the automatic linking of hundreds of peptides to sequences with software such as SEQUEST (Yates *et al.*, 1995). Some of the most advanced applications of this to date concern work undertaken by Amgen Inc., with the analysis of human urine (S.D. Patterson, unpublished). A single experiment identified peptides from approximately 100 different proteins. An advantage of this approach is that it is of very high throughput, and that it has built-in redundancy insofar as that many peptides from each protein are usually seen. However, the technique is highly data-intensive, where considerable effort in data analysis is required before one can assess the outcomes of an experiment. Also, as all work is carried out on peptides rather than proteins, it is likely that the technique will not be

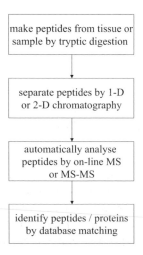

Figure 3. Proteomic analysis can be automated by digesting entire samples to peptides, then separating all peptides from the sample via HPLC and finally analysing them by MS or MS-MS. This approach, however, may give less detail of a sample than 2-D gel based techniques.

useful for distinguishing closely related isoforms, and neither will the technique be useful where detailed characterization of a single protein is required.

3. Informatic solutions for proteomics

Proteomics is an information science at many levels. For the proteins themselves, proteomics involves sequence information, processing and modification information, expression information, functional information, structural information and information concerning how the proteins interact with other proteins to form small or large protein complexes. Importantly, proteomics can also consider the effects of time and environment, so the above parameters might be assessed at different developmental time points or under different environmental conditions. On a more mundane level, sample tracking and cataloging of results is another important requirement of proteomic research. High-throughput screening requires laboratory information management systems (LIMS) to track how samples have been analyzed and the results of these analyses. Relevant results need to be databased effectively for future reference. Pivotal in the analysis of large amounts of proteomic data are proteomic tools, which are essential for jobs including protein identification and characterization. Remembering that even simple organisms such as *S. cerevisiae* have more than 6000 proteins, and that more complex organisms might have 200 different tissue types with 10 000 proteins each, the task proteomics faces in this regard is massive. In the remainder of this review, we will describe the types of proteome informatic resources that are available to researchers in the field. These resources will be divided into proteomic databases and proteomic tools. Currently available resources will be described and some future developments will be discussed.

3.1 Proteomic databases

To a certain degree, a genome is easy to define, it being the base-by-base sequence of a genome. Open reading frames can then be delineated, as can introns and exons, promoter and enhancer regions, and any structural regions of a chromosome (e.g. centromere and telomere). By contrast, a proteome may be seen as more diverse due to the different data types involved and the dynamic nature of proteins in a biological system. Therefore, a database that is a mere translation of nucleic acid sequences, whilst useful, is by no means a complete proteomic resource. A consequence of this is that there is no single database that currently addresses all areas of proteomics, but instead there are a series of databases that cover different proteomics-related fields.

The most comprehensive database for proteomics is currently SWISS-PROT (Bairoch and Apweiler, 2000; http://www.expasy.ch/). This is a curated and annotated database of proteins, maintained at the Swiss Institute for Bioinformatics in Geneva, Switzerland and the European Bioinformatics Institute. Apart from containing protein sequences, the database contains large numbers of annotations (*Figure 4*). These include information about protein expression, function, domain

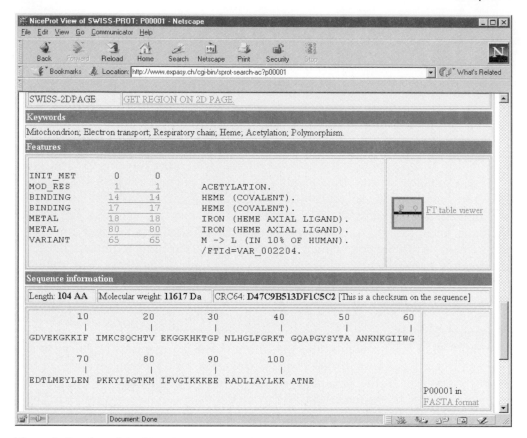

Figure 4. Portion of the SWISS-PROT entry for human cytochrome C (accession number P00001), showing the feature table and protein sequence. Note that the feature table has information on protein post-translational modifications, metal binding, and a known variant. Information on the function of the protein is earlier in the entry, in the comments field.

structures, variants, and protein processing and post-translational modifications. It also contains a very high level of cross-references to other databases, such as Medline, GenBank, the PDB structural database, and a series of cross-links to other specialized databases such as pattern and profile databases, metabolic databases or organism-specific databases. Whilst this database is approximately 84 000 entries, and thus small in comparison to the large sequence repositories such as GenBank, the information is of very high quality, and remains the gold standard of protein databases.

A challenge faced by SWISS-PROT is that the speed of genome sequencing and scientific publication has made it impossible to comprehensively cover all relevant published information for any particular organism, let alone the thousands of species for which SWISS-PROT contains some information. The company Proteome Inc. has taken a complementary approach to SWISS-PROT, concentrating on annotating only a few model or commercially significant organisms. One database from Proteome Inc. is the Yeast Proteome Database (YPD; see http://www.proteome.com/), which is currently the most comprehensive database for the budding yeast *S. cerevisiae*. All proteins in the proteome are annotated with information like that in SWISS-PROT; however there is additional information provided (e.g. codon bias, chromosome position, gene structure). The database features extensive reviewing and referencing of the yeast literature, providing highly detailed information about protein functions and the experiments that were undertaken to determine this information. There are also features such as transcript profile displays, expression profile searching and analysis, and sequence similarity reports for each protein. Proteome Inc. has similar databases for the organism *Candida albicans* and the worm *Caenorhabditis elegans*, and has received funding to develop databases for the pathogens *Pneumocystis carinii* and *Aspergillus fumigatus* as well as fungal model organisms such as *Aspergillus nidulans* and *Schizosaccharomyces pombe*. In the public sector, there are other organism-specific databases, whose completeness and usefulness as a resource are generally a reflection of how well an organism's research community is organized. Some of the better public organism databases include FlyBase (see http://flybase.bio.indiana.edu/) and the *Arabidopsis thaliana* database AAtDB (see http://genome-www.stanford.edu/Arabidopsis/).

A different type of proteomic database is the 2-DE protein expression database (*Figure 5*). These databases are based on 2-DE gel separations of an organism, tissue or cell subfraction, and contain images of these 2-DE gels and associated text information concerning the sample under study and proteins identified on the gels. Internet addresses for most 2-DE protein expression databases are cataloged at http://www.expasy.ch/ch2d/2d-index.html. Mass spectrometry has enabled large numbers of proteins to be identified from 2-DE gels, and many 2-DE databases now contain up to hundreds of identified proteins. In some cases (e.g. the SWISS-2DPAGE database), there is considerable textual information associated with each identified protein spot on a 2-DE gel, including information about the protein mass and pI, method of identification, details of other tissues where the protein is found (if applicable) and any pathological information. Recently, 2-DE protein expression databases have started to reflect the necessity for more than one gel for any sample, and some databases are now showing different separations (e.g. narrow pI range analyses) of samples. However, a major challenge that is faced by 2-DE gel databases remains that of consistency, both at the level of the 2-DE separation and the way in which data is formatted and presented. Whilst considerable effort has been placed in recent years on standardizing 2-DE sample preparation and electrophoresis conditions, the explosion of interest in proteomics has meant that there is now greater diversity than ever in the techniques being adopted to produce reference 2-DE maps. These techniques may be highly reproducible for a certain laboratory; however it may be challenging to precisely reproduce these separations elsewhere. There has been a level of consistency applied to the way 2-DE database images, text and crosslinks should be stored and presented (Appel *et al.*, 1996), which has been useful. However, as long as 2-DE gel separations in databases are performed by different methods and thus remain difficult or impossible to compare from lab to lab, the utility of 2-DE databases as a general resource may be limited. Nevertheless it is

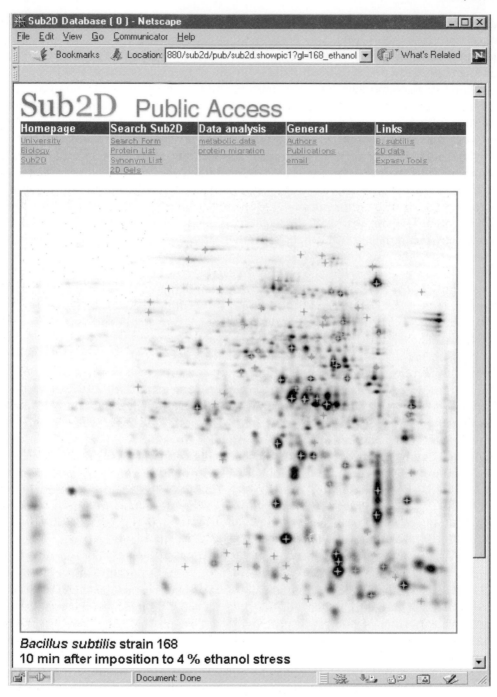

Figure 5. An annotated 2-De gel from the *Bacillus subtilis* Sub2D database. The crosses on the image represent clickable links to pages that give details of a particular protein, such as protein sequence, expression levels etc. Most 2-D gel databases use a similar system of clickable links.

notable that peptide mass information, generated by the widespread application of mass spectrometry to the identification of 2-DE-separated proteins, is providing parameters for proteins that are absolute, laboratory-independent, and thus very useful in databases.

There are many other database types that are important to proteomics. It is impossible to review all of these databases here, as they are growing in scope and number every day. However they can logically be organized into databases that address the areas of: nucleic acid sequences; protein sequences; protein pattern, profile and domains; post-translational modifications; three-dimensional protein structures; and metabolic databases. Arguably the best single internet site which lists and categorizes these resources is organized by Amos Bairoch, and is kept at the site http://www.expasy.ch/alinks.html. Over 1000 relevant sites are now listed.

3.2 Proteomics tools

Along with the construction of many proteomic-related databases, there has been a large number of informatic tools made in recent years to assist in various aspects of proteomic data analysis. These tools range from programs to aid in the interpretation and annotation of 2-DE gel images, through to tools for the *de novo* sequencing of peptides from their MS-MS spectra. We will briefly discuss these tools below and comment on how they contribute to the implementation of high-throughput approaches in proteomics.

Two-dimensional gel image analysis software. Since 1981, software has existed to assist the analysis of gel images (Anderson *et al.*, 1981). Such software has a number of core tasks to perform. These include correction of background, detection and quantitation of spots, alignment of one gel image to another, gel-to-gel matching via images and statistics, merging of one or more gels to create synthetic or master gel images, and the construction of databases where information concerning protein spots can be stored and queried. Whilst most 2-DE gel image analysis packages began in academic laboratories (e.g. ELSIE, TYCHO, MELANIE and QUEST), most of these packages have now been greatly enhanced and are available as commercial products that operate as standalone applications on PC platforms (e.g. KEPLAR, MELANIE II and PDQUEST). Internet- or intranet-based packages are also being developed by some groups (e.g. Lemkin *et al.*, 1999). However, it is worth noting that, even with advanced commercial packages, automatic gel analysis is rarely satisfactory and the complexity of 2-DE gels can mean that detailed image analysis is a very time- and labor-intensive task. But many hours of analysis time can be saved if effort is first placed into creating gels that are well controlled in the areas of sample preparation, electrophoretic separation and staining. A detailed explanation of the issues concerning computer analysis of 2-DE gel images can be found in Appel and Hochstrasser (1999).

Protein identification software. There is a large variety of protein parameters that can be generated analytically and subsequently used for protein identification. These include protein mass and isoelectric point, protein sequence tags, peptide mass fingerprints, peptide sequence tags and protein amino acid composition. There is software available for matching these parameters alone or in combination against databases (for review see Wilkins and Gooley, 1997), usually as Internet-based tools, but in some cases also as standalone packages (e.g. the ProteinLynx software available on mass spectrometers from the Micromass corporation). With the increased use of mass spectrometry for protein analysis in proteomics, numerous software packages have focused on providing the tools necessary for this type of research. Generally, these packages include software for: protein identification by peptide mass fingerprinting with variables including species, protein mass and pI; protein identification by MS/MS, where MS/MS information may result from CID or MALDI-TOF PSD fragmentation techniques; tools for the prediction of fragment masses after endoproteinase digestion and/or the MS/MS fragmentation of peptides; and in some cases, tools for the *de novo* prediction of peptide sequences from MS-MS data. Examples of software packages with some or all of these tools include Protein Prospector from UCSF (http://prospector.ucsf.edu/), PROWL from Rockefeller University, (http://prowl.rockefeller.edu/), MASCOT from Matrix Science Ltd (http://www.matrixscience.com/), SEQUEST from the University of Washington (Yates *et al.*, 1995), and the proteomic tools on the ExPASy server (http://www.expasy.ch/tools/#proteome; Wilkins *et al.*, 1999a). All of these packages work effectively as tools to identify proteins one at a time, but the greater challenge is in how they may be applied to deal with large numbers of samples with a minimum of human intervention. This is more complicated than it seems, as it is desirable that spectra from mass spectrometers can have peaks picked automatically, pipelined through protein identification tools automatically, and that results be ultimately stored in databases in association with other information such as 2-DE gel images. Some groups have begun to address this in part or whole (e.g. Berndt *et al.*, 1999; Gras *et al.*, 1999; Link *et al.*, 1999; Traini *et al.*, 1998) and it is apparent that this software, in combination with the advances in hardware automation, means that the industrialization of protein identification is almost complete.

Protein characterization software. Whilst protein identification is becoming increasingly industrialized, the process of characterizing protein isoforms and examining protein processing and post-translational modifications remains difficult and less popular. However, researchers are beginning to understand that the resolving power of 2-DE gels and the accuracy and resolution of mass spectrometers can be applied to detailed protein characterization, as well as identification. The wealth of information on protein processing and modifications in databases such as SWISS-PROT can also be used for the characterization task. The PeptideMass program, a tool for predicting the masses of peptides generated from the endoproteinase digestion of a protein (Wilkins *et al.*, 1997b), considers all protein processing and modification information available for a protein in SWISS-PROT and shows positions of modified amino acids and appropriate masses for modified peptides where known. The PeptIdent tool (http://www.expasy.ch/tools/ peptident.html; Wilkins *et al.*, unpublished) is a peptide mass fingerprinting identification engine which matches user-entered peptides against unmodified and modified peptides in the SWISS-PROT database and therefore identifies a protein and, where appropriate, shows peptides from that protein that may carry modifications. Other peptide mass fingerprinting engines have a similar approach but search only for mass differences rather than considering what modifications are known in current databases. A tool for predicting new post-translational modifications in peptide mass data has recently been described (Wilkins *et al.*, 1999b). The user specifies the protein sequence and its peptide mass data, so that the program can search for mass differences that correspond to a certain modification. Potentially modified peptides are then examined with a series of rules to predict which amino acids in that peptide, if any, might carry the modification. Rules exist for more than 20 different modifications; some rules are very simple, whilst others are complex sequence motifs. A further protein characterization tool that has been released on the ExPASy server is GlycoMod (N. Packer, personal communication), which predicts the composition of a protein-associated glycan from its mass. As this will be linked to a database of known glycosylation structures (Cooper *et al.*, 1999), it will present a comprehensive means of studying protein glycosylation when masses of the glycoforms can be generated.

3.3 Looking into the bioinformatic future

The above sections have given an overview of the informatic resources that are available for proteomics. Clearly, these resources are impressive and have evolved in a relatively short period of time. However, the major challenge that researchers face is how the many databases and tools required in proteomics can be integrated to work in a seamless and meaningful fashion, and above all in a way that allows us to pose involved biological questions. For example, as disease states are often epigenetic, we need to track the expression levels, modification and processing of large numbers of proteins at a time. This may be in one tissue of an organism (e.g. liver), but might also be in more than one tissue (e.g. liver and kidney). Alternatively, we may need to screen thousands of isolates of a certain microbe to check for proteins associated with pathogenesis. This may require interfacing with chip-based mRNA studies to see if novel proteins are expressed, or if an always-expressed protein is differentially processed or otherwise modified. The data-handling issues in the above examples are massive, and are undoubtedly the subject of much current work. Another major challenge is how large proteomic datasets will best be visualized. Clearly, the 10 000 or so proteins in the 200 different human tissues can not be displayed on a composite 2-DE gel. Instead, multidimensional visualization tools will be required. The use of color and ordered arrays will certainly be essential, as shown by Weinstein *et al.* (1997) in their proteomic screening of large numbers of compounds against a library of human cell lines. Increasing standardization of data types and tools will be required to achieve all of this and to remain competitive with functional genomics groups, who may advocate that large-scale mRNA analysis can answer questions that are essentially proteomic.

4. Conclusions

This review has covered much of the diversity of proteomics, but in particular the areas of automation and proteome informatics. By necessity, this has focused on areas that are technical. However, we must bear in mind that any set of tools is only as good as the results they yield when applied to a research task. Shotgun sequencing in genomics has shown that a combination of technologies, none of which were revolutionary but which were nevertheless judiciously applied, can produce results that have changed biology forever. Proteomics is now in the limelight, and the question we might ask is will proteomic technologies also have an impact of such magnitude? For this, one presumes, we'll have to wait and see.

References

Anderson, N.G. and Anderson, L. (1982) The human protein index. *Clin. Chem.* **28:** 739–748.

Anderson, N.L., Taylor, J., Scandora, A.E., Coulter, B.P. and Anderson, N.G. (1981) The TYCHO system for computer analysis of two-dimensional gel electrophoresis patterns. *Clin. Chem.* **27:** 1807–1820.

Anderson, N.L., Giere, F.A., Nance, S.L., Gemmell, M.A., Tollaksen, S.L. and Anderson, N.G. (1987) Effects of toxic agents at the protein level: quantitative measurement of 213 mouse liver proteins following xenobiotic treatment. *Fundam. Appl. Toxicol.* **8:** 39–50.

Appel, R.D. and Hochstrasser, D.F. (1999) Computer analysis of 2-D images. *Methods Mol. Biol.* **112:** 363–381.

Appel, R.D., Bairoch, A., Sanchez, J.-C., Vargas, J.R., Golaz, O., Pasquali, C. and Hochstrasser, D.F. (1996) Federated two-dimensional electrophoresis database: a simple means of publishing two-dimensional electrophoresis data. *Electrophoresis* **17:** 540–546.

Bairoch, A. and Apweiler, R. (2000) The SWISS-PROT protein sequence database and its supplement TrEMBL in 2000. *Nucleic Acids Res.* **28:** 45–48.

Berndt, P., Hobohm, U. and Langen, H. (1999) Reliable automatic protein identification from matrix-assisted laser desorption/ionization mass spectrometric peptide fingerprints. *Electrophoresis* **20:** 3521–3526.

Bienvenut, W.V., Sanchez, J.-C., Karmime, A., Rouge, V., Rose, K., Binz, P.-A. and Hochstrasser, D.F. (1999) Toward a clinical molecular scanner for proteome research: parallel protein chemical processing before and during western blot. *Anal. Chem.* **71:** 4800–4807.

Binz, P.-A., Muller, M., Walther, D., Bienvenut, W., Gras, R., Hoogland, C., Bouchet, G., Gasteiger, E., Fabbretti, R., Gay, S., Palagi, P., Wilkins, M.R., Rouge, V., Tonella, L., Paesano, S., Rossellat, G., Karmime, A., Bairoch, A., Sanchez, J.-C., Appel, R.D. and Hochstrasser, D.F. (1999) A molecular scanner to automate proteomic research and to display proteome images. *Anal. Chem.* **71:** 4981–4988.

Bjellqvist, B., Sanchez, J.C., Pasquali, C., Ravier, F., Paquet, N., Frutiger, S., Hughes, G.J. and Hochstrasser, D. (1993) Micropreparative two-dimensional electrophoresis allowing the separation of samples containing milligram amounts of proteins. *Electrophoresis* **14:** 1375–1378.

Brewer, J., Grund, E., Hagerlid, P., Olsson, I. and Lizana, J. (1986) *Electrophoresis '86* (ed. M.J. Dunn). VCH, Weinheim, pp. 226–229.

Briggs, M.M. and Capaldi, R.A. (1977) Near-neighbor relationships of the subunits of cytochrome c oxidase. *Biochemistry* **16:** 73–77.

Burnette, W.N. (1981) "Western Blotting": electrophoretic transfer of proteins from sodium dodecyl sulfate–polyacrylamide gels to unmodified nitrocellulose and radiographic detection with antibody and radioiodinated protein A. *Anal. Biochem.* **112:** 195–203.

Cavalcoli, J.D., VanBogelen, R.A., Andrews, P.C. and Moldover, B. (1997) Unique identification of proteins from small genome organisms: theoretical feasibility of high throughput proteome analysis. *Electrophoresis* **18:** 2703–2708.

Cooper, C.A., Wilkins, M.R., Williams, K.L. and Packer, N.H. (1999) BOLD—a biological O-linked glycan database. *Electrophoresis* **20:** 3589–3598.

Corbett, J.M., Dunn, M.J., Posch, A. and Gorg, A. (1994) Positional reproducibility of protein spots in two-dimensional polyacrylamide gel electrophoresis using immobilised pH gradient isoelectric focusing in the first dimension: an interlaboratory comparison. *Electrophoresis* **15:** 1205–1211.

Eckerskorn, C., Strupat, K., Karas, M., Hillenkamp, F. and Lottspeich, F. (1992) Mass spectrometric analysis of blotted proteins after gel electrophoretic separation by matrix-assisted laser desorption/ionization. *Electrophoresis* **13:** 664–665.

Eckerskorn, C., Strupat, K., Schleuder, D., Hochstrasser, D., Sanchez, J.C., Lottspeich, F. and Hillenkamp, F. (1997) Analysis of proteins by direct-scanning infrared-MALDI mass spectrometry after 2-D PAGE separation and electroblotting. *Anal. Chem.* **69:** 2888–2892.

Edman, P. and Begg, G. (1967) A protein sequenator. *Eur. J. Biochem.* **1:** 80–91.

Figeys, D., Ning, Y. and Aebersold, R. (1997) A microfabricated device for rapid protein identification by microelectrospray ion trap mass spectrometry. *Anal. Chem.* **69:** 3153–3160.

Figeys, D., Lock, C., Taylor, L. and Aebersold, R. (1998) Microfabricated device coupled with an electrospray ionization quadrupole time-of-flight mass spectrometer: protein identifications based on enhanced-resolution mass spectrometry and tandem mass spectrometry data. *Rapid Commun. Mass Spectrom.* **12:** 1435–1444.

Fountoulakis, M., Langen, H., Evers, S., Gray, C. and Takacs, B. (1997) Two-dimensional map of *Haemophilus influenzae* following protein enrichment by heparin chromatography. *Electrophoresis* **18:** 1193–1202.

Fountoulakis, M., Takacs, B. and Langen, H. (1998a) Two-dimensional map of basic proteins of *Haemophilus influenzae*. *Electrophoresis* **19:** 761–766.

Fountoulakis, M., Juranville, J.F., Roder, D., Evers, S., Berndt, P. and Langen, H. (1998b) Reference map of the low molecular mass proteins of *Haemophilus influenzae*. *Electrophoresis* **19:** 1819–1827.

Garrels, J.I. and Franza, B.R. (1989) The REF52 protein database. Methods of database construction and analysis using the QUEST system and characterizations of protein patterns from proliferating and quiescent REF52 cells. *J. Biol. Chem.* **264:** 5283–5298.

Gras, R., Muller, M., Gasteiger, E., Gay, S., Binz, P.-A., Bienvenut, W., Hoogland, C., Sanchez, J.-C., Bairoch, A., Hochstrasser, D.F. and Appel, R.D. (1999) Improving protein identification from peptide mass fingerprinting through a parameterized multi-level scoring algorithm and an optimized peak detection. *Electrophoresis* **20:** 3535–3550.

Gygi, S.P., Rochon, Y., Franza, B.R. and Aebersold, R. (1999) Correlation between protein and mRNA abundance in yeast. *Mol. Cell. Biol.* **19:** 1720–1730.

Harrington, M.G., Lee, K.H., Yun, M., Zewert, T., Bailey, J.E. and Hood, L.E. (1993) Mechanical precision in two-dimensional electrophoresis can improve spot positional reproducibility. *Appl. Theor. Electrophor.* **3:** 347–353.

Henzel, W.J., Billeci, T.M., Stults, J.T., Wong, S.C., Grimley, C. and Watanabe, C. (1993) Identifying proteins from two-dimensional gels by molecular mass searching of peptide fragments in protein sequence databases. *Proc. Natl Acad. Sci. USA* **90:** 5011–5015.

Herbert, B. (1999) Advances in protein solubilisation for two-dimensional electrophoresis. *Electrophoresis* **20:** 660–663.

Hochstrasser, D.F. (1989) Pre-cast gel systems for two-dimensional electrophoresis. US patent no. US4874490.

Hochstrasser, D.F. (1998) Two-dimensional electrophoresis device. US patent no. US5773645.

Houthaeve, T., Gausepohl, H., Ashman, K., Nillson, T. and Mann, M. (1997) Automated protein preparation techniques using a digest robot. *J. Protein Chem.* **16:** 343–348.

Isobe, T., Uchida, K., Taoka, M., Shinkai, F., Manabe, T. and Okuyama, T. (1991) Automated two-dimensional liquid chromatographic system for mapping proteins in highly complex mixtures. *J. Chromatogr.* **588:** 115–123.

Kenrick, K.G. and Margolis, J. (1970) Isoelectric focusing and gradient gel electrophoresis: a two-dimensional technique. *Anal. Biochem.* **33:** 204–207.

Klose, J. (1975) Protein mapping by combined isoelectric focusing and electrophoresis of mouse tissues. A novel approach to testing for induced point mutations in mammals. *Humangenetik* **26:** 231–243.

Klose, J. (1999) Genotypes and phenotypes. *Electrophoresis* **20:** 643–652.

Kuster, B. and Mann, M. (1998) Identifying proteins and post-translational modifications by mass spectrometry. *Curr. Opin. Struct. Biol.* **8:** 393–400.

Langen, H., Gray, C., Roder, D., Juranville, J.F., Takacs, B. and Fountoulakis, M. (1997) From genome to proteome: protein map of *Haemophilus influenzae*. *Electrophoresis* **18:** 1184–1192.

Langen, H., Fountoulakis, M. and Berndt, P. (1998) Modified in-gel protein digestion procedure allows high throughput MS analysis following 2-D gel gridding. Abstract from the *3rd Siena 2-D Electrophoresis Meeting 'From Genome to Proteome'*, 31 August–3 September, 1998.

Latham, K.E., Garrels, J.I., Chang, C. and Solter, D. (1991) Quantitative analysis of protein synthesis in mouse embryos. I. Extensive reprogramming at the one- and two-cell stages. *Development* **112:** 921–932.

Lemkin, P.F., Myrick, J.M., Lakshmanan, Y., Shue, M.J., Patrick, J.L., Hornbeck, P.V., Thornwal, G.C. and Partin, A.W. (1999) Exploratory data analysis groupware for qualitative and quantitative electrophoretic gel analysis over the Internet-WebGel. *Electrophoresis* **20:** 3492–3507.

Link, A.J., Hays, L.G., Carmack, E.B. and Yates, J.R. III (1997) Identifying the major proteome components of *Haemophilus influenzae* type-strain NCTC 8143. *Electrophoresis* **18:** 1314–1334.

Link, A.J., Eng, J., Schieltz, D.M., Carmack, E., Mize, G.J., Morris, D.R., Garvik, B.M. and Yates, J.R. III (1999) Direct analysis of protein complexes using mass spectrometry. *Nature Biotechnol.* **17:** 676–682.

Loo, J.A., Brown, J., Critchley, G., Mitchell, C., Andrews, P.C. and Ogorzalek Loo, R.R. (1999) High sensitivity mass spectrometric methods for obtaining intact molecular weights from gel-separated proteins. *Electrophoresis* **20:** 743–748.

Lopez, M.F. and Patton, W.F. (1997) Reproducibility of polypeptide spot positions in two-dimensional gels run using carrier ampholytes in the isoelectric focusing dimension. *Electrophoresis* **18:** 338–343.

Mann, M. and Wilm, M. (1994) Error tolerant identification of peptides in sequence databases by peptide sequence tags. *Anal. Chem.* **66:** 4390–4399.

Matsudaira, P. (1987) Sequence from picomole quantities of proteins electroblotted onto polyvinylidene difluoride membranes. *J. Biol. Chem.* **262:** 10 035–10 038.

Nokihara, K., Morita, N. and Kuriki, T. (1992) Applications of an automated apparatus for two-dimensional electrophoresis, Model TEP-1, for microsequence analysis of proteins. *Electrophoresis* **13:** 701–707.

O'Farrell, P.H. (1975) High resolution two-dimensional electrophoresis of proteins. *J. Biol. Chem.* **250:** 4007–4021.

O'Farrell, P.Z. and Goodman, H.M. (1976) Resolution of simian virus 40 proteins in whole cell extracts by two-dimensional electrophoresis: heterogeneity of the major capsid protein. *Cell* **9:** 289–298.

O'Farrell, P.H. and Ivarie, R.D. (1979) The glucocorticoid domain of response: measurement of pleiotropic cellular responses by two-dimensional gel electrophoresis. *Monogr. Endocrinol.* **12:** 189–201.

Ogorzalek Loo, R.R., Mitchell, C., Stevenson, T.I., Martin, S.A., Hines, W.M., Juhasz, P., Patterson, D.H., Peltier, J.M., Loo, J.A. and Andrews, P.C. (1997) Sensitivity and mass accuracy for proteins analyzed directly from polyacrylamide gels: implications for proteome mapping. *Electrophoresis* **18:** 382–390.

Ogorzalek Loo, R.R., Loo, J.A. and Andrews, P.C. (1999) Obtaining molecular weights of proteins and their cleavage products by directly combining gel electrophoresis with mass spectrometry. *Methods. Mol. Biol.* **112:** 473–485.

Oliver, S.G., van der Aart, Q.J., Agostoni-Carbone, M.L., Aigle, M., Alberghina, L., Alexandraki, D., Antoine, G., Anwar, R., Ballesta, J.P. and Benit, P. (1992) The complete DNA sequence of yeast chromosome III. *Nature* **357:** 38–46.

Opiteck, G.J., Ramirez, S.M., Jorgenson, J.W. and Moseley, M.A. III (1998) Comprehensive two-dimensional high-performance liquid chromatography for the isolation of overexpressed proteins and proteome mapping. *Anal. Biochem.* **258:** 349–361.

Phillips, T.A., Bloch, P.L. and Neidhardt, F.C. (1980) Protein identifications on O'Farrell two-dimensional gels: locations of 55 additional *Escherichia coli* proteins. *Bacteriology* **144:** 1024–1033.

Prober, J.M., Trainor, G.L., Dam, R.J., Hobbs, F.W., Robertson, C.W., Zagursky, R.J., Cocuzza, A.J., Jensen, M.A. and Baumeister, K. (1987) A system for rapid DNA sequencing with fluorescent chain-terminating dideoxynucleotides. *Science* **238:** 336–341.

Qin, J., Herring, C.J. and Zhang, X. (1998) De novo peptide sequencing in an ion trap mass spectrometer with ^{18}O labeling. *Rapid Commun. Mass Spectrom.* **12:** 209–216.

Reeh, S., Pedersen, S. and Friesen, J.D. (1976) Biosynthetic regulation of individual proteins in relA + and relA strains of *Escherichia coli* during amino acid starvation. *Mol. Gen. Genet.* **149:** 279–289.

Rigaut, G., Shevchenko, A., Rutz, B., Wilm, M., Mann, M. and Seraphin, B. (1999) A generic protein purification method for protein complex characterization and proteome exploration. *Nature Biotechnol.* **17:** 1030–1032.

Saiki, R.K., Scharf, S., Faloona, F., Mullis, K.B., Horn, G.T., Erlich, H.A. and Arnheim, N. (1985) Enzymatic amplification of beta-globin genomic sequences and restriction site analysis for diagnosis of sickle cell anemia. *Science* **230:** 1350–1354.

Sanger, F., Nicklen, S. and Coulson, A.R. (1977) DNA sequencing with chain-terminating inhibitors. *Proc. Natl Acad. Sci. USA* **74:** 5463–5467.

Scheele, G.A. (1975) Two-dimensional gel analysis of soluble proteins. Characterization of guinea pig exocrine pancreatic proteins. *J. Biol. Chem.* **250:** 5375–5385.

Schena, M., Shalon, D., Davis, R.W. and Brown, P.O. (1995) Quantitative monitoring of gene expression patterns with a complementary DNA microarray. *Science* **270:** 467–470.

Schleuder, D., Hillenkamp, F. and Strupat, K. (1999) IR-MALDI-mass analysis of electroblotted proteins directly from the membrane: comparison of different membranes, application to on-membrane digestion, and protein identification by database searching. *Anal. Chem.* **71:** 3238–3247.

Shevchenko, A., Jensen, O.N., Podtelejnikov, A.V., Sagliocco, F., Wilm, M., Vorm, O., Mortensen, P., Shevchenko, A., Boucherie, H. and Mann, M. (1996) Linking genome and proteome by mass spectrometry: large-scale identification of yeast proteins from two dimensional gels. *Proc. Natl Acad. Sci. USA* **93:** 14 440–14 445.

Steinberg, R.A., O'Farrell, P.H., Friedrich, U. and Coffino, P. (1977) Mutations causing charge alterations in regulatory subunits of the cAMP-dependent protein kinase of cultured S49 lymphoma cells. *Cell* **10:** 381–391.

Stoeckli, M., Farmer, T.B. and Caprioli, R.M. (1999) Automated mass spectrometry imaging with a matrix-assisted laser desorption ionization time-of-flight instrument. *J. Am. Soc. Mass Spectrom.* **10:** 67–71.

Tonella, L., Walsh, B.J., Sanchez, J.C., Ou, K., Wilkins, M.R., Tyler, M., Frutiger, S., Gooley, A.A., Pescaru, I., Appel, R.D., Yan, J.X., Bairoch, A., Hoogland, C., Morch, F.S., Hughes, G.J., Williams, K.L. and Hochstrasser, D.F. (1998) '98 *Escherichia coli* SWISS-2DPAGE database update. *Electrophoresis* **19:** 1960–1971.

Towbin, H., Staehlin, T. and Gordon, J. (1979) Electrophoretic transfer of proteins from polyacrylamide gels to nitrocellulose sheets: procedure and some applications. *Proc. Natl Acad. Sci. USA* **76:** 4350–4354.

Traini, M., Gooley, A.A., Ou, K., Wilkins, M.R., Tonella, L., Sanchez, J.-C., Hochstrasser, D.F. and Williams, K.L. (1998) Towards an automated approach for protein identification in proteome projects. *Electrophoresis* **19:** 1941–1949.

VanBogelen, R.A., Hutton, M.E. and Neidhardt, F.C. (1990) Gene-protein database of *Escherichia coli* K-12: edition 3. *Electrophoresis* **11:** 1131–1166.

Vestling, M.M. and Fenselau, C. (1994) Polyvinylidene difluoride (PVDF): an interface for gel electrophoresis and matrix-assisted laser desorption/ionization mass spectrometry. *Biochem. Soc. Trans.* **22:** 547–551.

Weinstein, J.N., Myers, T.G., O'Connor, P.M., Friend, S.H., Fornace, A.J. Jr, Kohn, K.W., Fojo, T., Bates, S.E., Rubinstein, L.V., Anderson, N.L., Buolamwini, J.K., van Osdol, W.W., Monks, A.P., Scudiero, D.A., Sausville, E.A., Zaharevitz, D.W., Bunow, B., Viswanadhan, V.N.,

Johnson, G.S., Wittes, R.E. and Paull, K.D. (1997) An information-intensive approach to the molecular pharmacology of cancer. *Science* **275**: 343–349.

Wilkins, M.R. and Gooley, A.A. (1997) Protein identification in proteome projects. In: *Proteome Research: New Frontiers in Functional Genomics* (eds M.R. Wilkins, R.D. Appel, K.L. Williams and D.F. Hochstrasser). Springer, Berlin, pp. 35–64.

Wilkins, M.R., Sanchez, J.-C., Gooley, A.A., Appel, R.D., Humphrey-Smith, I., Hochstrasser, D.F. and Williams, K.L. (1995) Progress with proteome projects: why all proteins expressed by a genome should be identified and how to do it. *Biotechnol. Genet. Eng. Rev.* **13**: 19–50.

Wilkins, M.R., Sanchez, J.-C., Williams, K.L. and Hochstrasser, D.F. (1996) Current challenges and future applications for protein maps and post-translational vector maps in proteome projects. *Electrophoresis* **17**: 830–838.

Wilkins, M.R., Williams, K.L., Appel, R.D. and Hochstrasser, D.F. (eds) (1997a) *Proteome Research: New Frontiers in Functional Genomics.* Springer, Berlin.

Wilkins, M.R., Lindskog, I., Gasteiger, E., Bairoch, A., Sanchez, J.C., Hochstrasser, D.F. and Appel, R.D. (1997b) Detailed peptide characterization using PEPTIDEMASS – a World-Wide-Web-accessible tool. *Electrophoresis* **18**: 403–408.

Wilkins, M.R., Gasteiger, E., Sanchez, J.-C., Bairoch, A. and Hochstrasser, D.F. (1998a) The limitations of two-dimensional gel electrophoresis for proteome projects: the effects of protein copy number and hydrophobicity. *Electrophoresis* **19**: 1501–1505.

Wilkins, M.R., Gasteiger, E., Tonella, L., Ou, K., Sanchez, J.-C., Tyler, M., Gooley, A.A., Williams, K.L., Appel, R.D. and Hochstrasser, D.F. (1998b) Protein identification with *N*- and *C*-terminal sequence tags in proteome projects. *J. Mol. Biol.* **278**: 599–608.

Wilkins, M.R., Gasteiger, E., Bairoch, A., Sanchez, J.-C., Williams, K.L., Appel, R.D. and Hochstrasser, D.F. (1999a) Protein identification and analysis tools in the ExPASy server. *Methods Mol. Biol.* **112**: 531–552.

Wilkins, M.R., Gasteiger, E., Gooley, A.A., Herbert, B.R., Molloy, M.P., Binz, P.A., Ou, K., Sanchez, J.-C., Bairoch, A., Williams, K.L. and Hochstrasser, D.F. (1999b) High-throughput mass spectrometric discovery of protein post-translational modifications. *J. Mol. Biol.* **289**: 645–657.

Wilm, M. and Mann, M. (1996) Analytical properties of the nanoelectrospray ion source. *Anal. Chem.* **68**: 1–8.

Wilm, M., Shevchenko, A., Houthaeve, T., Breit, S., Schweigerer, L., Fotsis, T. and Mann, M. (1996) Femtomole sequencing of proteins from polyacrylamide gels by nano-electrospray mass spectrometry. *Nature* **379**: 466–469.

Wodicka, L., Dong, H., Mittmann, M., Ho, M.H. and Lockhart, D.J. (1997) Genome-wide expression monitoring in *Saccharomyces cerevisiae*. *Nat. Biotechnol.* **15**: 1359–1367.

Yates, J.R. III (1998a) Mass spectrometry and the age of the proteome. *J. Mass. Spectrom.* **33**: 1–19.

Yates, J.R. III (1998b) Database searching using mass spectrometry data. *Electrophoresis* **19**: 893–900.

Yates, J.R. III., Eng, J.K. and McCormack, A.L. (1995) Mining genomes: correlating tandem mass spectra of modified and unmodified peptides to sequences in nucleotide databases. *Anal. Chem.* **67**: 3202–3210.

Chapter 9

Message in a bottleneck: a complete approach to automating proteome analysis

Mary F. Lopez

1. Introduction

Proteome (Wasinger *et al.*, 1995) analysis is a rapidly developing field that has become a focus of pharmaceutical and biotechnology companies involved in drug discovery (Blackstock and Weir, 1999; Lopez 1998, 1999a). This observation is reflected in the exponentially growing number of proteomics publications (*Figure 1*). Much of the interest in this area has been spurred by recent studies indicating a lack of correlation between transcriptional profiles and actual protein levels in cells (Anderson and Anderson, 1998; Gygi *et al.*, 1999). These studies make it clear that cellular protein analysis is complementary to genomic analysis, and that no drug discovery program can be successful without the incorporation of a proteomics platform. Another incentive for monitoring protein profiles is that many diseases are correlated with dysfunctional post-translational modifications (such as glycosylation) of proteins. These biological errors cannot be tracked at the genomic level, since they occur at the metabolic level (Orntoft and Vestergaard, 1999; Packer and Harrison, 1998).

2. 2-D gels coupled with mass spectrometry: A paradigm for proteomics?

Comprehensive proteome analysis can be undertaken using variety of methods. However, the currently preferred approach involves combining several techniques including two-dimensional gel electrophoresis (2-DE) (O'Farrell, 1975; Patton *et al.*, 1990) with subsequent comparison and subtractive image analysis of the 2-DE gel maps to identify of potential drug or diagnostic targets. The target proteins are then identified by excision from the gels, proteolysis and extraction of the resulting peptides and identification by mass spectrometry-based peptide mass fingerprinting (Henzel *et al.*, 1993; James *et al.*, 1993; Mann *et al.*, 1993; Yates *et al.*, 1993), or tandem mass spectrometry sequencing of individual peptides (Hunt *et al.*, 1986; Jonsher *et al.*, 1997; Yates *et al.*, 1999) (Chapter 5).

Proteomics, edited by S.R. Pennington and M.J. Dunn
© 2001 BIOS Scientific Publishers, Oxford.

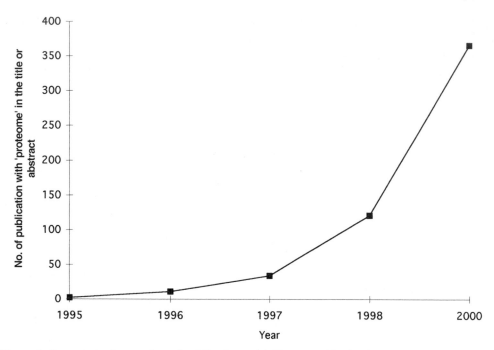

Figure 1. Increase in the number of publications with 'Proteome' in the title or abstract since 1995.

Unfortunately, because of their inherent complexity, proteins have not been as amenable to high throughput analysis as nucleic acids. Methods for studying complex mixtures of proteins, such as 2-DE, are difficult or impossible to automate successfully (Lopez 1999a; Nakihara *et al.*, 1992; Quadroni and James, 1999). Substitutes for gel electrophoresis in proteome projects have been avidly sought without much success (Lopez 1999a; Skipp, 1999). The critical issue is the exceedingly parallel nature of the 2-DE gel separation. When executed in a large format and with sensitive stains, a 2-DE gel can easily resolve 1000 and as many as 10 000 different proteins at one time (Klose, 1999; Klose and Kobalz, 1995). No other technique or combination of techniques has to date been developed that can deliver this level of detail in a similar time frame and with similar amounts of starting material. In addition to high resolution, 2-DE gels are also quantitative. Although chip based methods may offer a potential alternative to 2-DE gels (see Chapter 10), there are numerous issues with unnatural binding and phosphorylation events that remain to be resolved (Carr, 1999).

3. The automation concept

Surprisingly, although running 2-DE gels is a labor intensive, mostly manual procedure, it is typically not the bottleneck in a high-throughput proteome analysis facility (Table 1 in Lopez, 2000). Analysis of the 2-DE gels is still somewhat subjective and requires human intervention for editing. Bioinformatic processing of gigabytes of acquired data is time consuming and not yet efficient. Nevertheless, many of the procedures in the process of proteome analysis can be fully or semi-automated. This can lead to a dramatic increase in efficiency and productivity, which in turn increases the likelyhood of target identification (Lopez, 2000). Although some core proteomics facilities are now incorporating automation into some of the analysis procedures, commercially available complete systems for proteome analysis have become available only recently. Many companies are actively developing components and proteomics platforms to support drug and target discovery (BioRad; APB; PE Biosystems; GSI; Proteome Systems; Teca; and others).

An ideal high-throughput complete system for proteomics would be likely to include sample fractionation technology, high resolution electrophoresis equipment, automated gel staining instrumentation, semi-automated image analysis software, robotic modules for protein spot processing, MALDI-TOF and tandem mass spectrometers with integrated bioinformatic software to link the individual modules and expedite sample handling, tracking, data analysis and data archiving. An example of such a system assembled from currently commercially available components is illustrated in *Figure 2*. Ideally, each component should be compatible with the others allowing for efficient transfer of data and sample information along the discovery path. The components could theoretically be interfaced to a relational database for integration of the tracking information (such as barcodes) from each step in the process. In this way, all the information that is acquired from each sample would eventually be linked back to the original spot on the gel.

The individual features of some system components designed to meet the demands of high-throughput proteome analysis are outlined in the following sections.

4. 2-D gel electrophoresis and visualization

4.1 2-D gels

The quality of the 2-DE gels directly affects the results of any proteomics project. The ideal system will deliver large format, highly reproducible gels. The exquisite sensitivity of isoelectric focusing to ionic and other contaminants necessitates the use of very high quality reagents and pre-cast gels to minimize variables (Lopez *et al.*, 1991; Patton *et al.*, 1991a; Walsh *et al.*, 1998). A typical large format system is dedicated to 2-DE gels and can be moderate to high-throughput. Each unit should be capable of running at least 10–12 large format gels simultaneously. Multiple system modules can be added to accommodate higher throughput labs. Minigels can also be used successfully for 2-DE separations if prefractionation to reduce sample complexity is performed prior to the 2-DE run (Harry *et al.*, 2000). Because of convenience in handling, minigel electrophoresis is much more amenable to automation than large format gels.

The availability of pre-cast gels for the first and second dimension enhances the consistency of separations. A choice of immobilized or carrier ampholyte first dimension chemistries allows for flexibility in isoelectric focusing. Carrier

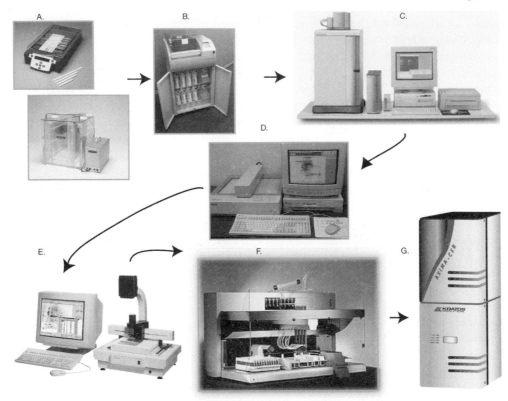

Figure 2. Schematic of a complete proteomic system assembled from currently commercially available components. (a) IPGphor, Hoefer DALT SDS PAGEgel System, APB. (b) Investigator Gel Processor, GSI. (c) LAS-1000 Digital Image acquisition system, Fujimed. (d) Investigator HT 2-D Image Analysis software System, GSI. (e) Protean 2-D Spot Cutter, BioRad. (f) Genesis RSP robotic system for digestion and MALDI spotting, Tecan. (g) Axima-CFR MALDI-TOF mass spectrometer, Kratos.

ampholyte tube gels may be preferable for hydrophobic or membrane proteins and immobilized gradients offer convenience and ease of use for routine applications (Lopez, 1999b; Patton *et al.*, 1999b). The combination of all these parameters results in gels that have very consistent spot positions from gel to gel, varying less than 1% in either the IEF (isoelectric focusing) or molecular weight dimensions (Fishmann, 1999; Lopez and Patton, 1997). *Figure 3* shows a series of close-ups from the same area of six different human serum 2-DE gels illustrating the appearance of novel spots in the disease samples versus the control samples. Comparisons of this type are not possible unless the gel to gel pattern reproducibility is excellent.

4.2 Enrichment and fractionation techniques

Although 2-DE is at present the highest resolution technique available for separating protein mixtures, it still falls short of resolving all (> 10 000) the expressed proteins

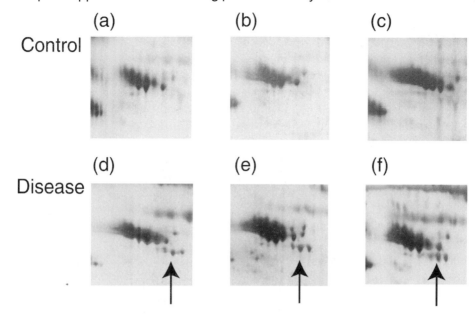

Figure 3. Close-ups of matching regions from 2-DE gel images of serum proteins. (a–c) control samples. (d–f) disease samples. Gels were run and analyzed as described in (Lopez and Patton, 1997) except using version 6.2 of the BioImage 2-DE Analyzer software (GSI). Arrows indicate differentially expressed proteins in the disease samples. (Samples courtesy of S. Rasheed, UCLA).

in any given cell. Most researchers have sought to solve this problem by using enrichment or prefractionation techniques such as affinity chromatography, charge fractionation in solution or subcellular fractionation prior to 2-DE (Harry *et al.*, 2000; Lopez, 1999a; Lopez *et al.*, 2000a; Quadroni and James, 1999). Another approach to fractionation has been the development of narrow range (1–2 pH units) immobilized pH gradient gels for the isoelectric focusing dimension (Harry *et al.*, 2000; APB, BioRad, Proteome Systems, GSI).

4.3 Visualization techniques

There currently exist few options for visualizing proteins on 2-DE gels. By far the most sensitive and quantitative technique, radioactive labeling, has become much less attractive as more and more laboratories are unable to use radioactive isotopes. Radioactive labeling also requires that live cells or tissue under investigation be incubated with the radioisotope prior to analysis and this clearly precludes its use for a number of important samples such as human tissue and fluids. The most commonly used colorimetric stains are Coomassie blue and silver stain (Merril, 1987; Patton *et al.*, 1999; Ramsby and Makowski, 1998; Wirth and Romano, 1995). The poor sensitivity of Coomassie blue makes it of very limited value for looking at any but the most abundant housekeeping proteins. Silver stains are frequently the best option since they can be extremely sensitive (to nanogram quantities), quantitative in some instances (Patton, 1995) and compatible with mass spectrometry and sequencing technologies, although the peptide coverage for peptide mass fingerprinting from in-gel digests of silver stained proteins is relatively poor (Gharahdaghi *et al.*, 1999). Unfortunately, there are several other serious disadvantages to silver stains. The dynamic range of silver stains is typically not very large and higher abundance proteins tend to saturate easily. The fact that silver stain is not an end-point stain means that quantitation can be inaccurate. Another alternative is fluorescent staining. There are several very good fluorescent stains on the market with sensitivity that parallels that of silver stain but they have not found wide acceptance. Recently, a new fluorescent stain, SYPRO® Ruby Protein Gel Stain (Molecular Probes), was introduced for proteomics applications. SYPRO Ruby Protein Gel Stain has increased sensitivity, a wider dynamic range and better peptide recovery from in-gel digests for mass spectrometry than silver stain (Berggren *et al.*, 2000; Lopez *et al.*, 2000a; Patton 2000, 2000a). *Figure 4* shows an example of pixel intensity surface plots taken through similar regions of identically loaded silver and SYPRO Ruby stained 2-DE fibroblast gels. These plots graphically illustrate the difference in dynamic range of the two stains and suggest the advantages of the fluorescent stain for quantitative comparisons.

4.4 *Automated staining*

Regardless of the staining method chosen, an automated staining system will reduce manual labor and increase consistency. There currently exist very few commercially available options for fully automated staining of 2-D gels (APB; GSI). A system compatible with high-throughput proteomics facilities should be capable of simultaneously staining multiple electrophoresis gels using different user definable protocols.

5. Image analysis

Image analysis is perhaps the most crucial element of a high-throughput proteomics program since it determines the quality of the data extracted from the collection of 2-DE gel images in the database, and guides the search trajectory for disease-specific protein markers (Wirth and Romano, 1995; see also Chapter 6). The image analysis step in the proteomics discovery path is also most likely to be a significant bottleneck in the procedure (see Lopez, 2000, Table 1). The first step is capture of the gel image in a digital format. There are many systems available on the market that offer high resolution CCD camera systems for image capture (Perkin Elmer, Fuji, APB, BioRad, Boehringer Mannheim). One advantage of CCD camera systems is flexibility and compatibility with many staining procedures including fluorescent stains (Patton, 2000, 2000a). Laser scanners can also successfully be used to capture images from gels with white light visible stains. Several reviews have been published on this subject (Patton, 1995, 2000).

Because of the computationally intense nature of 2-DE gel pattern analysis, unix-based software is typically preferred. The software must be capable of analyzing complex gels containing anywhere from 1500 to 10 000 protein spots and databases of hundreds of gels. Powerful database queries should allow the filtering of

(a)   (b)

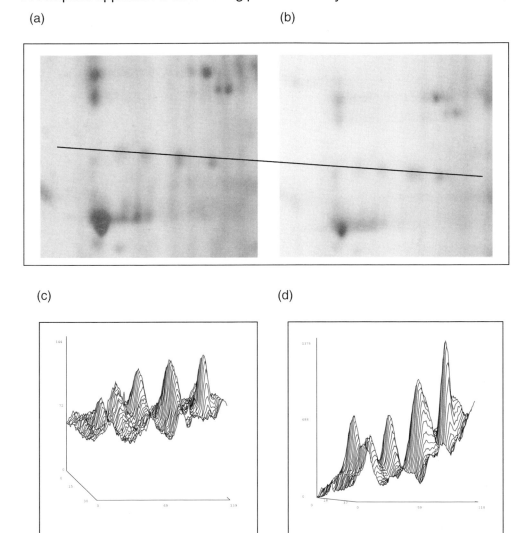

(c)   (d)

Figure 4. Pixel intensity surface plots of similar sections from 2-DE gels stained with silver stain and SYPRO Ruby protein gel stain. (a) Close-up section of silver stained 2-DE gel. (b) Close-up section of a 2-DE gel matched to (a) but stained with SYPRO Ruby protein gel stain. The line indicates the cluster of protein spots that the surface plots were taken from. (c) Surface plot of the area indicated by the line through (a). (d) Surface plot of the area indicated by the line through (b). Gels were run and analyzed as described in (Lopez and Patton, 1997) except using version 6.2 of the BioImage 2-D Analyzer software and stained with silver stain or SYPRO Ruby protein gel stain (Molecular Probes).

information based on the existence of spots, quantitative ratios of matched spots, spot integrated intensities, molecular weight, isoelectric point, area and user-defined spot or image characteristics. The multi-threading capabilities of the unix platform result in robust and very fast processing of complex, multi-level queries and very

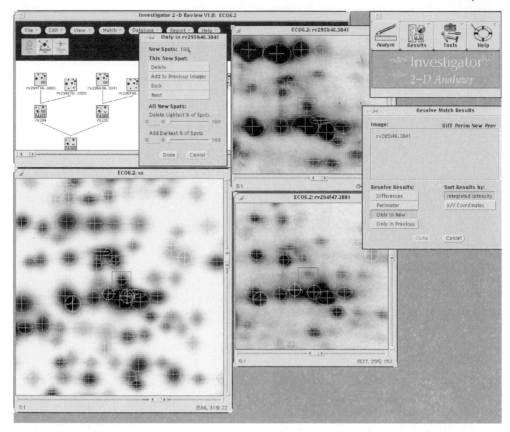

Figure 5. Screen capture of Investigator 2-D Analyzer software (GSI) illustrating match resolution option. This option can increase the efficiency and therefore the throughput of match editing, a significant bottleneck in 2-DE gel image analysis.

large gel databases. A strategy for increasing the efficiency and speed of analysis is to incorporate decision points in the match editing process that resolve differences between each new gel image added to the database (*Figure 5*). Several advanced 2-DE gel image analysis software packages are commercially available (Imagemaster, APB; Advanced 2-D Software, Phoretix; MELANIE II, Swiss Institute of Bio-informatics, GenBio; PDQUEST, BioRad, HT Investigator, GSI). A necessary feature of the image analysis software is integration with the other components of the complete proteomic system facilitating accurate excision and processing of designated protein spots (see following sections).

6. Robotics

6.1 Spot excision

Once potentially interesting protein spots on 2-DE gels have been identified by image analysis, they must be characterized. Typically, the spots are excised from the gel and

(a)   (b)

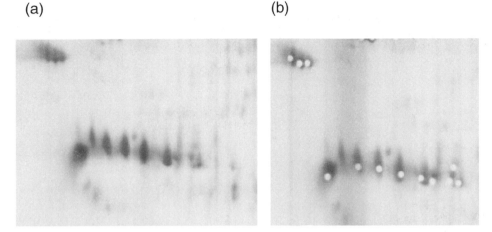

Figure 6. Close-ups of a section of a silver stained 2-DE gel, before and after processing with the Flexys spot excision robot (GSI). (a) Gel before processing. (b) Gel after processing to excise spots. The locations of the excised spots were chosen in a report generated by the 2-D Analyzer software for picking by the Flexys robot. A 2 mm plastic picking tip was used for spot excision. Gels were run and analyzed as described in Lopez and Patton (1997) except using version 6.2 of the BioImage 2-D Analyzer software (GSI).

subjected to in-gel protease (i.e. trypsin) digestion. Clearly, this procedure is laborious and fraught with potential contamination from skin keratins if done entirely manually for a large number of spots. Several robots for automated spot excision from 2-DE gels have been developed (BioRad; APB; GSI; Proteome Systems) and are commercially available. Some companies are developing all-in-one integrated platforms for spot excision, digestion and spotting onto MALDI-TOF targets (Proteome Systems).

A fully automated spot excision system may include a CCD camera for detection and excision of spots from 2-DE gels. The excision robot should be integrated with the 2-DE analysis software to allow either directed excision of target spots or of all spots in the 2-DE gel. The robot should be compatible with silver, Coomassie Blue stained or fluorescent gels to allow the greatest degree of flexibility. The highest throughput systems currently excise approximately 200 spots per hour. Spots should be excised by nonreactive, plastic or stainless steel picking tips and placed in microtiter plates for delivery to a digestion robot (see below). *Figure 6* shows an example of a 2-DE gel of plasma proteins excised with an automated spot excision robot.

6.2 Spot digestion and peptide extraction

After excision from 2-DE gels, the protein spots are subjected to protease (typically trypsin) digestion and the resulting peptide mixture is extracted from the gel plug. Again, efforts have been made to automate this process. A fully automated digestion robot should automate all the steps in the procedure from digestion to extraction with very high throughput as in-gel trypsin digestion can take as long as 8 hours. All sample washing and extraction steps should be automatically carried out in microtiter plates for ease of handling and minimal cross contamination of samples. Once the procedure is complete, the extracted peptides should be ready for delivery either to the next robot for spotting onto a MALDI-TOF target or to an autosampler for tandem mass spectrometry analysis. It may be desirable for ultra high throughput, to be able to control several digestion robots remotely through a networked PC. Digestion protocols should be easily customizable and then down-loaded to each instrument before operation.

6.3 Peptide desalting, clean-up and spotting onto MALDI target stage

Once the peptides have been extracted from the gel plugs, contaminants such as salts and detergents must be removed and the peptides concentrated before application to a mass spectrometer. If the peptides are to be analyzed on a MALDI-TOF-MS, they must be mixed with matrix (typically α-cyanocinnamic acid) and spotted onto a MALDI target. There are several commercially available robots that can perform some or all of these tasks (PE BioSystems; GSI; BioRad; Tecan; Proteome Systems). Some of the robotic platforms allow the use of Zip Tips® (Millipore) for peptide purification using hydrophobic resins (Courchesne and Patterson, 1997; Jensen *et al.*, 1999). It is necessary that volumes of 0.5–1 μl or less be spotted accurately in arrays, and that the robotic platform can accommodate MALDI-TOF sample plates from most manufacturers. Remote control of the MALDI-TOF robot(s) from a networked PC is also a desirable feature. *Figure 7* shows a mass spectrum of a peptide mixture obtained from a spot excised from a 2-DE gel and processed in the manner described above. The spot was identified as 50S ribosomal protein L9 by searching the Swiss Prot database.

7. Interfacing mass spectrometry stategies

There are currently two main approaches to the integration of mass spectrometry into a proteomics program. In order to identify proteins that are not present in the various genomic and proteomic databases, the amino acid sequence must be determined. Previously, this was accomplished by using Edman chemistry in an automated protein sequencer. This procedure required relatively large amounts of protein and time (Tempst *et al.*, 1990). More recently, most researchers have been using quadrupole or electrospray tandem mass spectrometry (Micromass, SciEX/ Perkin Elmer, Finnigan, Bruker) to identify the amino acid sequences of individual peptides and using these short sequence tags to search public and private EST and sequence databases (Yates *et al.*, 1993, 1999). These sequencing techniques are accurate but still require relatively large (picomole) amounts of material and have a limited throughput since they can be slow. For high-throughput identification of proteins from 2-DE gels, peptide mass fingerprinting using MALDI-TOF mass spectrometry is a preferred technique (Henzel *et al.*, 1993; James *et al.*, 1993; Pappin *et al.*, 1993). MALDI-TOF is an extremely sensitive (femtomoles to attomoles), high-resolution technique, (although it is not quantitative). It is also

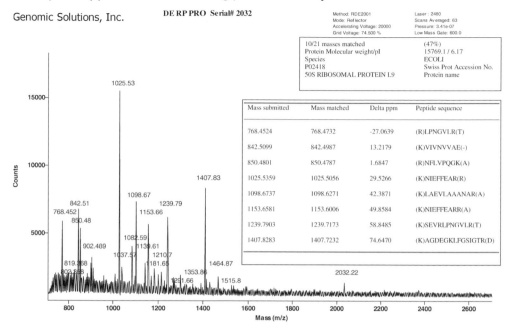

Figure 7. Mass spectrum of an in-gel peptide digest from an excised spot on a 2-DE gel. The spot (approximately 1 picomole of a Coomassie blue stained protein) was processed using the ProGest digestion and ProMS MALDI robots (GSI). The mass spectrum was acquired using a Voyager DE-RP Pro MALDI-TOF mass spectrometer (PE BioSystems) essentially as described in (Parker *et al.*, 1998). The peptide mass search was done using the Protein Prospector program on the world wide web (http://prospector.ucsf.edu) and searching the Swiss Prot database. The insets show the mass accuracy of the search as well as the sequence of each matched peptide. The probable identity of the protein is also indicated. Gels were run and analyzed as described in (Lopez and Patton, 1999) except using version 6. 2 of the BioImage 2-D Analyzer software (GSI).

well suited to automation and therefore high-throughput. In addition, since the sequencing of the genomes of many organisms is proceeding at a phenomenal rate, there is a wealth of sequence data available in public databases on the World Wide Web for mass profiling searches (Lopez 1998, 1999a). These techniques allow rapid identification of dozens and even hundreds of protein spots per day in an efficient and automated system. Numerous manufacturers (PE BioSystems, Bruker, Ciphergen, Micromass, Kratos,) have introduced accurate benchtop MALDI-TOF instruments in the past few years.

8. The link between robotics modules, image-analysis software and information databases

Processing large numbers of samples through the complete system described in the paragraphs above will generate large amounts of data and challenges for sample tracking. Hence, a critical issue in the automation of a proteomics program is the integration, at a software level, of the various instruments used in the analysis and the management of sample information and data. The development of a bioinformatic system to fulfill this need is a high priority for those involved in high-throughput proteomics projects. A bioinformatic solution might be based on a relational database to allow sample tracking and exchange of information and data between the various system components. Samples could be tracked with barcodes to eliminate manual tracking, often a source of error. The core component, the central database, would allow the other system components such as the 2-DE image analysis software, spot excision robots, digesters, mass spectrometers, etc., to deposit and access information. The database should also allows easy access to data for custom querying and reporting. In addition, sample preparation information and 2-DE gel running parameters should be recorded. Ultimately, the information obtained from mass spec analysis of the spots would be linked back to the gel images so that detailed preparation, processing and identification data for each spot would be available at a mouse click. An ideal user interface for this type of bioinformatic package would be a web browser. Currently, several companies have bioinformatic packages for proteomics under development (BioRad, APB, GSI, Proteome Systems).

9. Conclusions

Deciphering the complexity of biological systems requires an understanding of the expressed proteins as well as their genomic blueprint. Proteome analysis is the attempt to capture the protein profile of an organism or tissue at any given moment. Comparisons of expressed protein profiles can provide unique information pertinent to the identification of targets for potential drugs or diagnostic reagents. Automation and integration of the various procedures in proteome analysis is necessary to increase efficiency thereby facilitating target identification. Resolution of the bottlenecks in strategies for high-throughput proteome analysis is facilitating the drug and diagnostic marker discovery path. A future challenge of proteomics will be the integration with genomic and small molecule metabolic information. The eventual result of this integration will be a more comprehensive understanding of the metabolic processes in cells. It is evident that the advent of proteomics has encouraged many biologists to think out of the genomics box.

References

Anderson, N.L. and Anderson, N.G. (1998) Proteome and proteomics: New technologies, new concepts, and new words. *Electrophoresis*, **19:** 1853–1861.

Berggren, K, Steinberg, T.H., Lauber, W.M. *et al.* (1999) A luminescent ruthenium complex for ultrasensitive detection of proteins immobilized on membrane supports. *Anal. Biochem.* **276:** 129–143.

Blackstock, W. and Weir, M. (1999) Proteomics: quantitative and physical mapping of cellular proteins. *Trends in Biotech.* **17:** 121–127.

Carr, F. (1999) Proteins on chips. Presentation to IBC Conf., London, Feb. 17–19.

Courchesne, P.L. and Patterson, S.D. (1997) Manual microcolumn chromatography for sample clean-up prior to mass spectrometry. *Biotechniques* **22:** 244–250.

Fischmann, J. (1999) Advantages of immobilized pH gradients. In: *2-D Proteome Analysis Protocols* (ed. A.J. Link). Humana Press, Totowa, NJ, pp. 173–174.

Gharahdaghi, F., Weinberg, C.R., Meagher, D.A., Imal, B.S. and Mische, S.M. (1999) Mass spectrometric identification of proteins from silver-stained polyacrylamide gel: a method for the removal of silver ions to enhance sensitivity. *Electrophoresis* **20**: 601–605.

Gygi, S.P., Rochon, Y., Franza, R. and Aebersold, R. (1999) Correlation between protein and mRNA abundance in yeast. *Mol. Cell. Biol.* **19**: 1720–1730.

Harry, J.L., Wilkins, M.R., Herbert, B.R., Packer, N.H., Gooley, A.A. and Williams, K.L. (2000) Proteomics: Capacity *versus* utility. *Electrophoresis* **21**: 1071–1081.

Henzel, W.J., Billeci, T.M., Stults, J.T., Wong, S.C., Grimley, C. and Watanabe, C. (1993) Identifying proteins from two-dimensional gels by molecular mass searching of peptide fragments in protein sequence databases. *Proc. Natl. Acad. Sci. USA* **90**: 5011–5015.

Hunt, D.F., Yates, J.R. II, Shabanowitz, J., Winston, S. and Hauer, C.R. (1986) Protein sequencing by tandem mass spectrometry. *Proc. Natl. Acad. Sci. USA* **83**: 6233–8238.

James, P., Quadroni, M., Carafoli, E. and Gonnet, G. (1993) Protein identification by mass profile finger printing. *Biochem. Biophys., Res. Commun.* **195**: 58–64.

Jensen, O.N., Wilm, M., Shevchenko, A. and Mann, M. (1999) Sample preparation methods for mass spectrometric peptide mapping directly from 2-DE gels. In: *2-D Proteome Analysis Protocols* (ed. A.J. Link). Humana Press, Totowa, NJ, pp. 513–530.

Jonscher, K.R., Yates, I. and John R. (1997) The quadrupole ion trap mass spectrometer-a small solution to a big challenge. *Anal. Biochem.* **244**: 1–15.

Klose, J. (1999) Genotypes and phenotypes. *Electrophoresis* **20**: 643–652.

Klose, J. and Kobalz, U. (1995) Two-dimensional electrophoresis of proteins: an updated protocol and implications for a functional analysis of the genome. *Electrophoresis* **16**: 1034–1059.

Lopez, M.F. (1998) Proteomic databases: roadmaps for drug discovery. *Am. Biotechnol. Lab.* July.

Lopez, M.F. (1999a) Advantages of carrier ampholyte IEF. In: *2-D Proteome Analysis Protocols* (ed. A.J. Link). Humana Press, Totowa, NJ, pp. 109–110.

Lopez, M.F. (1999b) Proteome analysis I. Gene products are where the biological action is. *J. Chromatogr. B.* **722**: 191–202.

Lopez, M.F., (2000) Better approaches to finding the needle in a haystack: Optimizing Proteome analysis through automation. *Electrophoresis* **21**: 1082–1093.

Lopez, M.F. and Patton, W.F. (1997) Reproducibility of polypeptide spot positions in two-dimensional gels run using carrier ampholytes in the isoelectric focusing dimension. *Electrophoresis* **18**: 338–343.

Lopez, M.F., Patton, W.F., Utterback, B.L., Chung-Welch, N., Barry, P., Skea, W.M. and Cambria, R.P. (1991) Effect of various detergents on protein migration in the second dimension of two-dimensional gels. *Analytical Biochem.* **199**: 35–44.

Lopez, M.F., Kristal, B.S., Chernokalskaya, E., Lazarev, A., Shestopalov, A.I., Bogdanova, A. and Robinson, M. (2000a) High–throughput profiling of the mitochondrial proteome using affinity fractionation and automation. *Electrophoresis* (in press).

Lopez, M.F., Berggren, K., Chernokalskaya, E., Lazarev, A., Robinson, M. and Patton, W.F. (2000b) A comparison of silver stain and SYPRO Ruby Protein Gel Stain with respect to protein detection in two-dimensional gels and identification by peptide mass profiling. *Electrophoresis* (in press).

Lopez, M.F., Berggren, K., Chernokalskaya, E., Lazarev, A., Robinson, M. and Patton, W.F. (2000c) A comparison of silver stain and SYPRO Ruby Protein Gel Stain with respect to protein detection in two-dimensional gels and identification by peptide mass profiling. *Electrophoresis* (in press).

Mann, M., Hojrup, P. and Roepstorff, P. (1993) Use of mass specctrometric molecular weight information to identify proteins in sequence databases. *Biol Mass. Spectrom.* **22**: 338–345.

Merril, C. (1987) Detection of proteins separated by electrophoresis. In: *Advances in Electrophoresis, vol. I* (eds. A. Chrambach, M. Dunn and B. Radola). VCH, Federal Republic of Germany, pp. 111–139.

Nokihara, K., Morita, N. and Kuriki, T. (1992) Applications of an automated apparatus for two-dimensional electrophoresis. *Electrophoresis* **13**: 701–707.

O'Farrell, P.H. (1975) High resolution two-dimensional electrophoresis of proteins. *J. Biol. Chem.* **250:** 4007–4021.

Orntoft, T.F. and Vestergaard, E.M. (1999) Clinical aspects of altered glycosylation of glycoproteins in cancer. *Electrophoresis* **20:** 362–371.

Packer, N.H. and Harrison, M.J. (1998) Glycobiology and proteomics: is mass spectrometry the holy grail? *Electrophoresis* **19:** 1872–1882.

Parker, K.C., Garrels, J.I., Hines, W., Butler, E.M., McKee, A.H.Z., Patterson, D. and Martin, S. (1998) Identification of yeast proteins from two-dimensional gels: Working out spot cross-contamination. *Electrophoresis* **19:** 1920–1922.

Pappin, D.J.C., Hojrup, P. and Bleasby, A.J. (1993) Rapid identification of proteins by peptide-mass fingerprinting. *Curr. Biol.* **3:** 327–332.

Patton, W.F. (1995) Biologist's perspective on analytical imaging systems as applied to protein gel electrophoresis. *J. Chromatogr. A.* **698:** 55–87.

Patton, W.F. (2000) A thousand points of light: the application of fluorescence detection technologies to two dimensional gel electrophoresis and proteomics. *Biotechniques* **28:** 944–957.

Patton, W.F. (2000a) Making blind robots 'see'; the synergy between fluorescent dyes and imaging devices in automated proteomics. *Electrophoresis* **21:** 1123–1144.

Patton, W.F., Pluskal, M.G., Skea, W.M., Buecker, J.L., Lopez, M.F., Zimmermann, R., Belanger, L.M. and Hatch, P.D. (1990) Development of a dedicated two-dimensional gel electrophoresis system that provides optimal pattern reproducibility and polypeptide resolution. *Biotechniques* **8:** 518–524.

Patton, W.F., Lopez, M.F., Barry, P. and Skea, W.M. (1991a) A mechanically strong matrix for protein electrophoresis with enhanced silver staining properties. *Biotechniques* **12:**580–585.

Patton, W.F., Chung-Welch, N., Lopez, M.F., Cambria, R.P., Utterback, B.L. and Skea, W.M. (1991b) Tris-tricine and tris-borate buffer systems provide better estimates of human meso-thelial cell intermediate filament protein molecular weights than the standard tris-glycine system. *Anal. Biochem.* **197:** 24–29.

Patton, W.F., Lim, M.J. and Shepro, D. (1999) Protein detection: using reversible metal chelate stains. In: *2-D Proteome Analysis Protocols* (ed. A.J. Link). Humana Press, Totowa, NJ, pp. 331–339.

Quadroni, M. and James, P. (1999) Proteomics and automation. *Electrophoresis* **20:** 664–667.

Ramsby, M.L. and Makowski, G.S. (1999) Differential detergent fractionation of eukaryotic cells: analysis by two-dimensional gel electrophoresis. In: *2-D Proteome Analysis Protocols* (ed. A.J. Link). Humana Press, Totowa, NJ, pp. 53–66.

Scheler, C. (1998) Peptide mass fingerprint sequence coverage from differently stained proteins on two-dimensional electrophoresis patterns by matrix assisted laser desorption/ionization-mass spectrometry (MALDI-MS). *Electrophoresis* **19:** 918–927.

Skipp, P. (1999) Proteome analysis without gels: current status of capilllary isoelectric focusing-electrospray ionization (CIEF-ESI)-MS. *Biochemical Soc. Trans.* **27:** 68.

Tempst, P., Link, A.J., Riviere, L.R., Fleming, M. and Elicone, C. (1990) Internal sequence analysis of proteins separated on polyacrylamide gels at the sub-microgram level: improved methods, applications and gene cloning strategies. *Electrophoresis* **11:** 537–553.

Walsh, B.J., Molloy, M.P. and Williams, K.L. (1998) The Australian proteome analysis facility (APAF): assembling large scale proteomics through integration and automation. *Electrophoresis* **19:** 1883–1890.

Wasinger, V.C., Cordwell, S.J., Cerpa-Poljak, A., Yan, J.X., Gooley, A.A., Wilkins, M.R., Duncan, M.W., Harris, R., Williams, K.L. and Humphery-Smith, I. (1995) Progress with gene-product mapping of the Mollicutes: *Mycoplasma genitalium. Electrophoresis* **16:** 1090–1094.

Wirth, P. and Romano, A. (1995) Staining methods in gel electrophoresis, including the use of multiple detection methods. *J. Chromatogr. A.* **698:** 123–143.

Yates, J.R. III, Carmack, E., Hays, L., Link, A.J. and Eng, J.K. (1999) Automated protein identification using microcolumn liquid chromatography-tandem mass spectrometry. In: *2-D Proteome Analysis Protocols* (ed. A.J. Link). Humana Press, Totowa, NJ, pp. 553–558.

Yates, J.R. III, Speicher, S., Griffin, P.R. and Hunkapiller, T. (1993) Peptide mass maps: a highly informative approach to protein identification. *Anal. Biochem.* **214:** 397–408.

Chapter 10

Novel approaches to protein expression analysis

Rosalind E. Jenkins and Stephen R. Pennington

1. Introduction

The progress in genome sequencing projects has been dramatic and widely publicized (see for example http://www.tigr.org/tdb/ and http://www.ornl.gov/ TechResources/Human_Genome/home.html). The completed genomes provide a rich opportunity for advancing our understanding of the biology of these organisms. These projects have already spawned new approaches to mRNA expression analysis (Chee *et al.*, 1996; Eisen *et al.*, 1998; Gerhold *et al.*, 1999) and techniques to undertake comprehensive approaches to the analysis of protein–protein interactions using the two-hybrid assay (Fields and Song, 1989; Fromont-Racine *et al.*, 1997; Lecrenier *et al.*, 1998). Together they are producing enormous quantities of information on organisms as diverse as *Saccharomyces cerevisiae* (Dujon, 1996; Goffeau *et al.*, 1996; Uetz *et al.*, 2000), *Caenorhabditis elegans* (Chalfie, 1998; *C. elegans* consortium, 1998; Plasterk, 1999) and humans (Chee *et al.*, 1996). More recent advances in microarray technology have increased still further the speed at which sequence information and differential gene expression data may be gathered (Bowtell, 1999). The integration of these and other approaches will provide much information that is broadly classified as functional genomics and this is already transforming the study of biological processes (Alizadeh *et al.*, 2000; Alon *et al.*, 1999; DeRisi *et al.*, 1996; Perou *et al.*, 1999; Pollack *et al.*, 1999; Wang *et al.*, 1999). However, as such methods become more routine and more widespread in use, it is increasingly becoming accepted that they will have to be complemented by the direct analysis of the products encoded by the genes and mRNAs – the proteins. Ultimately the analysis of the expression level, localization, structure and function of the protein products of the genome will define the activity of a cell or organism (Dove, 1999; Parekh, 1999; Pennington *et al.*, 1997; Wilkins *et al.*, 1995). However, proteins are inherently more complex than their corresponding mRNA or DNA sequences and it is not surprising, therefore, that the challenge associated with the development of techniques to support genome-scale analysis of protein function currently lags some way behind the impressive nucleic acid-based methods.

2. The scope of functional proteomics

There are several important and complementary components to functional proteo-mics. To date, much attention has been focused on the attempt to undertake global analysis of protein expression and in particular to analyze the changes in protein expression associated with, for example, disease states, knockout of individual genes, drug treatments and changing extracellular conditions. This analysis has often been justified on the basis that the expression level of individual mRNAs often does not reflect the level of expression of their protein product(s); and this is of course a good justification. However, it is also apparent that the level of expression of individual proteins within a cell, organism or tissue does not necessarily reflect the activity of the protein. The latter is a complex integration of the post-translational modifications to which the protein may be subjected, its tissue, cell and subcellular distribution and its interaction with other proteins. To analyze protein function therefore requires that each of these be examined and the data amalgamated. Clearly, detailed discussion of the current strategies for each of these approaches is beyond the scope of a single chapter and so here we focus on: (i) an introduction to some of the technical limitations of the major current approach to protein expression analysis; (ii) describing alternative methods that are actively under investigation; and (iii) discussing potential novel approaches for protein expression profiling. Before this, however, it is important to reiterate that significant advances are being made in the development of high-throughput techniques for protein localization (see for example Chapter 12), protein-structure analysis (Brünger, 1997; Cusack et al., 1998; Gerstein and Hegyi, 1998; Robson, 1999; Sali and Kuriyan, 1999; Swindells et al., 1998; Wery and Schevitz, 1997), prediction of protein function (Bork et al., 1998; Marcotte et al., 1999a,b; Pellegrini et al., 1999) and in mapping protein–protein interactions both by practical methods such as the two-hybrid approach (Uetz et al., 2000) and by computational methods (Enright et al., 1999). Other impressive approaches to genome-wide analysis of gene function have also been reported (see for example, RossMacdonald et al., 1999).

3. Proteome analysis: the 2-DE based strategy

At present, most proteome analysis projects begin with the separation of proteins by two-dimensional electrophoresis (2-DE) (see Chapter 3). This involves a first-stage separation of the complex protein mixture based on the isoelectric point, followed by a perpendicular separation based on the molecular weight. Up to 10 000 individual protein spots may be resolved on a single 2-DE gel (Klose, 1999) with multiple gels providing comparative data on protein expression. Notably, a single protein may be present as several spots on a 2-DE gel and at least some of this diversity reflects different post-translationally modified forms of the protein. 2-DE therefore has the capacity to support analysis of protein expression and post-translational modifica-tion. For some time, the enormous potential of 2-DE was severely restricted by the difficulty of identifying the protein(s) within the protein spots. Despite this a few

Current approach

- 2-DE for protein separation
- Protein detection
- Protein identification
  - Matching
    - pI, Mr, peptide mass fingerprinting, Amino acid composition, sequence tag
  - *De novo*
    - Tandem MS, microsequencing
- 2-D maps and databases

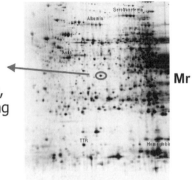

Figure 1. Protein expression analysis and protein characterization by 2-DE.

investigators, most notably Celis and colleagues, made remarkable in-roads in the identification of 2-DE-resolved proteins and the development of 2-DE maps and databases (Celis *et al.*, 1994, 1995), but these were isolated cases. The development of mass spectrometry (MS)-based methods to characterize 2-DE-resolved protein spots and of course proteins from other sources (see *Figure 1* and Chapter 5) has led to a striking increase in the efforts now being made to exploit 2-DE as a protein expression mapping tool (see for example, Celis *et al.*, 1998; Futcher *et al.*, 1999; Li *et al.*, 1999; Shevchenko *et al.*, 1996).

There is much continued debate about the sustainability of 2-DE as a platform for protein expression mapping. This debate is fueled by its technical demands and limitations. The most obvious of these are summarized in *Figure 2*. Thus, even in the most advanced laboratories, the quality of gels both in terms of spatial reproducibility of individual protein spots, and the quantitative reproducibility of the intensity of the stained protein spots, are difficult to achieve. There have been a number of studies on the uniformity of gels both within and between laboratories and there is no doubt that with appropriate care and experience, good reproducibility can be achieved (see Corbett *et al.*, 1994). However, it is also obvious that many still underestimate the difficulty in achieving such reproducibility and the need for replicate gels if statistically significant differences in the expression of individual protein spots are to be revealed. At present the methods in routine use for the detection of proteins on 2-DE gels are such that only moderate- to high-abundance proteins are displayed (see Chapter 4). As with all the limitations outlined here, there are ways of overcoming this shortcoming. For example, enrichment of samples prior to 2-DE analysis is one potential method to enable lower-abundance proteins to be detected. As might be anticipated, efforts are being made to develop new fluorescent

2-DE as a platform for proteomics

- Reproducibility
- Sensitivity
- Range of display
  - Numbers
  - Control
  - 'Difficult' proteins
- Data analysis
- Skill, effort and time required
- Lack of automation

Figure 2. Technical limitations of 2-DE as a platform technology for protein expression analysis.

dyes for protein staining. Of particular interest is the possible use of fluorescent dyes to label proteins prior to 2-DE resolution and the use of dyes that permit two samples to be stained with different fluorophores and then mixed prior to 2-DE (Unlü *et al.*, 1997). The resultant fluorescence emission of the individual protein spots provides a measure of the relative abundance of the protein in the two samples. This approach has some inherent limitations, but affords the important advantage that it overcomes problems in reproducibility between gels.

A single gel can potentially display up to 10 000 protein spots (Klose, 1999), although most gels presented in the literature show two to five times fewer spots. Given that individual proteins are present as more than one protein spot, the total number of proteins resolved is fewer still. Importantly, each gel affords little control over which proteins within the sample are displayed and some proteins, including membrane proteins, are poorly represented on 'standard condition' gels. Again, there are methods to overcome such limitations: multiple (overlapping) gels covering narrower pI and molecular weight ranges can be run; alternative detergents can be used to improve the solubilization of membrane proteins; and the protein samples can themselves be subjected to a hieratical extraction process. All of this increases the workload (number of gels that have to be run and analyzed) and for some samples, including human biopsy samples, there may be insufficient material for multiple gels. Of course, 2-DE seems particularly well suited to the analysis of simpler organisms and in this application there is considerable scope to analyze a significant proportion of the total 'proteome'.

The images obtained from 2-DE gels contain complex 3-D information that represents a significant analytical challenge (see Chapter 6). Effective image analysis of gels is therefore still time-consuming and in need of further development. Overall, as anyone who has used the method will testify, the process of 2-DE and gel analysis requires skill, and is labor-intensive and time-consuming. Automation of the process would alleviate many of these problems and has long been a goal for researchers in the field. Some success in the automation of the gel running and image processing has apparently been achieved but it remains conspicuously 'low profile'.

Automation of the process of 2-DE gel image analysis and automated protein spot excision and characterization by MS-based methods are actively under development. Several commercial organizations are producing and/or developing improved gel analysis packages, protein-spot picking robots and instruments to automate MALDI-TOF-based peptide mass fingerprinting and the more sophisticated electrospray tandem MS approaches (see Chapters 5 and 8). This has already improved the ability to collect data from 2-DE gels by automated selection of spots for mass spectrometric analysis (Traini *et al.*, 1998) and automated searching of sequence databases (Jensen *et al.*, 1997; Nicola *et al.*, 1998; see also Chapter 8).

4. Alternatives to 2-DE for protein expression analysis

The potential impact of proteome analysis on our understanding of biology is such that it is driving the rapid development of new technologies with which to study protein expression and function. These new approaches fall broadly into two categories – those that, like 2-DE, are separation-dependent, and others that seek to measure protein expression in a separation-independent manner.

4.1 Separation-dependent methods

The methods described in this section all require that the complex protein mixtures be subjected to a degree of purification/separation prior to analysis. The power of 2-DE lies in its orthogonal application of two separation steps that resolve proteins on the basis of distinct physiochemical properties – pI and M_r. However, before describing some of the alternative orthogonal approaches to protein separations that are being developed, two interesting 2-DE-based methods that have been explored should be mentioned. Infrared MALDI-TOF-MS has been used as an alternative to the more common UV MS and has been adapted to allow a high throughput of samples. The proteins separated by 2-DE have been electroblotted onto PVDF membranes, incubated with matrix solution and then scanned by an IR-MALDI-MS (Eckerskorn *et al.*, 1997a). A conceptually similar approach seeks to use a 'molecular scanner' to directly characterize 2-DE-resolved proteins (Bienvenut *et al.*, 1999; Binz *et al.*, 1999). In this method the entire 2-DE gel is electroblotted onto a PVDF membrane with the proteins being subjected to proteolytic cleavage during the transfer. The membrane is then scanned by MS. This avoids the need to stain the proteins, analyze the gel images and subject the protein spots to individual digestion reactions and MS, but more importantly exploits the sensitivity and analytical power of MS. The approach is in its infancy but if the recently reported commercial interest in the method (Buter, 2000) materializes then its development could be accelerated dramatically.

'Simple' separations. It has become apparent that electrospray tandem MS affords the potential to analyze protein mixtures, and there is much interest in exploiting this capability. The idea of rapidly analyzing protein complexes following fractionation and relatively limited protein separation was proposed by Lamond and Mann (1997) and then very successfully applied to the analysis of protein complexes (Neubauer *et al.*, 1998; Rappsilber *et al.*, 2000; Rigaut *et al.*, 1999; Rout *et al.*, 2000; Wigge *et al.*, 1998; Zachariae *et al.*, 1998). In this approach, complex protein assemblies such as spliceosomes are purified, subjected to 1-D gel electrophoresis and the resulting protein bands (which may contain several proteins) proteolytically digested and analyzed by MS. This approach has proved very powerful for the rapid and detailed characterization of such protein complexes and has resulted in the identification of a number of proteins not previously recognized as components of the complexes. A similar strategy is also applicable to the analysis of less complex protein assemblies that have, for example, been recovered by immunoaffinity methods. Ciphergen produces a protein analysis system that with use of appropriate antibodies is able to capture femtomole quantities of target proteins and to measure their mass using surface-enhanced laser desorption/ionization (SELDI) MS (www.ciphergen.com). The company also produces a range of protein 'chips' which are MALDI targets that have been coated with various chromatographic resins (e.g. ion exchange) to support protein expression profiling experiments. Briefly, complex protein mixtures may be applied to the targets, washed under different conditions and then the retained proteins characterized on the basis of their intact mass by MS. This approach has many limitations but an interesting panel of 'chips' are available.

Orthogonal separations. In order to eliminate the requirement for 2-DE, methods that combine two or more established methods of fractionation or separation to achieve equivalent resolving power and/or throughput are being examined. In particular, methods that use liquid chromatography or capillary electrophoresis are attractive as these are more readily amenable to automation, thereby increasing the potential for development of a high-throughput environment.

Link and colleagues have used a two-dimensional chromatographic separation approach to characterize yeast ribosome complex proteins (Link *et al.*, 1999). Purified yeast 80S ribosomes were subjected to proteolytic digestion and then automated chromatography on a cation exchange (SCX) column; 12 step-wise eluates from the column were passed onto a reversed-phase column and the eluate from this analyzed by tandem MS. In a single run, 75 of the 78 predicted ribosomal proteins were identified. Interestingly, when the same process was applied to a total yeast cell extract, 71 of the predicted 78 ribosomal proteins were identified. The sensitivity of the method was improved to a detection limit of 10 fmol of trypsin-digested protein by using a novel biphasic microcapillary column (Link *et al.*, 1999).

High-performance liquid chromatography (HPLC) has been used as a means to separate proteins and peptides for some time and recent improvements have seen modifications that allow the method to be both miniaturized, using capillary columns, and automated. Attempts have also been made to combine reversed-phase capillary HPLC with automatic fraction collection directly onto MALDI-

TOF-MS targets, Edman sequencing membranes or nitrocellulose membranes (Grimm and Grasser, 1998). In applying this orthogonal approach to the analysis of proteins from maize kernels the authors were able to characterize proteins from nanogram quantities of starting material. Ultrahigh-pressure capillary liquid chromatography potentially affords the opportunity to increase the speed of HPLC analysis by two to three times whilst also increasing the number of analytes that may be characterized by a similar order of magnitude (MacNair et al., 1999).

An interesting modification of the IR-MALDI-MS scanning method described above has been reported (Eckerskorn et al., 1997b). Protein mixtures were separated by HPLC and the eluate collected directly onto PVDF strips as an elution time course. The PVDF membrane was then incubated with matrix solution prior to scanning IR-MALDI-MS. It has been reported that adequate spectra may be obtained when using ice as the matrix for IR-MALDI-MS, possibly opening the way to analyzing proteins in situ on frozen tissue sections by scanning IR-MALDI-MS (Berkenkamp et al., 1996).

Opiteck et al. (1998) have sought to improve earlier attempts to undertake 2D-HPLC and have developed a fully automated method of 2D-HPLC that parallels 2-DE. In their approach proteins are first separated by size exclusion chromatography and fractions are then automatically separated by reversed-phase liquid chromatography, thereby generating 2D-chromatograms. The analytes are transferred to 96-well plates from which fractions can be automatically loaded for MALDI-MS identification. A total of 576 fractions are produced from a typical separation run. The MS results may also be analyzed on-line so that the manual involvement in such analyses is kept to a minimum. This method was used to successfully identify two gene products in bacterial lysates, the src homology domain of a nonreceptor tyrosine kinase and β-lactamase, the sequences for which had been inserted into E. coli (Opiteck et al., 1998). This approach affords a number of advantages: the separations are highly reproducible, sample loading, running and fraction collecting are all readily automated and the proteins remain in solution throughout, making subsequent analysis by MS very straightforward. In its present form the system falls a long way short of the resolving power of 2-DE, but with further modification and enhanced capabilities for automated electrospray MS/MS the multiple proteins present in each of the final fractions could be readily characterized.

Capillary electrophoresis (CE) has proved to be an extremely effective method of analyzing clinical protein samples as it is rapid, potentially highly reproducible and does not require large sample volumes for high-sensitivity detection. For example, it has been applied to the diagnosis of conditions such as adenylosuccinase deficiency, nephrotic syndrome and cancer cachexia (Choudhary et al., 1999). In the latter, a combination of CE and MALDI-TOF-MS was used to identify a glycoprotein present in urine samples that may play a role in the severe muscle proteolysis associated with cachexia. The sensitivity of MS makes it readily amenable to such CE separations.

Others have investigated the low-abundance protein components of human cerebrospinal fluid (CSF) by two-dimensional liquid phase electrophoresis followed by MALDI-TOF-MS (Davidsson et al., 1999). This method allows large quantities of protein mixture to be loaded on the first dimension so that sufficient quantities of low-abundance proteins can be recovered at elution for MALDI analysis. This is particularly relevant for serum and CSF samples as they are dominated by the presence of albumin and immunoglobulin that in many analyses may mask the presence of other potentially more important proteins.

Exciting developments are being made in the miniaturization of CE through the use of 'chips' (see Becker *et al.*, 1998; Kopp *et al.*, 1997; Yao *et al.*, 1999). For example, Yao *et al.* (1999) have scaled down the process of capillary electrophoresis by generating a chip with a 40 μm-deep by 4.5 cm long channel etched into it, with eluates detected by absorbance at 214 nm or by laser-induced fluorescence (for fluorescein-derivatized samples). The authors speculate that microcapillary arrays and 2-DE in chip format may soon be possible (Yao *et al.*, 1999).

A major hurdle to linking such microseparations to MS-based protein characterization lies in the adequate delivery of the separated proteins to the mass spectrometer. Small amounts of proteins present at low concentrations are particularly prone to sample losses by adsorption onto surfaces and are susceptible to contamination if too much human intervention is required. For these reasons microfabricated fluidic systems are being developed to deliver samples directly to the ESI-MS apparatus via modules designed to minimize sample loss through adsorption to surfaces and which avoid manipulation of the small sample volumes (Figeys and Aebersold, 1999). Modules are also being developed that concentrate the samples through small C-18 reversed-phase cartridges before they are delivered to the ES-MS: such devices provide sensitivity below the fmol/μl level (Figeys and Aebersold, 1999).

4.2 Separation-independent methods: towards 'protein chips'?

Despite the advances in miniaturized and combined approaches to protein identification, the separation-dependent techniques described above still require relatively large quantities of protein sample and, where more than one step is required, operate in a sequential process. A powerful way to undertake highly parallel analyses and reduce the sample required for analysis is to exploit the power of 'molecular recognition', that is to use the specific recognition between two molecules to isolate, detect and/or identify the target molecule.

Perhaps the most dramatic example of the power of molecular recognition is in the application of 'DNA chips' that utilize specific base-pairing between an immobilized target and fluorescently labeled nucleic acid probes to provide rapid simultaneous analysis of thousands of different sequences (Bowtell, 1999; Cheung *et al.*, 1999). The cDNA or synthetic oligonucleotide sequences are arrayed on glass chips such that up to 10 000 genes may be represented on a chip that is 3.6 cm^2 in area (Gerhold *et al.*, 1999). By labeling probe cDNA with different color labels, comparisons between RNA expression profiles in healthy and diseased tissue, for example, may be rapidly performed through simultaneous hybridization to an appropriate chip. This methodology has now been applied to a number of detailed investigations (Alizadeh *et al.*, 2000; Alon *et al.*, 1999; DeRisi *et al.*, 1996; Perou *et al.*, 1999; Pollack *et al.*, 1999; Wang *et al.*, 1999).

There is much interest in the development of 'protein chips' and clearly the application of molecular recognition in a manner analogous to 'DNA chips' would make a significant impact on protein expression mapping (see Abbott, 1999). Such chips would require two key components: (i) spatially addressed molecules that recognize individual protein moieties; and (ii) a method to detect the interaction of the individual proteins in the mixture with their corresponding recognition molecules. As might be anticipated, attempts to develop such an analogous approach have been initiated in several laboratories. However, attempts to directly mimic the DNA chip approach may be fraught with difficulties. For example, fluorescent labeling of proteins with different fluorophores in a manner similar to the labeling of cDNAs has several inherent difficulties. Firstly, it is unlikely that the incorporation of the fluorophore into the proteins will be as well controlled as the process used for nucleic acids. Relatively small differences in efficiency of incorporation of the different fluorophores into the same protein present in different samples will compromise the ability to undertake quantitative comparisons. Site-specific labeling at a single site would overcome this difficulty but would result in severely reduced sensitivity of detection. Furthermore, whilst incorporation of fluorescent moieties into nucleic acids has a relatively predictable effect on the hybridization reactions, this is unlikely to be the case for proteins. Thus, the ability of the immobilized molecule to recognize the protein target is likely to be influenced by the incorporation of fluorophore in a poorly predictable or controllable manner. It seems likely that producing protein chip arrays with spatially defined sites for the recognition of individual proteins will require the integration of new approaches to generate the recognition molecules with those for detecting the interaction of proteins with the recognition molecules. Below we outline just some of the alternative approaches currently being explored. Some of these methods are well developed but may not be applicable to 'protein chips' for highly parallel protein expression analysis (i.e. hundreds or thousands of proteins analyzed simultaneously), but rather support more modest but nonetheless important protein expression analysis. Others are more prototypical but potentially applicable to 'protein chips'.

Protein recognition molecules. The most obvious molecule that can effectively be used for the recognition of a protein is an antibody. This form of molecular recognition has been widely used in a diverse range of applications. Both polyclonal and monoclonal antibodies have been used. However, if one envisages the use of antibodies for 'protein chips', several important criteria have to be fulfilled. Firstly, one has to have access to a process that can support the generation and selection of antibodies to individual proteins and possibly each of their post-translationally modified forms. The antibodies must be highly specific, that is they must be capable of recognizing and distinguishing the individual protein when present in a complex mixture. Finally, the binding of the individual proteins within the mixture to their cognate antibodies must occur under similar conditions. This latter objective could, in part, be achieved by careful control over the screening step in antibody production. Clearly, current approaches to the generation of antibodies are unlikely to be able to support the development of genome-scale 'protein chips'.

However, it seems likely that many will want to undertake more modest protein-expression mapping. For example, diagnostic protein chips that could monitor the expression and/or post-translational modifications of tens of different proteins would be a significant enhancement of current capabilities.

The production of synthetic phage-displayed antibodies provides considerable scope for the generation of protein recognition molecules to support the development of 'protein chips'. At present a range of different applications for these antibodies are being exploited (see Chapter 12).

There are a number of different methods for arraying proteins (including antibodies) and these have recently been reviewed (Blawas and Reichert, 1998). One potential method is described in the recent work of Rowe *et al.* (1999), who have developed a method for arraying capture antibodies on microscope slides with sufficient spatial separation to allow specific detection of multiple-bound antigens. Briefly, they silanized and esterized a glass surface (microscope slide) and then coated it with NeutrAvidin. A flow chamber module then allowed biotinylated capture antibodies to be laid down in six vertical channels. Application of the samples at right angles to the capture antibodies enabled the presence of six analytes to be measured in six samples simultaneously. The captured analytes were detected by a sandwich method using a fluorescently labeled detector antibody and CCD-based optical readout (Rowe *et al.*, 1999).

Alternative approaches use a hydrogel stamper to apply a molecular film of protein onto a gold-coated surface (Gaber *et al.*, 1999) or immobilization of protein within polyacrylamide gel pads arrayed on silanized glass (Guschin *et al.*, 1997). In the latter, a 3% solution of acrylamide containing glycerol, methylene blue and *N,N,N',N'*-tetramethylethylenediamine (TEMED) was photo-polymerized, the grid being generated by masking with a nontransparent chromium film. The gel matrix was activated by treatment with glutaraldehyde and a solution of mouse IgG1, rabbit IgG or bovine serum albumin was applied by loading approximately 1 nl onto each gel pad using a gold-plated fiber-optic pin. The authors showed that mouse IgG1 arrayed on a glass slide was selectively identified by a fluorescein-labeled anti-mouse IgG1 antibody (Guschin *et al.*, 1997).

Antibodies are not the only protein recognition molecules that might be used for 'protein chips'. Any group of molecules that can specifically recognize individual proteins with appropriate affinity and avidity and can do so in an immobilized form would be suitable. Indeed, there are disadvantages to the use of antibody proteins as protein recognition molecules, the most obvious ones being that the chips might suffer from poor storage properties and that the presence of any protease activity within the protein mixtures to be analyzed may 'degrade' the chip. The large libraries of compounds generated by combinatorial chemistry could provide a good source of protein recognition molecules, but again an effective method for screening the individual molecules for their capacity to bind individual proteins would be required.

Aptamers are single-stranded oligonucleotides possessing high affinity for conformational biomolecules, such as proteins. Libraries of aptamers up to a few hundred nucleotides in length may be immobilized to an appropriate surface for screening and may provide information on conformation, position of hydrogen bonds and other data which help to build up a three-dimensional model of the target protein (Brody *et al.*, 1999; Ellington and Szostak, 1990; Hermann and Patel, 2000; Romig *et al.*, 1999). This methodology has potential for development. Artificial receptors or antigen-binding sites have also recently been generated using molecularly imprinted polymers.

Molecularly imprinted polymers. Molecular imprinting involves the synthesis of artificial recognition sites on a surface by mimicking the shape of the template molecule in a polymeric film, thereby forming a molecularly imprinted polymer (MIP). It is proposed that any biological situation in which shape plays a part, such as antibody–antigen interactions, substrate–enzyme binding and receptor–ligand binding, may be mimicked by MIPs (Ansell and Mosbach, 1998; Haupt and Mosbach, 1998; Ye *et al.*, 1998).

There appear to be two general methods of constructing MIPs. The first involves mixing the template molecule with the polymer reagents, allowing the matrix to harden and then removing the template with a specific solvent. This leaves holes in the polymer that mimic the shape of the template molecule and may be employed to capture the target from a sample. At present the suitability of various monomers and cross-linkers for the manufacture of a polymer with the required specificity and affinity whilst retaining the advantageous properties of durability, resistance to high temperatures and pressures, and lack of reaction to acids, bases and solvents are being actively explored (Owens *et al.*, 1999). For example, Takeuchi and colleagues prepared a combinatorial library of triazine herbicides by varying the ratio of the monomers methacrylic acid and 2-trifluoromethylacrylic acid in polymers whilst maintaining a constant level of herbicide: the molecular imprint was completed when the herbicide was released by incubation with acetonitrile (Takeuchi *et al.*, 1999). They then assessed the MIP by comparing the concentration of free herbicide before and after incubation with the MIP library. Similarly, others have prepared an MIP from methacrylic acid polymerized in the presence of theophylline (Mullett and Lai, 1999). The resulting polymer was then ground to a powder, the theophylline was removed by extraction with methanol-acetic acid and the smallest MIP particles used to prepare a column. The column was shown to bind theophylline relatively specifically. It has been suggested that further modification of such MIPs to incorporate magnetic iron oxide would allow them to be withdrawn from solution by application of a magnetic field (Ansell and Mosbach, 1998).

The second method of generating MIPs uses mica, a substance that is atomically flat, as the adsorption substrate (Shi *et al.*, 1999). Protein is coated onto the mica and then covered with a disaccharide layer that mimics the surface conformation of the protein. A polymeric film of hexafluoropropylene is subsequently deposited over the sugar layer by glow discharge plasma deposition, after which the mica and protein are removed leaving a sugar-coated imprint of the protein (Shi *et al.*, 1999). In a demonstration of this approach, an MIP was generated by imprinting with streptavidin: the MIP preferentially bound biotin from a bovine serum albumin/ biotin mixture (Shi *et al.*, 1999).

Clearly, much further work is required if such MIPs are to evolve into an appropriate recognition surface for protein chips. Three issues appear to us to be significant. The first is the extent to which the MIP surface could contain discrete regions that would recognize individual proteins. It is also not apparent how the binding of protein to the MIP could be detected. Finally, and perhaps most significantly, by being generated to mimic the intact protein, the MIPs are directed against all the conformational epitopes within the protein, and it is increasingly apparent that proteins are modular in their 'design'. It is not clear therefore whether the MIPs could be sufficiently specific for an individual protein.

Detection methods. Detection of the interaction of proteins with recognition molecules on a 'protein chip' in a rapid, highly parallel and sensitive manner is likely to prove very challenging. Many of the methods currently in development are very effective for less complex devices but are probably not applicable to 'protein chips', at least as envisaged here. For example, methods (like the one described above) that require a second fluorescently labeled recognition molecule to detect the captured protein are unlikely to prove applicable. Despite the potential drawbacks, development of a fluorescence-based fiber-optic biosensor in which the capture antibodies are coated directly onto a fiber-optic probe is interesting (Tempelman *et al.*, 1996). In this approach detection is achieved by application of a detection antibody labeled with cyanine dye and laser excitation of the fiber-optic probe to enable quantification of the bound analyte by fluorimetry. The system developed by Tempelman and colleagues was able to analyze four samples simultaneously for the presence of a single analyte and had an optimum detection range of 5–200 ng/ml for Staphylococcal enterotoxins (Tempelman *et al.*, 1996). Another interesting fluorescence-based detection method has also been described in which an antibody-coated matrix is saturated with fluorescently labeled target antigen. The fluorescent target antigen is displaced by unlabeled target present in the test sample and the released fluorescent material measured. The method has been used by the US Federal Drugs Administration to quantify cocaine metabolites in urine (Yu *et al.*, 1996) and by the Office of Naval Research to detect explosives (Narang *et al.*, 1998). Both of these studies reported detection limits in the femtomole–picomole range. Although in the present form this would not be translatable to a 'protein chip', one could foresee a system in which the protein recognition molecules arrayed on the surface are designed such that they all bind a generic fluorescent moiety that would be released on binding of the individual proteins. The fluorescence remaining at an individual site would then be inversely related to the amount of relevant protein in the sample mixture.

Biosensors, including surface plasmon resonance (SPR)/resonant mirror biosensors, allow real-time detection of protein interactions with other proteins or ligands (see Gizeli and Lowe, 1996). These devices comprise a glass surface onto which a thin layer of metal, usually gold, has been deposited and is subsequently covered by a self-assembled monolayer (SAM) of matrix such as carboxymethylated hydrogel (Laricchia-Robbio *et al.*, 1996), carboxymethylated dextran (Vikinge *et al.*, 1998; Watts *et al.*, 1994) or a mixture of alkanethiolates (Dubrovsky *et al.*, 1999; Lahiri *et al.*, 1999). The chemistry of the matrix is such that proteins and ligands may be immobilized on the sensor and when the target protein binds, the refractive index of the matrix changes. The change in refractive index may be measured by a diode array detector (Lahiri *et al.*, 1999). Clearly, if such devices could be made more parallel, that is with a larger number of protein recognition sites, and miniaturized, they could possess considerable potential.

Quartz crystal microbalance (QCM) devices are an interesting method for detecting antibody–antigen interactions. In these, antibodies are bound to quartz crystals through amino groups incorporated into ethylenediamine plasma-polymerized films (Nakanishi *et al.*, 1996). The QCM is then exposed to the fluid to be analyzed, binding of target antigen results in small changes in mass on the surface of the crystal, and these changes are measured by detecting changes in resonance frequency via two gold electrodes. The sensitivity of these devices is such that they have the potential to detect the interaction of a single protein molecule with a single antibody molecule.

Other devices that have the potential to detect single molecule interactions include the force-amplified biological sensor being developed by Colton and colleagues (Baselt *et al.*, 1997). This is a modification of the atomic force microscope in which micron-sized magnetic particles are used in combination with micromachined piezoresistive cantilevers to measure antibody–antigen interactions. Finally, the possibility of incorporating biosensors into microelectronic circuits has been explored using molecular ion channels as sensing devices (Cornell *et al.*, 1997). In this approach, a gold electrode was attached to gramicidin molecules in the lower layer of a synthetic lipid membrane with the ion channel being formed by a second gramicidin molecule present in mobile form in the upper membrane layer. The mobile gramicidin molecules were linked to an antibody such that binding of antigen to antibody altered the conductance properties of the channel. There are many advantages offered by such a device, including the amplification that occurs as part of the detection event; thus, a single antigen–antibody interaction results in fluxes of millions of ions per second (see Turner, 1997).

5. Conclusions

Despite the technical limitations associated with 2-DE, there is little doubt that, at present, it is unrivalled as a method to resolve up to several thousand proteins simultaneously (see Abbott, 1999). Furthermore, powerful methods to characterize 2-DE-resolved proteins in a semi-automated and high-throughput fashion are well advanced and are currently in 'continuous improvement'. It seems likely, therefore, that 2-DE and these supporting techniques will remain in widespread use for some time yet. The technical limitations are sufficient to have motivated many to attempt to develop alternative methods of protein expression mapping. We have reviewed some of these developments and, in doing so, it has become evident that any future technology for high-throughput protein expression analysis will almost certainly require a multidisciplinary approach and the further development of novel methods.

The rapidly growing academic and commercial interest in proteomics seems likely to catalyze such developments.

References

Abbott, A. (1999) A post-genomic challenge: learning to read patterns of protein synthesis. *Nature* **402**: 715–720.

Alizadeh, A.A., Eisen, M.B., Davis, R.E., Ma, C., Lossos, I.S., Rosenwald, A., Boldrick, J.G., Sabet, H., Tran, T., Yu, X., Powell, J.I., Yang, L.M., Marti, G.E., Moore, T., Hudson, J., Lu, L.S., Lewis, D.B., Tibshirani, R., Sherlock, G., Chan, W.C., Greiner, T.C., Weisenburger, D.D., Armitage, J.O., Warnke, R., Levy, R., Wilson, W., Grever, M.R., Byrd, J.C., Botstein, D., Brown, P.O. and Staudt, L.M. (2000) Distinct types of diffuse large B-cell lymphoma identified by gene expression profiling. *Nature* **403**: 503–511.

Alon, U., Barkai, N., Notterman, D.A., Gish, K., Ybarra, S., Mack, D. and Levine, A.J. (1999) Broad patterns of gene expression revealed by clustering analysis of tumor and normal colon tissues probed by oligonucleotide arrays. *Proc. Natl Acad. Sci. USA* **96**: 6745–6750.

Ansell, R.J. and Mosbach, K. (1998) Magnetic molecularly imprinted polymer beads for drug radioligand binding assay. *Analyst* **123**: 1611–1616.

Baselt, D.R., Lee, G.U., Hanse, K.M., Chrisey, L.A. and Colton, R.J. (1997) A high-sensitivity micromachined biosensor. *Proc. IEEE* **85**: 672–680.

Becker, H., Lowack, K. and Manz, A. (1998) Planar quartz chips with submicron channels for two-dimensional capillary electrophoresis applications. *J. Micromech. Microengng* **8**: 24–28.

Berkenkamp, S., Karas, M. and Hillenkamp, F. (1996) Ice as a matrix for IR-matrix-assisted laser desorption/ionisation: mass spectra from a protein single crystal. *Proc. Natl Acad. Sci. USA* **93**: 7003–7007.

Bienvenut, W.V., Sanchez, J.C., Karmine, A., Rouge, V., Rose, K., Binz, P.A. and Hochstrasser, D.F. (1999) Toward a clinical molecular scanner for proteome research: parallel protein chemical processing before and during western blot. *Anal. Chem.* **71**: 4800–4807.

Binz, P.A., Muller, M., Walther, D., Bienvenut, W.V., Gras, R., Hoogland, C., Bouchet, G., Gasteiger, E., Fabbretti, R., Gay, S., Palagi, P., Wilkins, M.R., Rouge, V., Tonella, L., Paesano, S., Rossellat, G., Karmine, A., Bairoch, A., Sanchez, J.C., Appel, R.D. and Hochstrasser, D.F. (1999) A molecular scanner to automate proteomic research and to display proteome images. *Anal. Chem.* **71**: 4981–4988.

Blawas, A.S. and Reichert, W.M. (1998) Protein patterning. *Biomaterials* **19**: 595–609.

Bork, P., Dandekar, T., Diaz-Lazcoz, Y., Eisenhaber, F., Huynen, M. and Yuan, Y. (1998) Predicting function: from genes to genomes and back. *J. Mol. Biol.* **283**: 707–725.

Bowtell, D.D.L. (1999) Options available – from start to finish – for obtaining expression data by microarray. *Nature Genet.* **21**: 25–32.

Brody, E.N., Willis, M.C., Smith, J.D., Jayasena, S., Zichi, D. and Gold, L. (1999) The use of aptamers in large arrays for molecular diagnostics. *Mol. Diagnostics* **4**: 381–388.

Brünger, A.T. (1997) X-ray crystallography and NMR reveal complementary views of structure and dynamics. *Nature Struct. Biol.* **4** (Suppl.): 862–865.

Buter, D. (2000) Celera in talks to launch private sector human proteome project. *Nature* **403**: 815–816.

C. elegans consortium (1998) Genome sequence of the nematode *C. elegans*: a platform for investigating biology. The *C. elegans* Sequencing Consortium. *Science* **282**: 2012–2018.

Celis, J.E., Rasmussen, H.H., Olsen, E., Madsen, P., Leffers, H., Honoré, B., Dejgaard, K., Gromov, P., Vorum, H., Vassilev, A. *et al.* (1994) The human keratinocyte two-dimensional protein database (update 1994): towards an integrated approach to the study of cell proliferation, differentiation and skin diseases. *Electrophoresis* **15**: 1349–1458.

Celis, J.E., Rasmussen, H.H., Gromov, P., Olsen, E., Madsen, P., Leffers, H., Honoré, B., Dejgaard, K., Vorum, H., Kristensen, D.B. *et al.* (1995) The human keratinocyte two-dimensional gel protein database (update 1995): mapping components of signal transduction pathways. *Electrophoresis* **16**: 2177–2240.

Celis, J.E., Ostergaard, M., Jensen, N.A., Gromova, I., Rasmussen, H.H. and Gromov, P. (1998) Human and mouse proteomic databases: novel resources in the protein universe. *FEBS Lett.* **430:** 64–72.

Chalfie, M. (1998) Genome sequencing. The worm revealed. *Nature* **396:** 620–621.

Chee, M., Yang, R., Hubbell, E., Berno, A., Huang, X.C., Stern, D., Winkler, J., Lockhart, D.J., Morris, M.S. and Fodor, S.P.A. (1996) Accessing genetic information with high-density DNA arrays. *Science* **274:** 610–614.

Cheung, V.G., Morley, M., Aguilar, F., Massimi, A., Kucherlapati, R. and Childs, G. (1999) Making and reading microarrays. *Nature Genet.* **21:** 15–19.

Choudhary, G., Chakel, J., Hancock, W., Torres-Duarte, A., McMahon, G. and Wainer, I. (1999) Investigation of the potential of capillary electrophoresis with off-line matrix-assisted laser desorption/ionization time-of-flight mass spectrometry for clinical analysis: examination of a glycoprotein factor associated with cancer cachexia. *Anal. Chem.* **71:** 855–859.

Corbett, J.M., Dunn, M.J., Posch, A. and Görg, A. (1994) Positional reproducibility of protein spots in two-dimensional polyacrylamide gel electrophoresis using immobilised pH gradient isoelectric focusing in the first dimension: an interlaboratory comparison. *Electrophoresis* **15:** 1205–1211.

Cornell, B.A., Braach-Maksvytis, V.L.B., King, L.G., Osman, P.D.J., Raguse, B., Wieczorek, L. and Pace, R.J. (1997) A biosensor that uses ion-channel switches. *Nature* **387:** 580–583.

Cusack, S., Belrhali, H., Bram, A., Burghammer, M., Perrakis, A. and Riekel, C. (1998) Small is beautiful: protein micro-crystallography. *Nature Struct. Biol.* **5** (Suppl.): 634–637.

Davidsson, P., Westman, A., Puchades, M., Nilsson, C.L. and Blennow, K. (1999) Characterization of proteins from human cerebrospinal fluid by a combination of preparative two-dimensional liquid-phase electrophoresis and matrix-assisted laser desorption/ionization time-of-flight mass spectrometry. *Anal. Chem.* **71:** 642–647.

DeRisi, J., Penland, L., Brown, P.O., Bittner, M.L., Meltzer, P.S., Ray, M., Chen, Y., Su, Y.A. and Trent, J.M. (1996) Use of a cDNA microarray to analyse gene expression patterns in human cancer. [See comments.] *Nature Genet.* **14:** 457–460.

Dove, A. (1999) Proteomics: translating genomics into products? *Nature Biotechnol.* **17:** 233–236.

Dubrovsky, T.B., Hou, Z., Stroeve, P. and Abbott, N.L. (1999) Self-assembled monolayers formed on electroless gold deposited on silica gel: a potential stationary phase for biological assays. *Anal. Chem.* **71:** 327–332.

Dujon, B. (1996) The yeast genome project: what did we learn? *Trends Genet.* **12:** 263–270.

Eckerskorn, C., Strupat, K., Schleuder, D., Hochstrasser, D., Sanchez, J.C., Lottspeich, F. and Hillenkamp, F. (1997a) Analysis of proteins by direct scanning infrared-MALDI mass spectrometry after 2D PAGE separation and electroblotting. *Anal. Chem.* **69:** 2888–2892.

Eckerskorn, C., Strupat, K., Kellermann, J., Lottspeich, F. and Hillenkamp, F. (1997b) High sensitivity peptide mapping by micro-LC with on-line membrane blotting and subsequent detection by scanning IR-MALDI mass spectrometry. *J. Protein Chem.* **16:** 349–362.

Eisen, M.B., Spellman, P.T., Brown, P.O. and Botstein, D. (1998) Cluster analysis and display of genome-wide expression patterns. *Proc. Natl Acad. Sci. USA* **95:** 14863–14868.

Ellington, A.D. and Szostak, J.W. (1990) In vitro selection of RNA molecules that bind specific ligands. *Nature* **346:** 818–822.

Enright, A.J., Iliopoulos, I., Kyrpides, N.C. and Ouzounis, C.A. (1999) Protein interaction maps for complete genomes based on gene fusion events. *Nature* **402:** 86–90.

Fields, S. and Song, O. (1989) A novel genetic system to detect protein–protein interactions. *Nature* **340:** 245–246.

Figeys, D. and Aebersold, R. (1999) Microfabricated modules for sample handling, sample concentration and flow mixing: application to protein analysis by tandem mass spectrometry. *J. Biomech. Engng* **121:** 7–12.

Fromont-Racine, M., Rain, J.C. and Legrain, P. (1997) Toward a functional analysis of the yeast genome through exhaustive two-hybrid screens. *Nature Genet.* **16:** 277–282.

Futcher, B., Latter, G.I., Monardo, P., McLaughlin, C.S. and Garrels, J.I. (1999) A sampling of the yeast proteome. *Mol. Cell. Biol.* **19:** 7357–7368.

Gaber, B.P., Martin, B.D. and Turner, D.C. (1999) Create a protein microarray using a hydrogel 'stamper'. *Chemtech* **29:** 29–24.

Gerhold, D., Rushmore, T. and Caskey, C.T. (1999) DNA chips: promising toys have become powerful tools. *Trends Biochem. Sci.* **24:** 168–173.

Gerstein, M. and Hegyi, H. (1998) Comparing genomes in terms of protein structure: surveys of a finite parts list. *FEMS Microbiol. Rev.* **22:** 277–304.

Gizeli, E. and Lowe, C.R. (1996) Immunosensors. *Curr. Opin. Biotechnol.* **7:** 66–71.

Goffeau, A., Barrell, B.G., Bussey, H., Davis, R.W., Dujon, B., Feldmann, H., Galibert, F., Hoheisel, J.D., Jacq, C., Johnston, M., Louis, E.J., Mewes, H.W., Murakami, Y., Philippsen, P., Tettelin, H. and Oliver, S.G. (1996) Life with 6000 genes. *Science* **274:** 562–567.

Grimm, R. and Grasser, K.D. (1998) Nanogram scale separations of proteins using capillary high-performance liquid chromatography with fully-automated on-line microfraction collection followed by matrix-assisted laser desorption/ionisation time-of-flight mass spectrometry, protein sequencing and Western blot analysis. *J. Chromatogr. A* **800:** 83–88.

Guschin, D., Yershov, G., Zaslavsky, A., Gemmell, A., Shick, V., Proudnikov, D., Arenkov, P. and Mirzabekov, A. (1997) Manual manufacturing of oligonucleotide, DNA, and protein microchips. *Anal. Biochem.* **250:** 203–211.

Haupt, K. and Mosbach, K. (1998) Plastic antibodies: developments and applications. *Trends Biotechnol.* **16:** 468–475.

Hermann, T. and Patel, D.J. (2000) Biochemistry-adaptive recognition by nucleic acid aptamers. *Science* **287:** 820–825.

Jensen, O.N., Mortensen, P., Vorm, O. and Mann, M. (1997) Automation of matrix-assisted laser desorption/ionization mass spectrometry using fuzzy logic feedback control. *Anal. Chem.* **69:** 1706–1714.

Klose, J. (1999) Large-gel 2D electrophoresis. *Methods Mol. Biol.* **112** (2-D Proteome Analysis Protocols): 147–172.

Kopp, M.U., Crabtree, H.J. and Manz, A. (1997) Developments in technology and applications of microsystems. *Curr. Opin. Chem. Biol.* **1:** 410–419.

Lahiri, J., Isaacs, L., Tien, J. and Whitesides, G.M. (1999) A strategy for the generation of surfaces presenting ligands for studies of binding based on an active ester as a common reactive intermediate: a surface plasmon resonance study. *Anal. Chem.* **71:** 777–790.

Lamond, A.I. and Mann, M. (1997) Cell biology and the genome projects – a concerted strategy for characterizing multiprotein complexes by using mass spectrometry. *Trends Cell Biol.* **7:** 139–142.

Laricchia-Robbio, L., Liedberg, B., Platou-Vikinge, T., Rovero, P., Beffy, P. and Revoltella, R.P. (1996) Mapping of monoclonal antibody- and receptor-binding domains on human granulocyte-macrophage colony-stimulating factor (rhGM-CSF) using a surface plasmon resonance based biosensor. *Hybridoma* **15:** 343–350.

Lecrenier, N., Foury, F. and Goffeau, A. (1998) Two-hybrid systematic screening of the yeast proteome. *BioEssays* **20:** 1–5.

Li, X.P., Pleissner, K.P., Scheler, C., Regitz-Zagrosek, V., Salnikow, J. and Jungblut, P.R. (1999) A two-dimensional electrophoresis database of rat heart proteins. *Electrophoresis* **20:** 891–897.

Link, A.J., Eng, J., Schieltz, D.M., Carmack, E., Mize, G.J., Morris, D.R., Garvik, B.M. and Yates, J.R. (1999) Direct analysis of protein complexes using mass spectrometry. *Nature Biotechnol.* **17:** 676–682.

MacNair, J.E., Patel, K.D. and Jorgenson, J.W. (1999) Ultrahigh-pressure reversed-phase capillary liquid chromatography: isocratic and gradient elution using columns packed with 1.0 μm particles. *Anal. Chem.* **71:** 700–708.

Marcotte, E.M., Pellegrini, M., Ng, H.L., Rice, D.W., Yeates, T.O. and Eisenberg, D. (1999a) Detecting protein function and protein–protein interactions from genome sequences. *Science* **285:** 751–753.

Marcotte, E.M., Pellegrini, M., Thompson, M.J., Yeates, T.O. and Eisenberg, D. (1999b) A combined algorithm for genome-wide prediction of protein function. *Nature* **402:** 83–86.

Mullett, W.M. and Lai, E.P.C. (1999) Molecularly imprinted solid phase extraction microcolumn with differential pulsed elution for theophylline determination. *Microchem. J.* **61:** 143–155.

Nakanishi, K., Muguruma, H. and Karube, I. (1996) A novel method of immobilising antibodies on a quartz crystal microbalance using plasma-polymerized films for immunosensors. *Anal. Chem.* **68:** 1695–1700.

Narang, U., Gauger, P.R., Kusterbeck, A.W. and Ligler, F.S. (1998) Multianalyte detection using a capillary-based flow immunosensor. *Anal. Biochem.* **255:** 13–19.

Neubauer, G., King, A., Rappsilber, J., Calvio, C., Watson, M., Ajuh, P., Sleeman, J., Lamond, A. and Mann, M. (1998) Mass spectrometry and EST-database searching allows characterization of the multi-protein spliceosome complex. *Nature Genet.* **20:** 46–50.

Nicola, A.J., Gusev, A.I., Proctor, A. and Hercules, D.M. (1998) Automation of data collection for matrix assisted laser desorption/ionization mass spectrometry using a correlative analysis algorithm. *Anal. Chem.* **70:** 3213–3219.

Opiteck, G.J., Ramirez, S.M., Jorgenson, J.W. and Moseley, M.A. III (1998) Comprehensive two-dimensional high-performance liquid chromatography for the isolation of overexpressed proteins and proteome mapping. *Anal. Biochem.* **258:** 349–361.

Owens, P.K., Karlsson, L., Lutz, E.S.M. and Andersson, L.I. (1999) Molecular imprinting for bio- and pharmaceutical analysis. *Trends Anal. Chem.* **18:** 146–154.

Parekh, R. (1999) Proteomics and molecular medicine. *Nature Biotechnol.* **17** (Suppl.): BV19–BV20.

Pellegrini, M., Marcotte, E.M., Thompson, M.J., Eisenberg, D. and Yeates, T.O. (1999) Assigning protein functions by comparative genome analysis: protein phylogenetic profiles. *Proc. Natl Acad. Sci. USA* **96:** 4285–4288.

Pennington, S.R., Wilkins, M.R., Hochstrasser, D.F. and Dunn, M.J. (1997) Proteome analysis: from protein characterisation to biological function. *Trends Cell Biol.* **7:** 168–173.

Perou, C.M., Jeffrey, S.S., van de Rijn, M., Rees, C.A., Eisen, M.B., Ross, D.T., Pergamenschikov, A., Williams, C.F., Zhu, S.X., Lee, J.C., Lashkari, D., Shalon, D., Brown, P.O. and Botstein, D. (1999) Distinctive gene expression patterns in human mammary epithelial cells and breast cancers. *Proc. Natl Acad. Sci. USA* **96:** 9212–9217.

Plasterk, R.H. (1999) The year of the worm. *BioEssays* **21:** 105–109.

Pollack, J.R., Perou, C.M., Alizadeh, A.A., Eisen, M.B., Pergamenschikov, A., Williams, C.F., Jeffrey, S.S., Botstein, D. and Brown, P.O. (1999) Genome-wide analysis of DNA copy-number changes using cDNA microarrays. *Nature Genet.* **23:** 41–46.

Rappsilber, J., Siniossoglou, S., Hurt, E.C. and Mann, M. (2000) A generic strategy to analyze the spatial organization of multi-protein complexes by cross-linking and mass spectrometry. *Anal. Chem.* **72:** 267–275.

Rigaut, G., Shevchenko, A., Rutz, B., Wilm, M., Mann, M. and Seraphin, B. (1999) A generic protein purification method for protein complex characterisation and proteome exploration. *Nature Biotechnol.* **17:** 1030–1032.

Robson, B. (1999) Beyond proteins. *Trends Biotechnol.* **17:** 311–315.

Romig, T.S., Bell, C. and Drolet, D.W. (1999) Aptamer affinity chromatography: combinatorial chemistry applied to protein purification. *J. Chromatogr. B* **731:** 275–284.

RossMacdonald, P., Coelho, P.S.R., Roemer, T., Agarwal, S., Kumar, A., Jansen, R., Cheung, K.H., Sheehan, A., Symoniatis, D., Umansky, L., Heldtman, M., Nelson, F.K., Iwasaki, H., Hager, K., Gerstein, M., Miller, P., Roeder, G.S. and Snyder, M. (1999) Large-scale analysis of the yeast genome by transposon tagging and gene disruption. *Nature* **402:** 413–418.

Rout, M.P., Aitchison, J.D., Suprapto, A., Hjertaas, K., Zhao, Y.M. and Chait, B.T. (2000) The yeast nuclear pore complex: composition, architecture, and transport mechanism. *J. Cell Biol.* **148:** 635–651.

Rowe, C.A., Scruggs, S.B., Feldstein, M.J., Golden, J.P. and Ligler, F.S. (1999) An array immunosensor for simultaneous detection of clinical analytes. *Anal. Chem.* **71:** 433–439.

Sali, A. and Kuriyan, J. (1999) Challenges at the frontiers of structural biology. *Trends Genet.* **15:** M20–M24.

Shevchenko, A., Jensen, O.N., Podtelejnikov, A.V., Sagliocco, F., Wilm, M., Vorm, O., Mortensen, P., Shevchenko, A., Boucherie, H. and Mann, M. (1996) Linking genome and proteome by mass spectrometry: large-scale identification of yeast proteins from two dimensional gels. *Proc. Natl Acad. Sci. USA* **93:** 14 440–14 445.

Shi, H., Tsai, W.-B., Garrison, M.D., Ferrari, S. and Ratner, B.D. (1999) Template-imprinted nanostructured surfaces for protein recognition. *Nature* **398:** 593–597.

Swindells, M.B., Orengo, C.A., Jones, D.T., Hutchinson, E.G. and Thornton, J.M. (1998) Contemporary approaches to protein structure classification. *BioEssays* **20:** 884–891.

Takeuchi, T., Fukuma, D. and Matsui, J. (1999) Combinatorial molecular imprinting: an approach to synthetic polymer receptors. *Anal. Chem.* **71:** 285–290.

Tempelman, L.A., King, K.D., Anderson, G.P. and Ligler, F.S. (1996) Quantitating Staphylococcal enterotoxin B in diverse media using a portable fibre-optic biosensor. *Anal. Biochem.* **233:** 50–57.

Traini, M., Gooley, A.A., Ou, K., Wilkins, M.R., Tonella, L., Sanchez, J.C., Hochstrasser, D.F. and Williams, K.L. (1998) Towards an automated approach for protein identification in proteome projects. *Electrophoresis* **19:** 1941–1949.

Turner, A.P.F. (1997) Switching channels makes sense. *Nature* **387:** 555–557.

Uetz, P., Giot, L., Cagney, G., Mansfield, T.A., Judson, R.S., Knight, J.R., Lockshon, D., Narayan, V., Srinivasan, M., Pochart, P., QureshiEmili, A., Li, Y., Godwin, B., Conover, D., Kalbfleisch, T., Vijayadamodar, G., Yang, M.J., Johnston, M., Fields, S. and Rothberg, J.M. (2000) A comprehensive analysis of protein–protein interactions in *Saccharomyces cerevisiae*. *Nature* **403:** 623–627.

Unlü, M., Morgan, M.E. and Minden, J.S. (1997) Difference gel electrophoresis: a single gel method for detecting changes in protein extracts. *Electrophoresis* **18:** 2071–2077.

Vikinge, T.P., Askendal, A., Liedberg, B., Lindahl, T. and Tengvall, P. (1998) Immobilized chicken antibodies improve the detection of serum antigens with surface plasmon resonance (SPR). *Biosens. Bioelectron.* **13:** 1257–1262.

Wang, K., Gan, L., Jeffery, E., Gayle, M., Gown, A.M., Skelly, M., Nelson, P.S., Ng, W.V., Schummer, M., Hood, L. and Mulligan, J. (1999) Monitoring gene expression profile changes in ovarian carcinomas using cDNA microarray. *Gene* **229:** 101–108.

Watts, H.J., Lowe, C.R. and Pollard-Knight, D.V. (1994) Optical biosensor for monitoring microbial cells. *Anal. Chem.* **66:** 2465–2470.

Wery, J.P. and Schevitz, R.W. (1997) New trends in macromolecular X-ray crystallography. *Curr. Opin. Chem. Biol.* **1:** 365–369.

Wigge, P.A., Jensen, O.N., Holmes, S., Souès, S., Mann, M. and Kilmartin, J.V. (1998) Analysis of the Saccharomyces spindle pole by matrix-assisted laser desorption/ionization (MALDI) mass spectrometry. *J. Cell Biol.* **141:** 967–977.

Wilkins, M.R., Sanchez, J.-C., Gooley, A.A., Appel, R.D., Humphery-Smith, I., Hochstrasser, D.F. and Williams, K.L. (1995) Progress with proteome projects: why all proteins expressed by a genome should be identified and how to do it. *Biotechnology* **13:** 19–50.

Yao, S., Anex, D.S., Caldwell, W.B., Arnold, D.W., Smith, K.B. and Schultz, P.G. (1999) SDS capillary gel electrophoresis of proteins in microfabricated channels. *Proc. Natl Acad. Sci. USA* **96:** 5372–5377.

Ye, L., Ramstrom, O., Mansson, M.O. and Mosbach, K. (1998) A new application of molecularly imprinted materials. *J. Mol. Recog.* **11:** 75–78.

Yu, H., Kusterbeck, A.W., Hale, M.J., Ligler, F.S. and Whelan, J.P. (1996) Use of the USDT flow immunosensor for quantification of benzoylecgonine in urine. *Biosens. Bioelectron.* **11:** 725–734.

Zachariae, W., Shevchenko, A., Andrews, P.D., Ciosk, R., Galova, M., Stark, M.J., Mann, M. and Nasmyth, K. (1998) Mass spectrometric analysis of the anaphase-promoting complex from yeast: identification of a subunit related to cullins. *Science* **279:** 1216–1219.

Chapter 11

Application of proteome analysis to drug development and toxicology

Stefan Evers and Christopher P. Gray

1. Introduction

Good pharmaceuticals should be efficacious and specific. In general, they should interact primarily with their target while exerting only minimal effects on any other molecule. One of the major requirements in drug development is the confirmation that a substance is inhibiting the intended target (target validation) and does not have any undesired effects. A common shortcoming of assays and studies in drug development and toxicology is their focus on phenotypic observations, single macromolecules or enzymatic activity. Observed toxicity or pharmacological activity often remain poorly understood at the molecular level, and the direct or indirect effects of a substance on potential target molecules other than the one tested remain undetected. In order to bridge this gap between the observation of a pharmacological effect or toxicity and the understanding of the effects at the molecular level, there is an obvious need for techniques that address biological questions on a broader scale. These techniques should provide a deeper insight into the molecular processes of disease and of pharmacological or toxicological effects than hitherto achieved by conventional tests. Understanding these mechanisms can improve decision-making at critical stages in drug development and contribute considerably to increasing the efficiency of the process. An important advance in the development of such techniques is the progress in DNA sequencing methodology, which has revealed the entire repertoire of genes and corresponding proteins of prokaryotic and eukaryotic organisms. The advent of fully assembled genome sequences has opened up new avenues in biological research. However, the genome sequence alone of a specific organism is only of limited value. The term 'functional genomics' was coined for a relatively new discipline of research that utilizes technologies such as proteome analysis and high-density DNA arrays to utilize the wealth of information contained within the genomic sequences.

Proteome analysis has the potential to detect and quantify hundreds or even up to thousands of proteins in a single experiment. This provides the opportunity of making qualitative and quantitative assessments of changes in the protein synthesis patterns between different tissues, cells with different phenotype or genotype, healthy versus diseased, and cells/tissues in the presence or in the absence of drugs.

Proteomics, edited by S.R. Pennington and M.J. Dunn
© 2001 BIOS Scientific Publishers, Oxford.

Depending on the model system used, the response to different treatments or disease stages can be followed over time, allowing the study of the kinetics of the effect of a particular drug or the identification of features that change consistently with the progression of disease.

Superficially, transcriptional imaging and proteomics appear to be redundant technologies. Studying gene expression at the mRNA level seems to be an easier and more sensitive approach than proteomics. However, there are a number of arguments that underscore the importance of proteome analysis in the characterization of biological samples:

(i)   The study of gene regulation by the determination of mRNA concentrations is indirect and can only represent a measurement of an intermediate, leaving many phenomena undetected.

(ii) Regulation of gene expression at the level of translation has been documented in many cases (e.g. ribosomal proteins, codon usage, and attenuation mechanisms for the regulation of amino acid biosynthesis enzymes in *E. coli*) and remains undetected by quantification of mRNA.

(iii) A frequently neglected factor that determines the concentration of a protein in a cell is its rate of degradation that may also be subject to regulation (Tobias *et al.*, 1991).

(iv) An important point is that co- and post-translational covalent modifications (e.g. phosphorylation, acylation or glycosylation) play a significant biological role, often changing the properties of a protein completely (see Chapters 5 and 13). Several studies have shown that toxic effects can manifest themselves at the level of protein modification (Anderson *et al.*, 1992; Witzmann *et al.*, 1998). Many of the modified protein isoforms are resolved by two-dimensional gel electrophoresis (2-DE), such as, for example, the 'trains' of spots representing proteins with different states of glycosylation. Proteomics thus provides the researcher with a unique and powerful tool for the detection and study of the role of these modifications, for example in disease processes.

With respect to drug development, it should be realized that the functional macromolecules encoded by a gene are, in the vast majority of cases, proteins, and that the targets for pharmaceuticals are also almost exclusively proteins. A recent study illustrates how only a combination of transcriptional imaging with proteomics can lead to a comprehensive understanding of the molecular processes within a cell. In a proteomics experiment it was shown that, as the result of transfection with an oncogene, the unprocessed, inactive form of the ICE3 caspase accumulated in transformed malignant cells (*Figure 1*). Subsequently, mRNA quantification revealed that the synthesis of the protease that activates ICE3 (Wang *et al.*, 1996) was strongly downregulated in these cells, thus explaining the effect observed on the 2-DE gels (Certa *et al.*, 1999).

The adjustment of protein synthesis rates is an integral part of the cellular response to changing environmental conditions. The maturation of the cell through the cell cycle and differentiation are also dependent upon the regulation of gene

(a)

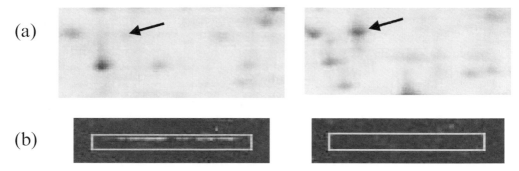

(b)

Figure 1. Selected regions of 2DE-PAGE gels (a) and Affymetrix GeneChips™ (b) obtained with protein and mRNA extracts from normal mouse mast cells (left panel) and ras-oncogene-transfected cells (right panel), respectively. An arrow indicates the position of the unprocessed form of the ICE3 caspase on the 2-DE gel. The region on the chip representing the granzyme B gene is boxed. Granzyme B proteolytically activates the ICE3 caspase (Wang *et al.*, 1996). These results indicate that, in the presence of the H-ras oncogene, granzyme B production is repressed, which leads to accumulation of unprocessed ICE3 pro-caspase. It is likely that the protease is involved in the apoptotic pathway and that, in its absence, cells undergo malignant, uncontrolled growth (Certa *et al.*, 1999).

expression. Complex regulatory networks control the concentration and enzymatic activities of proteins at more than one level (e.g. mRNA synthesis, translation, modification, transport, proteolytic processing, and degradation). Thus, the proteome reflects the environmental conditions. The comparative study of the proteome helps in gaining an understanding of the regulatory networks of a cell. As a consequence, it is a valuable tool for the identification and validation of drug targets, the study of the mechanisms of drug action, and the detection and rationalization of pharmacological or toxic effects.

2. Making use of the data

The amount of data generated by proteomics experiments requires computational analysis. In order to validate the observations and make reliable predictions, replicate gels of replicate samples have to be processed, stored and analyzed. The experimental variation can be very high, especially in cases where genetic variations come into play, as for example in toxicological studies using outbred animals (Steiner *et al.*, 1995). Clearly, statistics are crucial for the interpretation of the large amount of (often error-prone) data obtained from multiple 2-DE gels. Beyond a classical statistical analysis (mean values, standard deviations, significance estimations) provided by the standard 2-DE gel analysis tools, there are two fundamental ways to examine the data from these experiments:

(i)   The proteome as a whole can be regarded as a biosensor reflecting environmental changes. Questions such as 'Does compound A act in a similar way to compound B?' can be addressed in this way.

(ii) A more refined analysis addresses the effects of compounds on single proteins or
      sets of proteins implicated in metabolic pathways or belonging to different
      functional classes. This allows a rationalization of the observed effects of a drug,
      and the pinpointing of potential diagnostic markers for disease, pharmaco-
      logical activity or toxicity.

In the following, examples for both approaches will be discussed and illustrated in
more detail, with the focus on drug development and toxicology.

2.1 Comparative analysis

It can be assumed that two compounds with the same properties trigger similar
responses in a cell. The relative amounts of protein present under conditions of
induction (response) compared to the amount present without induction (control)
can be represented as a series of numerical values ('induction ratios'). The resulting
profiles can then be evaluated with respect to different conditions, providing
valuable information about the properties of the compound under study. An
example from antibacterial research illustrates this approach.

    A screen for dihydrofolate reductase (DHFR) inhibitors resulted in the compound
Ro-64-1874, a 2,4-diaminopyrimidine derivative, like trimethoprim (Tmp). Proteome analysis using *Haemophilus influenzae* as the model system resulted in a set of
data that could be compared to a database of effects of well-characterized
antimicrobials. The responses to Ro-64-1874 and to the standard antibiotics were
then compared pairwise as follows: the spots were partitioned into three groups –
(i) spots that were increased (induction ratio > 1.25); (ii) decreased (induction ratio
< 0.8); and (iii) did not change ($0.8 <$ induction ratio < 1.25) upon induction. Only
spots that were increased or decreased were selected for further analysis. The degree
of similarity between the two responses was estimated by comparing the number of
spots that were induced/repressed under both conditions with the total number of
spots and the number of spots induced/repressed by each one of the conditions
(*Figure 2*). The ratio between these values is a good measure for the degree and
significance of similarity between two effects. A comparison of the expression profile

---

Figure 2. Results of the pairwise comparison of the effects of 8 mg/l and 150 mg/l of Ro-64-1874
on *H. influenzae* cultures with the effects of other antibiotics [inhibitors of THF synthesis (2
compounds), cell wall synthesis (2), transcription (2), translation (7), tRNA synthesis (1), DNA
gyrase (2) and fatty acid synthesis (1)]. ($+$) and ($-$) designate sets of spots synthesized at an
increased or decreased rate (induction ratio > 1.25 and < 0.8, respectively). Abbreviations for
antibiotics and respective concentrations used are: Cam, chloramphenicol (2 mg/l); Fus, fusidic
acid (32 mg/l); Kan, kanamycin (32 mg/l); Smz, sulfamethoxazole (120 mg/l); Str, streptomycin
(24 mg/l); Tet, tetracycline (2 mg/l); and Tmp, trimethoprim (1 mg/l). The cultures were exposed to
Ro-64-1874 for 60 min. Other exposure times are indicated. The calculation of significance values
is illustrated by the comparison of the sets of spots increased under induction with 8 mg/l of Ro-64-
1874 and Tmp (column highlighted with bold lines): 58% of all spots induced by Tmp were also
induced by Ro-64-1874, while only 31% of all spots were induced by Ro-64-1874. The ratio
between these two percentages was 1.84 and was used as a measure of significance for the similarity
of the two responses. Only pairs with ratios above 1.1 are shown.

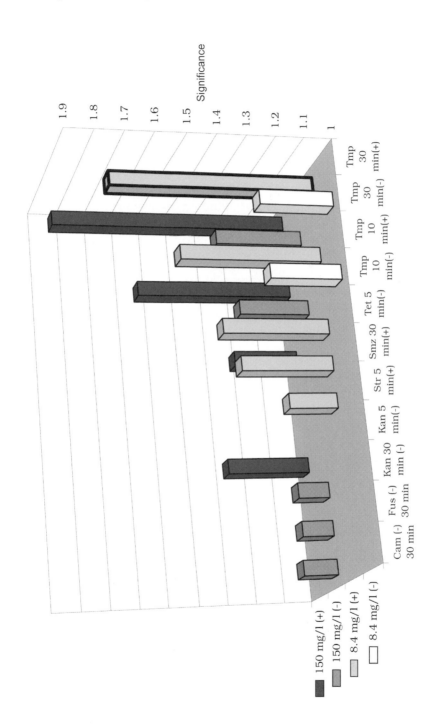

resulting from exposure to Ro-64-1874 with the other profiles in the database (including responses to inhibitors of transcription, translation, tetrahydrofolate synthesis, tRNA synthesis, fatty acid synthesis and DNA gyrase) yielded a good correlation only with those to Tmp and sulfamethoxazole (Smz). The expected mode of action of Ro-64-1874 could thus be confirmed. We recently encountered a case where a compound selected for its *in vitro* activity against gyrase triggered a secondary *in vivo* response that, from a comparison of the patterns, indicated inhibition of protein synthesis. This activity could be confirmed in an *in vitro* transcription/translation assay. The response to some compounds has been shown to be time- and concentration-dependent (VanBogelen *et al.*, 1987). Therefore, it is desirable to examine the response to compounds with an unclear mode of action at more than one concentration and at different time points after addition of the antibiotic.

This approach is amenable to a number of different applications in the fields of antibacterial research, pharmacology and toxicology, particularly in the study of structure–function relationships (e.g. Myers *et al.*, 1997). For this kind of comparison, the identity of the protein spots is unimportant. However, the success of this comparison is dependent upon having a database of responses derived from compounds for which the mode of action is understood, which is not always possible. Compounds with a novel mode of action will not yield significant correlation values with any of the standard substances. For such compounds a more detailed analysis is required to gain a better understanding of their properties. A comparison of the changes in the gene expression profile resulting from treatment with a drug with the changes resulting from the inactivation or downregulation of the gene encoding the drug target is an interesting approach to target validation. Using DNA microarrays, the effects of cyclosporine A (CyA) and FK506 on yeast cells could be correlated to the effects of disruption of the calcineurin and FPR1 genes which encode their target proteins, demonstrating the usefulness of this strategy (Marton *et al.*, 1998).

2.2 Detection of biomarker proteins

Proteome databases can be queried for changes in protein spot intensity that are indicative of a disease or the response to a group of pharmaceuticals. Such proteins are useful as surrogate markers for particular responses or for disease. Probably the best example for the successful application of proteomics in the field of toxicology is the study of the effects of CyA. By examining the induction pattern, a mechanism for the toxicity of CyA could be hypothesized. CyA is a widely used immunosuppressant, the application of which is compromised by a number of adverse side effects, the most important being nephrotoxicity. The most significant finding resulting from treatment with CyA was the disappearance of one particular spot, which was subsequently identified as representing calbindin-D 28 kDa, from the rat kidney protein pattern (Benito *et al.*, 1995; Steiner *et al.*, 1996). Further studies using an enzyme-linked immunosorbent assay confirmed a continuous decrease in the levels of calbindin-D 28 kDa in the rat kidney (Steiner *et al.*, 1996). This decrease was associated with increased urine calcium excretion and corticomedullary intratubular calcifications (Aicher *et al.*, 1997). The observed effects can be explained if calbindin-D 28 kDa acts as a facilitator of calcium transport and as a calcium buffer, as has been previously suggested (Feher, 1983). The successful application of proteomics in conjunction with other techniques resulted in the identification of a marker of CyA toxicity and provided a plausible explanation for its toxic effects. A second example for the successful identification of biomarker proteins comes from a long-term study of the progression of bladder squamous cell carcinoma (reviewed in Celis *et al.*, 1998). In this study, protein markers that define the degree and differentiation of the cancerous lesions could be detected (Celis *et al.*, 1996; Ostergaard *et al.*, 1997).

This approach to biomarker identification has enormous potential in different fields of the pharmaceutical industry, ranging from preclinical research to diagnostics, pharmacology and toxicology. The power of proteomics in detecting the distinguishing diagnostic features of a protein expression profile is largely dependent on the size and the quality of the database used to extract the information. For this purpose, several publicly available and proprietary databases have been constructed to archive large numbers of expression patterns and mass spectrometry data derived from 2-DE gels (see Chapter 8). Different types of multivariate statistical analysis, such as principal component analysis or predictive modeling, can be used to define biomarker proteins.

2.3 Detailed analysis of the regulation of gene expression

As more of the spot coordinates are correlated with their respective proteins, a more detailed analysis of the experimental data becomes possible. Protein levels of the majority of proteins involved in a particular pathway or, alternatively, in a regulon can be quantified. At present, only in the case of microbes has a sufficient number of the proteins been identified to allow for a detailed analysis of the effects of growth conditions and antibiotics on gene expression. F.C. Neidhardt's group was the first to establish a 2-DE map for *Escherichia coli* and to begin a systematic study and cataloging of the responses of these bacteria to different environmental conditions (reviewed in VanBogelen *et al.*, 1997). In this manner, a number of different regulons have been discovered and characterized (VanBogelen *et al.*, 1997). Proteome maps were published for several microorganisms, such as for example *Bacillus subtilis* (Antelmann *et al.*, 1997), *Haemophilus influenzae* (Cash *et al.*, 1997; Langen *et al.*, 1997), and *Saccharomyces cerevisiae* (Boucherie *et al.*, 1996; Hodges *et al.*, 1999). The availability of total genome sequences, which greatly facilitates the identification of the proteins and allows the construction of comprehensive metabolic tables, has had a major impact on this approach. It must, however, be borne in mind that the assignment of function is, in most cases, based on sequence homology and that traditional biochemistry remains crucial for the completion and validation of these pathways. An example where the detailed analysis of a biochemical pathway was possible has recently been published by our group (Evers *et al.*, 1998). As mentioned above, we have studied the response of *H. influenzae* to the tetrahydrofolate synthesis inhibitors, Tmp and Smz. Among the most significant changes was the increase in intensity of spots representing the enzymes involved in methionine biosynthesis. These results would indicate that the genes encoding these proteins are regulated in a similar manner to their counterparts in *E. coli* (Weissbach and Brot, 1991). An investigation of the effects of Tmp and Smz on purine and thymidine biosynthetic enzymes, which are also affected by inhibition of tetrahydrofolate synthesis, is at present not possible, as many of the proteins involved have not yet been identified on the 2-DE gels. In a similar study in yeast, the regulation of the synthesis rates of purine and histidine biosynthetic proteins was investigated (Denis *et al.*, 1998), demonstrating the feasibility of this approach in eukaryotes.

This type of detailed analysis can help in establishing a hypothesis for the mode of action of compounds which result in a response for which there is no counterpart in the database. Thus, it is potentially useful for the discovery of novel targets for pharmacologically active substances. Secondary effects observed in these analyses could also identify potential targets for pharmaceutical intervention. In combination with other methodologies (e.g. recombinant gene inactivation) a more complete picture of the effects of a drug can be obtained. The usefulness of this approach will be extended to include more complex organisms using the data accumulated in the large number of ongoing sequencing projects.

2.4 Prediction of protein function

For many proteins, an assignation of protein function on the basis of sequence comparisons is impossible. Such proteins may represent potential unexploited targets for pharmaceuticals and are therefore of particular interest for the industry. The concept of 'regulatory homology' was recently proposed by Anderson and Anderson (1998): proteins for which the relative abundance changes in a similar manner when confronted with changing environmental conditions may be involved in a common process. This concept may be useful in the prediction of protein function. Moreover, the changes in the proteome of recombinant organisms that are devoid of or overexpress the 'orphan' gene may indicate possible functions for these proteins. In an experiment studying the effect of sulfate-limitation on the gene expression in *E. coli*, eight sulfate-regulated proteins were identified by comparison of 2-DE gels (Dainese *et al.*, 1997). As a consequence of the recombinant inactivation of the gene corresponding to one of these proteins, TauD, the bacteria became growth-deficient on taurine as the sole sulfur source (Dainese *et al.*, 1997). Proteomics thus identified proteins involved in sulfate assimilation, while the recombinant techniques used in conjunction demonstrated the role for one of them, TauD, in taurine desulfonation.

3. Conclusions and outlook

It is now widely accepted that proteomics is useful in pharmaceutical research and development. Most pharmaceutical companies have established proteome analysis groups to improve the characterization of biological material with reference to drug action, and the finding of relevant biomarkers and potential targets. Large Scale

Biology Corp., now part of Biosource Technologies Inc., pioneered the application of 2-DE in molecular pharmacology and toxicology, constructing a comprehensive Molecular Effects of Drugs™ database (Anderson and Anderson, 1998), offering 2-DE as a service to the pharmaceutical industry and providing access to their proprietary protein profile database and 2DE maps. Several other companies have been founded following this example, also offering these services or conducting proteomics programs for drug discovery and development (see Chapter 11 on 'commercial proteomics').

Proteomics is applied to various stages in the drug discovery and development process. As mentioned above, it is useful in the early phase for target identification and validation. Later in development, its usefulness has been demonstrated in pharmacology and toxicology. Particularly in the latter two fields, the significance of the findings and their correlation with results and observations obtained from classical physiological and histological tests require careful evaluation. These studies will hopefully lead to the identification of novel biomarkers that may be useful in the development of speedier and more economic diagnostic tests. The detection of these biomarkers may be applicable at earlier stages in drug development, reducing the need for experimentation in animals.

In order to fully exploit the potential of proteomics, technical developments are underway to counter the limitations that are, at present, still associated with this methodology. One of the more unsatisfactory features is the use of 2-DE, which is cumbersome and time-consuming, preventing the use of proteomics as a high-throughput technology. In an intelligent combination of techniques, 2-DE can be employed to identify marker proteins which are then detected by other methods more suitable for high-throughput testing (e.g. ELISA, reporter gene systems, biochemical tests). The automation of 2-DE and the improvement of the image analysis are areas of intense study so that the acquisition of the data can be accelerated. A higher throughput that meets the demands of the pharmaceutical industry may eventually be achieved by the replacement of 2-DE gels by other, more easily automatable separation systems. At present, only the study of the more abundant and soluble proteins is possible due to the relative lack of sensitivity and the difficulty in solubilizing large and hydrophobic proteins. Proteome studies are therefore necessarily biased towards these proteins, which are often structural proteins or metabolic enzymes (*Table 1*). The potential of proteomics to detect, identify and quantify protein modifications is, at present, underexploited. There are only few examples where the modifications have been thoroughly identified and characterized. Since technological development proceeds at a rapid pace, an increase in the sensitivity and accuracy of mass spectrometric measurements is to be expected. This, together with improved analysis software, will allow routine analysis of protein modification and widen the scope of proteome studies.

Most of the phenomena observed by proteome studies or mRNA analysis are difficult to interpret. Since the regulatory pathways are interconnected, an effect is often seen on many seemingly unrelated proteins belonging to different functional classes. Primary, secondary or tertiary effects overlap, creating a highly complex expression pattern. In biological research, these methodologies serve to generate

Table 1. Representation of different functional classes of proteins on preparative 2-DE gels from *H. influenzae* soluble protein extracts

Functional class[a]	Percentage of each class identificed
Amino acid biosynthesis	44
Biosynthesis of cofactors, prosthetic groups and carriers	27
Cell envelope	22
Cellular processes	30
Central intermediary metabolism	37
Energy metabolism	41
Fatty acid and phospholipid metabolism	40
Hypothetical	19
Other categories	23
Purines, pyrimidines, nucleosides and nucleotides	40
Regulatory functions	16
Replication	25
Transcription	33
Translation	49
Transport and binding proteins	20
Total	*29*

[a]According to Fleischmann *et al.* (1995).

hypotheses, which clearly must be validated using complementary techniques. Information from different sources such as the quantification of metabolites will also be necessary to obtain a basic understanding of the biochemistry of the cell. Bioinformatics plays a vital role in archiving and analyzing the data, integrating our present knowledge on gene regulation with the findings from proteomics and also other sources into a more comprehensive, 'holistic' view of an organism, and in presenting the data in a comprehendable mode.

Acknowledgments

We thank Hanno Langen und Uli Certa for supplying Figure 1 and W. Keck for a critical reading of the manuscript.

References

Aicher, L., Meier, G., Norcross, A., Jakubowski, J., Varela, M., Cordier, A. and Steiner, S. (1997) Decrease in kidney calbindin-D 28 kDa as a possible mechanism mediating cyclosporine A- and FK-506-induced calciuria and tubular mineralization. *Biochem. Pharmac.* **53:** 723–731.

Anderson, N.L. and Anderson, N.G. (1998) Proteome and proteomics: new technologies, new concepts, and new words. *Electrophoresis* **19:** 1853–1861.

Anderson, N.L., Copple, D.C., Bendele, R.A., Probst, G.S. and Richardson, F.C. (1992) Covalent protein modifications and gene expression changes in rodent liver following administration of methapyrilene: a study using two-dimensional electrophoresis. *Fundam. Appl. Toxicol.* **18:** 570–580.

Antelmann, H., Bernhardt, J., Schmid, R., Mach, H., Volker, U. and Hecker, M. (1997) First steps from a two-dimensional protein index towards a response-regulation map for *Bacillus subtilis*. *Electrophoresis* **18**: 1451–1463.

Benito, B., Wahl, D., Steudel, N., Cordier, A. and Steiner, S. (1995) Effects of cyclosporine A on the rat liver and kidney protein pattern, and the influence of vitamin E and C coadministration. *Electrophoresis* **16**: 1273–1283.

Boucherie, H., Sagliocco, F., Joubert, R., Maillet, I., Labarre, J. and Perrot, M. (1996) Two-dimensional gel protein database of *Saccharomyces cerevisiae*. *Electrophoresis* **17**: 1683–1699.

Cash, P., Argo, E., Langford, P.R. and Kroll, J.S. (1997) Development of a *Haemophilus* two-dimensional protein database. *Electrophoresis* **18**: 1472–1482.

Celis, J.E., Rasmussen, H.H., Vorum, H., Madsen, P., Honore, B., Wolf, H. and Orntoft, T.F. (1996) Bladder squamous cell carcinomas express psoriasin and externalize it to the urine. *J. Urol.* **155**: 2105–2112.

Celis, J.E., Ostergaard, M., Jensen, N.A., Gromova, I., Rasmussen, H.H. and Gromov, P. (1998) Human and mouse proteomic databases: novel resources in the protein universe. *FEBS Lett.* **430**: 64–72.

Certa, U., Hochstrasser, R., Langen, H., Buess, M. and Moroni, C. (1999) Biosensors in biomedical research: development and application of gene chips. *Chimia* **53**: 57–61.

Dainese, P., Staudenmann, W., Quadroni, M., Korostensky, C., Gonnet, G., Kertesz, M. and James, P. (1997) Probing protein function using a combination of gene knockout and proteome analysis by mass spectrometry. *Electrophoresis* **18**: 432–442.

Denis, V., Boucherie, H., Monribot, C. and Daignan-Fornier, B. (1998) Role of the myb-like protein bas1p in *Saccharomyces cerevisiae*: a proteome analysis. *Mol. Microbiol.* **30**: 557–566.

Evers, S., Di Padova, K., Meyer, M., Fountoulakis, M., Keck, W. and Gray, C.P. (1998) Strategies towards a better understanding of antibiotic action: folate pathway inhibition in *Haemophilus influenzae* as an example. *Electrophoresis* **19**: 1980–1988.

Feher, J.J. (1983) Facilitated calcium diffusion by intestinal calcium-binding protein. *Am. J. Physiol.* **244**: C303–307.

Fleischmann, R.D., Adams, M.D., White, O. *et al.* (1995) Whole-genome random sequencing and assembly of *Haemophilus influenzae* Rd. *Science* **269**: 496–512.

Hodges, P.E., McKee, A.H.Z., Davis, B.P., Payne, W.E. and Garrels, J.I. (1999) The Yeast Proteome Database (YPD): a model for the organization and presentation of genome-wide functional data. *Nucleic Acids Res.* **27**: 69–73.

Langen, H., Gray, C., Röder, D., Juranville, J.-F., Takacs, B. and Fountoulakis, M. (1997) From genome to proteome: protein map of *Haemophilus influenzae*. *Electrophoresis* **18**: 1184–1192.

Marton, M.J., DeRisi, J.L., Bennett, H.A., Vishwanath, I.R., Meyer, M.R., Roberts, C.J., Stoughton, R., Burchard, J., Slade, D., Dai, H., Bassett, D.E.J., Hartwell, L.H., Brown, P.O. and Griend, S.H. (1998) Drug target validation and identification of secondary drug target effects using DNA microarrays. *Nature Med.* **4**: 1293–1301.

Myers, T.G., Anderson, N.L., Waltham, M., Li, G., Buolamwini, J.K., Scudiero, D.A., Paull, K.D., Sausville, E.A. and Weinstein, J.N. (1997) A protein expression database for the molecular pharmacology of cancer. *Electrophoresis* **18**: 647–653.

Ostergaard, M., Rasmussen, H.H., Nielsen, H.V., Vorum, H., Orntoft, T.F., Wolf, H. and Celis, J.E. (1997) Proteome profiling of bladder squamous cell carcinomas: identification of markers that define their degree of differentiation. *Cancer Res.* **57**: 4111–4117.

Steiner, S., Wahl, D., Varela, M.C., Aicher, L. and Prieto, P. (1995) Protein variability in male and female Wistar rat liver proteins. *Electrophoresis* **16**: 1969–1976.

Steiner, S., Aicher, L., Raymackers, J., Meheus, L., Esquer-Blasco, R., Anderson, N. and Cordier, A. (1996) Cyclosporine A decreases the protein level of the calcium-binding protein calbindin-D 18 kDa in rat kidney. *Biochem. Pharmac.* **51**: 253–258.

Tobias, J.W., Shrader, T.E., Rocap, G. and Varshavsky, A. (1991) The N-end rule in bacteria. *Science* **254**: 1374–1377.

VanBogelen, R.A., Kelley, P.M. and Neidhardt, F.C. (1987) Differential induction of heat shock, SOS, and oxidation stress regulons and accumulation of nucleotides in *Escherichia coli*. *J. Bacteriol.* **169**: 26–32.

VanBogelen, R.A., Abshire, K.Z., Moldover, B., Olson, E.R. and Neidhardt, F.C. (1997) *Escherichia coli* proteome analysis using the gel-protein database. *Electrophoresis* **18**: 1243–1251.

Wang, S., Miura, M., Jung, Y., Zhu, H., Gagliardini, V., Shi, L., Greenberg, A.H. and Yuan, J. (1996) Identification and characterization of Ich-3, a member of the interleukin-1beta converting enzyme (ICE)/Ced-3 family and an upstream regulator of ICE. *J. Biol. Chem.* **271:** 20 580–20 587.

Weissbach, H. and Brot, N. (1991) Regulation of methionine synthesis in *Escherichia coli*. *Mol. Microbiol.* **5:** 1593–1597.

Witzmann, F.A., Daggett, D.A., Fultz, C.D., Nelson, S.A., Wright, L.S., Kornguth, S.E. and Siegel, F.L. (1998) Glutathione *S*-transferases: two-dimensional electrophoretic protein markers of lead exposure. *Electrophoresis* **19:** 1332–1335.

Chapter 12

Phage antibodies as tools for proteomics

Kevin Pritchard, Kevin S. Johnson and Ulla Valge-Archer

1. Approaches to proteomics

1.1. Protein data from EST data

Motivated by the desire to gain a more complete understanding of biological function, intense effort has been put into genome sequencing projects. While databases of curated, full-length sequences continue to grow, far more information is currently available in databases of short DNA sequences called 'expressed sequence tags' or ESTs (Adams *et al.*, 1991), which probably now represent most of the expressed human genome, albeit in a fragmented state. Interpreting the biological relevance of this data is a challenge that is proving to be far harder than generating the DNA sequences. Clearly, we need to understand the patterns of expression of the proteins that the ESTs encode as a clue to their function. Thus, the genome sequencing effort is now paralleled by more complex efforts to catalog and characterize the proteome.

Currently, proteomics relies heavily on the complementary techniques of two-dimensional gel electrophoresis (2-DE) and mass spectrometry to display, identify and catalog proteins comprising partial proteomes from selected tissues and organisms. In this approach, researchers work backwards from the fractionated proteins to the corresponding DNA sequences. A limitation of this approach is that the resulting databases are heavily biased towards abundant proteins. Unlike DNA or RNA, proteins cannot be amplified; functionally important proteins of low abundance are often not detected and a complete proteome cannot at present be defined.

An alternative approach would be to generate protein-specific molecular probes based on DNA sequence data, and to use these probes to identify individual proteins and to study their functions in cell extracts, intact cells, tissues, and in living organisms. It seems likely that this type of targeted approach will yield useful, comprehensive and interpretable data more rapidly than the existing approach. Proteomic analysis using targeted molecular probes depends on the generation of these specific probes at a rate and a scale capable of matching genomic projects. We describe here how this concept may be implemented in practice, by means of phage antibody libraries.

Proteomics, edited by S.R. Pennington and M.J. Dunn
© 2001 BIOS Scientific Publishers, Oxford.

Phage display (Barbas *et al.*, 1991; McCafferty *et al.*, 1990), is uniquely capable of providing specific antibodies in the numbers and at the rate needed to deal with the flow of information from genome-sequencing projects. Whereas classical approaches, such as 2-DE gels, use generic stains to detect proteins, specific probes, such as phage antibodies, can be used to reveal and quantify particular proteins in almost any biological context, from a Western blot to an intact organism. We will describe a high-throughput implementation of this concept, the ProAb^{TM} project. Uniquely, ProAb^{TM} analyzes protein expression not in extracts (as do dot-blots and 2-DE gels), but directly on histology sections, thereby accumulating details of tissue, cellular and subcellular distribution of each protein.

Proteins act not in isolation but through their interactions with other proteins and with smaller ligands. Information on associated proteins and their topology within multiprotein complexes is frequently a key to understanding function. We will describe a technique known as the ProxiMol^{TM} technique, whereby antibodies specific to a target recognized by a guide molecule, which may be a ligand of any type, can be obtained (Osbourn *et al.*, 1998). The ProxiMol^{TM} technique thus offers a means of obtaining specific antibodies to a protein, without having to purify or clone it from its native state. Furthermore, a single antibody, recognizing a protein epitope, can in principle be used as a guide molecule with which to obtain a broad panel of new antibodies, some of which should have a direct functional effect on the target, such as neutralization of ligand binding. This technique has great potential, in that initial data on the distribution of protein targets can be backed up rapidly by functional data from cell and tissue-based assays.

1.2 Antibodies are the logical choice as protein-specific molecular probes

Antibodies raised in animals have been indispensable to biochemical research for many decades, and are versatile reagents: they offer specific, sensitive and quantitative detection of proteins in tissues or extracts; they can reveal the subcellular location of individual proteins in histological material; and, by preventing their target proteins from interacting with their natural ligands, can thereby confirm the specific biological functions of those targets. Antibodies have great potential throughout functional genomics and proteomics, because new techniques of antibody generation based on phage display (McCafferty *et al.*, 1990) can be automated and run with a throughput high enough to supply high-quality reagents for the huge numbers of potential targets arising from genome projects. The next section is an overview of the methods that have become the building blocks of an antibody-based functional genomics platform.

1.3 Overview of key technologies

Phage display of antibodies. Functional antibody domains can be expressed on the surface of filamentous bacteriophages (McCafferty *et al.*, 1990; Winter and Milstein, 1991). Large phage-antibody libraries (now $> 10^{11}$ sequences) contain antibodies binding with high affinity to essentially any given antigen (Vaughan *et al.*, 1996). DNA encoding the antibody gene is packaged within the phage particle, linking antibody specificity to sequence. The desired antibody specificity is selected from the library by its affinity for target antigen. Thus, the phage library functions as an *in vitro* immune system. Conventional immunization involves whole animals (at least one per antigen) and requires weeks, and usually multiple immunizations, for the immune response to develop a useful reagent. In contrast, phage display is an *in vitro* system, and the antibody reagent can be obtained within days. Thus, phage display is the most efficient way of obtaining antibodies (McCafferty *et al.*, 1996).

Phage antibodies as reagents. The phage particles themselves, which display functional antibody domains, are used routinely as reagents in ELISA, immunocytochemistry, flow cytometry and other screening techniques. This greatly accelerates the screening process by removing the need for a purified protein reagent.

Antibody generation in high throughput. Using phage display, antibodies can be isolated within one working week. Antigens can be arrayed in microtiter plates, so that antibodies to hundreds of different antigens can be selected manually. Phage display lends itself well to automation, so that antibodies can now be generated at a rate capable of handling the diversity of sequences present in DNA sequence databases.

Antigens: linking gene sequences to protein in 'ProAb'^{TM}. Databases of ESTs are interrogated to identify sequences of interest. Peptide sequences are defined for the particular open reading frame, and corresponding peptide antigens are generated using a 96-well automated synthesizer. These antigens are then used to select antibodies of the desired specificities. Antibody probes for expressed sequences are thus made directly from information in DNA databases, requiring neither full-length DNA sequence nor protein antigen expression.

Antigens: probing uncloned proteins and their neighbors by the ProxiMol^{TM} technique. Phage antibodies specific for proteins on whole cells or tissues can be isolated, but how can the selection against an intact cell, or other multiprotein complex, be focused on a particular target molecule? Provided a guide molecule of any type exists, a technique called 'ProxiMol'^{TM} can be used to recover phages specific for the guide molecule's binding partner and its near neighbors (Osbourn *et al.*, 1998).

Antigens: solving the problem of self-tolerance. Phage display bypasses the immune system. Antibodies to 'self'-antigens, including sequences conserved completely, can be isolated from the library directly (Griffiths *et al.*, 1993). We have never failed to isolate antibodies to human proteins from our libraries of human antibodies.

Screening for expression and cellular distribution. Immunocytochemistry (ICC) gives a uniquely detailed picture of the distribution of proteins in tissues and in cells. The challenge has been to perform ICC at a high throughput, because interpreting the data requires human expertise, and the visual results and interpretation must be made accessible in a user-friendly database.

Bioinformatics drives the process. In the past, individual scientists could readily comprehend the flow of data from low-throughput techniques. Now, no one can handle the quantity and complexity of data being produced without sophisticated bioinformatics tools. Bioinformatics expertise must be involved throughout the discovery process, from EST to ICC pattern.

2. Construction and application of phage antibody libraries

2.1 Display of antibody fragments on bacteriophages

Functional antibody fragments can be displayed on the surface of filamentous bacteriophages as fusions to the *N*-terminus of the minor phage coat protein, g3p. The great majority of phage particles will not display more than one copy of the antibody (McCafferty *et al.*, 1990). The favored format of antibody fragment is single-chain Fv (SC-Fv), a protein composed of the variable regions only of the IgG heavy and light chains (V_H and V_L) connected by a flexible peptide linker. The DNA sequence encoding the antibody domains may be cloned into the phage genome (a phage vector); however, phagemid vectors, which are easier to manipulate and have higher transformation efficiencies, are favored generally. (A phagemid is a plasmid with both a plasmid and a phage origin of replication.) The work described in this chapter employed the phagemid vector pCANTAB6 (McCafferty *et al.*, 1994) and phage M13 (Yanisch *et al.*, 1985). Excellent practical handbooks are available (Kay *et al.*, 1996; McCafferty *et al.*, 1996).

2.2 Construction of antibody libraries

Human lymphoid cells are the ideal source of DNA sequences encoding the V_H and V_L regions of IgGs (*Figure 1*). The domains are amplified by polymerase chain reaction (PCR) using families of primers (Marks *et al.*, 1991). The V_H and V_L gene sequences are then linked together in an assembly PCR to give a complete scFv antibody DNA sequence. V_H and V_L domains are paired randomly, and these non-native pairings can increase functional diversity in the library. The scFv sequences are amplified by PCR using primers incorporating restriction sites that facilitate cloning into the phagemid vector. After ligation into the recipient phagemid, the library is transformed into *E. coli* by electroporation. The phage library proper is obtained by 'rescuing' the culture of the phagemid library in *E. coli* with a helper phage that will package the phagemid DNA (McCafferty *et al.*, 1996).

2.3 Library size and diversity

Specificity and affinity are desirable qualities in any antibody. The larger the library, the greater the likelihood that it will contain an antibody of very high affinity having the desired specificity (Perelson and Oster, 1979). Although the donors of cells from

Construction of 10^{11} Library

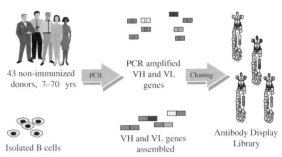

43 non-immunized donors, 7–70 yrs

PCR

PCR amplified VH and VL genes

Cloning

Isolated B cells

VH and VL genes assembled

Antibody Display Library

Figure 1. Principle of constructing a phage antibody library. The DNA encoding the antigen-binding domains of human antibodies is packaged within a bacteriophage particle that also displays the functional antibody protein at its tip. The binding specificity and affinity of the antibody can thus be used to isolate the corresponding antibody DNA.

which human antibody libraries have been constructed are described as 'nonimmunized', humans have large V-gene repertoires, are out-bred, and have been exposed to a huge range of antigens. Human B cells can make antibodies to self-antigens, that is normal human proteins, but those clones would normally be eliminated by the surveillance of the immune system – a restriction that cannot apply in phage display. Human antibody libraries readily yield clones recognizing any human protein tested so far. A library (Vaughan et al., 1996) of more than 10^9 clones was used as a source of scFv to a diverse range of antigens, and affinities were determined systematically. It was clear that the library contained scFv with affinities better than 10^8 M^{-1}, and into the 10^{10} M^{-1} range, for a large diversity of antigens. This work demonstrated that the phage technology delivered antibodies with affinities as high as those generally considered to be the best obtainable from immunization (Foote and Eisen, 1995). As libraries have increased in size, statistical proof of the expected increase in antibody quality has become more laborious, though experience seems to show that very high affinities are now obtained more readily than they were from smaller libraries. Moreover, measuring affinities tighter than 10^{10} M^{-1} becomes challenging technically. Currently, the largest library of human antibodies known to the authors exceeds 10^{11} different sequences. Construction of such a library is not a trivial task.

2.4 Selection of specific antibodies from a phage library

The basic principle is that of affinity capture. Since the DNA encoding the antibody sequence is packaged in the phage particle that displays the scFv specificity, the protein and its gene are coselected.

Selections on purified antigen. The most widely applicable selection technique is 'panning'. Antigen is coated onto a solid surface, such as the well of a microtiter plate, and the library of phage antibodies is incubated with the target. The surface is washed, leaving behind those phages that display antibodies binding to the target antigen. Specific phages are propagated by being infected into *E. coli*; the bacterial culture is spread upon a culture plate so as to produce bacterial colonies that have arisen from a single cell. Each colony then makes a monoclonal antibody.

Selections against complex antigens. Antibodies from phage libraries have frequently been isolated by selections on whole cells or subcellular fractions such as membrane preparations. Monoclonal scFv will still be obtained, but the antigen complexity will cause difficulty in determining the specificity of the output. The ProxiMol™ technique (see below) was invented to overcome this difficulty.

2.5 High-throughput selections

The procedures by which phage antibodies are selected from the library are simple and take place in small volumes of liquid ($100\,\mu l$) and can therefore be carried out in parallel on multiwell microtiter plates. A different antigen may be coated onto each well, and early manual experiments obtained specific antibodies to 83 of 93 antigens per plate. Most pharmaceutical research laboratories use generic robots to handle microtiter plate reactions in the standard 96- and 384-well formats. Similarly, commercial 'off-the-shelf' workstations (for example, from Tecan and GRI) can now be used to isolate specific antibodies to up to 200 different antigens per week.

2.6 Phage antibodies as reagents

Phage particles displaying scFv are highly useful research reagents. Discovery programs dealing with targets on a genomic scale clearly need reagents generated at the highest possible throughput, and phage antibodies meet this need. We will describe subsequently how soluble antibody proteins can readily be obtained from the phage display system; however we should emphasize that the phage format works well in all upstream screening assays. Phage antibodies have routinely been successful in: ELISA on protein or peptide antigens; ELISA on whole cells or subcellular fractions; flow cytometry; and immunocytochemistry. The phage particle has almost 3000 copies of the major coat protein g8p, so antiphage antiserum detects phage binding to its target with high sensitivity. Supernatants from *E. coli* cultures in 96-well plates, following inoculation with helper phage, can be used directly as reagents (see McCafferty *et al.*, 1996 for protocols).

2.7 Soluble antibodies from phage libraries

Applications of scFv. The fastest route to obtain data is to use phage antibodies throughout screening, and this is the normal practice. Nevertheless, it is often better to use soluble scFv protein in experiments downstream of the target discovery phase. Examples include: measurements of affinity and kinetics; cell-based neutralization assays; and *in vivo* studies. There may be other applications where the large size of the phage particle might impede its diffusion to antigen. Furthermore, the size of the phage limits the maximum antibody concentration to around 10 nM (6×10^{12} phage/ml). ScFv are well-proven reagents for proof-of-principle studies in cell biology either *in vitro* or *in vivo*. The following notes show how amounts adequate for experimental purposes (micrograms to milligrams) can be prepared simply.

Preparation of scFv in high throughput. Enough scFv for screening can readily be expressed without departing from the multiwell plate format. To induce soluble scFv expression into the bacterial periplasm, clones of cells containing the pCANTAB plasmid are incubated with IPTG. A proportion of scFv tends to leak out of the cells into the culture medium, especially after overnight induction. After centrifuging the plate, the supernants can be used in ELISA. This very simple protocol has been found to be adequate for screening assays, over many years of experience (McCafferty *et al.*, 1996).

More scFv can generally be obtained from periplasmic extracts, released from the cells by standard procedures. The periplasmic extract can be used directly in many assays. However, the pCANTAB vector adds a hexahistidine tag to the *C*-terminus of the scFv, allowing easy purification by immobilized metal affinity chromatography (IMAC). Magnetic IMAC beads (Qiagen) allow fast and easy purification of scFv in 96-well format.

Where 100 μg to milligram amounts of individual scFv are required (such as for *in vivo* studies) cultures are scaled up into shake flasks. The IMAC-purified product is adequate for most assays; when necessary, 90% purity can generally be achieved by simple gel filtration chromatography.

2.8 Expression as IgG

Whole IgG molecules have a much longer pharmacokinetic half-life than scFv, and are very stable proteins. Thus, IgG may be the appropriate format for *in vivo* or long-term *in vitro* studies. The phage display format allows facile cloning into IgG vectors, so that the research reagent could be directly on the path to manufacture of therapeutic human IgG. There are numerous well-proven methods for expressing micrograms to grams of human IgG from mammalian cell cultures. This makes a 'gene to clinic' program realistic.

3. Phage antibodies in functional genomics

3.1 The 'ProAb'^{TM} approach

Large resources supporting genomic sequencing projects and large proteomic databases have been committed mainly because it is believed that the investment will be repaid by the discovery of new targets for the next generation of drug discovery programs. From the point of view of the pharmaceutical industry and of medical research, the most interesting information about any new protein would be that, in a specific disease state, its abundance changed dramatically in particular tissues. Ideally, researchers would like to correlate gene expression with disease indications in high throughput. This is the theme underlying the ProAb project. A schematic of the process is shown in *Figure 2*. The strategy is to translate ESTs into synthetic peptide antigens, which are used to select phage antibodies that can detect protein antigens in immunocytochemistry.

ProAb ™ Summary

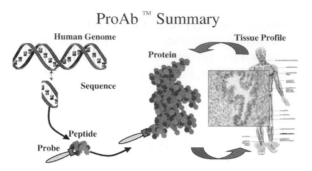

Figure 2. Principle underlying the 'ProAb'™ approach to functional genomics. Interesting segments of DNA, mined from a database of ESTs, are translated into protein. Peptides are synthesized chemically to represent the protein sequence. The peptides are then used as antigens against which phage antibodies are selected. The antibodies are used as detection reagents in immunocytochemistry, thereby revealing the distribution and abundance of the gene product in normal and diseased tissues.

Thus, the ProAb project takes the most direct path possible from EST to disease association. The account below describes the scientific and technical steps involved in implementing this approach in high throughput.

3.2 Making sense of ESTs by bioinformatics

ESTs are currently the richest source of DNA sequence information. There are now more than 10^6 ESTs in databases such as dbEST (Boguski *et al.*, 1993), so that (in theory) human open reading frames should statistically be represented in EST databases with several-fold redundancy. However high the throughput of the discovery process it will clearly be necessary to prioritize the choice of ESTs to enter that process. An excellent strategy is to focus on a known protein with interesting biological function (and, ideally, a known structure), and to search for family members (see Section 3.6). Having found the ESTs of interest, antigenic peptides must be chosen. Although it is possible to cover the entire sequence with overlapping peptides, a more selective approach can improve the success rate. A decision can be taken based on the weighted scores from algorithms designed to predict the probability: of adopting a native-like structure; of being in a solvent-accessible region of the native protein; and of having sequence features that might promote good antibody binding. The first step at which bioinformatics will be involved in the program, therefore, is in the identification of specific peptide sequences that should represent the protein of which the EST was a part.

3.3 Synthesizing antigens

Predicted antigenic peptide sequences are downloaded to an automated 96-well solid-phase peptide synthesizer (Advanced Chem Tech) that uses fmoc chemistry at micromole scale (5–100 mg product). At this stage, sequences may be searched to record the minority that (from experience) could be problematical to synthesize.

Usually, 15-residue antigenic peptides have been found suitable, but in fact much longer peptides (e.g. 30 amino acid residues) are also being made in high yield by automated synthesis. Some post-translational modifications such as phosphorylations can also be made in the synthetic peptides. At the conclusion of synthesis, peptides are deprotected, cleaved from the beads and cleaned free from reagents by precipitation. Washed peptide precipitates are redissolved and every peptide is subjected to quality control by reversed-phase liquid chromatography (LC) and electrospray mass spectrometry (Gilson HPLC/Micromass Platform-LC), to ensure that only those more than 85% pure will be used as antigens. The LC-MS instrument is linked to the bioinformatics system so that all peptide sequences are checked automatically against predicted masses. Antigenic peptides are often synthesized with an *N*-terminal cysteine residue, which is convenient for coupling to carrier protein. This helps to ensure that truncated products will not contaminate the antigen (peptides grow from the *C*-terminus during synthesis). *N*-terminal biotinylation is sometimes used for the same reason. The carrier protein is usually bovine serum albumin, activated at lysine residues using a bifunctional succinimide–maleimide reagent. SDS–PAGE and MALDI mass spectrometry are used on a proportion of conjugates to check the efficiency of coupling. Typically, a median coupling ratio of 20–25 peptides per protein molecule has been favored. It is worth noting that unprecedented numbers of high-quality peptides (up to 300 per week) can be synthesized using a single commercially available instrument. This has been achieved by software optimization, optimization of coupling times, careful selection of chemistries and careful choice of quality reagents.

3.4 High-throughput selection of phage antibodies to peptide antigens

The basic selection method is panning, as described above (Section 2.4). Peptide–protein conjugates are coated onto 96-well microtiter plates. The procedures for selection of specific antibodies are essentially as described above (Sections 2.4 and 2.5), but all the processes, through selection, elution of bound phage, infection spreading, colony-picking, arraying clones into 96-well plates, and screening for peptide-specific antibodies by ELISA, are run automatically on commercial robots. This means that the generation of antibodies will never be the rate-limiting step. Typically about 12 antibody clones are picked per peptide sequence. Connection of ELISA-positive clones to EST is maintained by the bioinformatics system by means of barcoding. At the end of the antibody selections, the *E. coli* carrying each ELISA-positive clone are archived as glycerol stocks at −70°C.

3.5 High-throughput immunocytochemistry

Immunocytochemistry should be the optimum technique for demonstrating association of antigen with disease. Unlike techniques such as dot-blots or even 2-DE gels, where the tissues must be ground up and solubilized, immunocytochemistry preserves the distribution of proteins within the tissue. This reveals protein distribution at cellular resolution and can demonstrate differences that would be missed altogether by the averaging effect of homogenizing together the various cell types within the tissue. The question might be asked, why is such an information-rich technique not used universally? The answer presumably lies in the perceived difficulty of operating the technique at high throughput. Indeed, many hurdles had to be overcome to achieve a rate of 1000 samples per month. Particular effort was put into the following areas.

Access to tissues. It is vital to have links with clinical centers maintaining diverse tissue banks; the tissues also need to be preserved promptly post-mortem. However, a protein-directed reagent could detect antigen at times post-mortem when little mRNA target sequence might remain.

Optimal processing. To increase the throughput, carefully chosen samples of a panel of different tissues are dissected and encapsulated together before sectioning. This greatly increases the parallelism of the process and, apart from increasing speed, facilitates cross-tissue comparisons by ensuring that the different samples within the block are all processed identically. To preserve antigenicity, cryosections have generally been used in preference to paraffin-embedded material; mild fixation using cold acetone has generally been preferred to cross-linking fixatives.

Use of phage antibodies in ICC. Phage antibodies have been found to give sensitive detection in ICC, with good spatial resolution. Supernatants of clones grown in 96-well format are used directly to label tissue sections.

Screening of ICC samples. Currently there is no substitute for human expertise when interpreting staining patterns. A team approach is essential; interesting patterns, picked out during prescreening, must be examined by experienced histochemists; there must also be a traceable system of cross-checking the interpretation.

Recording data. Images are best captured by digital cameras attached to the microscopes, and archived on an integrated database. Images by themselves are of little use without expert interpretation, which can only be provided by experienced cellular pathologists.

3.6 Bioinformatics

We have stressed throughout that bioinformatics has been fundamentally important at all stages of the ProAb™ program. No existing software was suitable, so a bespoke suite of software had to be written in-house. It may be useful to summarize the features that were found to be essential.

Bioinformatics in the discovery phase. We have indicated above, how each stage in the experimental strategy is controlled and monitored by the bioinformatics system. This is summarized in the flowchart below:

Choose and design peptide antigen
⇓
Control peptide synthesis
⇓
Log peptide–protein conjugates
⇓
Record peptide-positive clones from high-throughput selections
⇓
Archive antibody clones
⇓
Track antibody DNA sequences
⇓
Track samples into ICC
⇓
Record images and verbal interpretations from ICC

Choose and design peptide antigen. A statistical representation of a family of related sequences can be used as the template against which to search EST databases, by means of a search engine using a Hidden Markov Model. This is a particularly sensitive way of finding ESTs likely to represent new family members. The application used is 'Ptolemy'.

Record images and verbal interpretations from ICC. Images are archived using digital cameras. The histopathologist's interpretation is recorded and transcribed by voice-recognition software ('Plato') and then edited into a system of annotations that accompany the picture in the database.

Bioinformatics and data analysis. The product of the ProAb™ program is a database from which researchers can gain biological insights, based on the patterns of association of EST sequences with cellular pathology. This information must be organized into a database that is friendly to nonspecialists.

Applications framework. Java software (known as 'Continuity') ensures that information and procedures in all components within the database can be accessed on any hardware platform. Within the shell provided by Continuity, the components communicate using CORBA.

User interface. The interface to the ProAb™ database is called 'Voyager'. The general user can use a query form in Voyager to obtain specific data. Users wishing to explore the data in depth need a more interactive interface. One useful device is to reduce ICC data to a 'virtual tissue dot blot'. Staining is classified into four cellular regions: membrane; cytoplasmic; connective tissue; and nonconnective tissue. These regions are represented as color-coded dots, and the intensity of color is graded to reflect the histologists' evaluation of the intensity of staining.

ProAb^{TM} database. Within the Continuity framework and accessible via the user-friendly Voyager interface is the ProAb^{TM} database proper. This is an Oracle 8 database, which integrates all the experimental data including: digital image files; analysis of staining patterns; histological interpretations; data trail from EST through peptide design, synthesis and quality-control data for each individual sequence; antibody isolation data; and links to SWISS-PROT and dbEST databases.

3.7 *ProAb^{TM} results*

Several well-known proteins have been used as internal controls in the ProAb process. This validates the efficiency of the process in identifying correct expression patterns for ESTs.

Example 1. A segment of an EST (dbEST number 845032) was translated as VRSSSRTPSDKP; phage antibodies were selected against the corresponding synthetic peptide antigen; immunocytochemistry revealed strong staining in tonsil (*Figure 3a*). A BLAST search of dbEST matched the peptide sequence to a cDNA sequence from a library (id 595) made from human tonsillar cells, which had been enriched for germinal center B cells by flow sorting. The depositor had given this sequence a putative assignment as human tumor necrosis factor (TNFα) precursor. The antipeptide antibody was indeed found to react specifically with native human TNFα in ELISA.

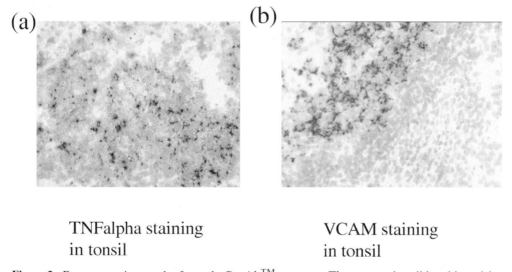

(a)

(b)

TNFalpha staining
in tonsil

VCAM staining
in tonsil

Figure 3. Representative results from the ProAb^{TM} program. The process is validated by raising phage antibodies against peptides representing known protein antigens. Immunocytochemistry demonstrates cell and tissue distribution of antigen consistent with data from the literature. (a) Specific staining of TNFα by phage antibodies raised against the peptide VRSSSRTPSDKP; (b) specific staining of VCAM by phage antibodies to the peptide ESRKLEKGIQVEIY.

Example 2. A peptide sequence ESRKLEKGIQVEIY was used to represent the human vascular cell adhesion molecule (VCAM-1), which corresponds to a sequence from EST 20523 from a spleen cDNA library; this EST was annotated as 'similar to vascular cell adhesion molecule 1'. Phage antibodies selected against this peptide stained germinal center of tonsil, consistent with the reported occurrence of VCAM-1 on dendritic cells (*Figure 3b*).

4. Phage antibodies in proteomics – the ProxiMol^{TM} technique

4.1 Conventional selection of phage antibodies to membrane proteins

Selection of antibodies that bind to membrane proteins has always proved challenging whether accomplished by phage display or by *in vivo* methods. This is largely because such proteins are particularly difficult to purify to use as immunogens. Simple transmembrane proteins can often be expressed as soluble molecules, which have successfully been used in phage antibody selections. In the absence of soluble cloned protein, native membrane protein must be purified, but this is notoriously difficult to accomplish without some protein degradation. Integral membrane proteins, such as 7-transmembrane (7-TM) chemokine receptors, represent another level of difficulty: they are impossible to express in a soluble secreted form, and are difficult to purify in quantity and without denaturation, whether from native or recombinant sources. Current approaches assume the availability of at least biochemical information about the target protein, if not its identity.

It has sometimes been possible to obtain useful antibodies using unpurified mixtures of antigens, and this has often been attempted. Whole cells can be used directly as immunogens; this transfers the problem downstream to screening. Very large numbers of antibodies have to be screened, with no guarantee of target specificity because the antibodies were selected randomly. One way of increasing the targeting of specific cell surface molecules has been to 'deselect' on common cell surface proteins. This may be accomplished by mixing a small amount of cells expressing the desired cell surface protein with a large number of cells that do not. The target cells of interest are tagged by adding cell-type specific antibodies. The phage antibody library is added to the cell mixture, and the target cells with attached phages are then isolated by means of the cell-type-specific antibody. This approach assumes the availability of antibody specific for the target cell population and is certainly applicable for well-characterized subpopulations of cells such as blood cells.

4.2 Guided selection to improve membrane protein selections

If the phage antibodies in the library could somehow be guided to the molecules of interest, then much more successful selections could be performed against proteins that are only available in complex mixtures, notably membrane proteins such as receptors. An ideal method would utilize a receptor's cognate ligand as the guide molecule, and also as a means of confirming the bioactive conformation of the target molecule. Phage binding in the proximity of the guide molecule would thus be isolated, thereby enriching the selected population with phage specific for the target receptor. This is the principle of the ProxiMol^{TM} method.

4.3 ProxiMol^{TM} method and applications

Catalyzed reporter deposition. In immunocytochemistry and other techniques, signals are often amplified by catalyzed reporter deposition (Bobrow *et al.*, 1989, 1992). Horseradish peroxidase-conjugated reagents are bound, either directly or indirectly, to a target antigen. In the presence of hydrogen peroxide and biotin tyramine, horseradish peroxidase (HRP) will catalyze the formation of short-lived biotin tyramine free radicals, which will covalently bind to and biotinylate proteins in close proximity to the horseradish peroxidase. The deposited biotin molecules can then be detected by streptavidin-coupled reagents, such as horseradish peroxidase, and the original signal is thus amplified since the newly deposited biotins outnumber the original target or horseradish peroxidase molecules.

Principle of the ProxiMol^{TM} method. Catalyzed reporter deposition has been adapted to allow the isolation of phage antibodies binding in close proximity to a biotinylated guide molecule in the process called the ProxiMol^{TM} method (Osbourn *et al.*, 1998). There are two requirements that must be met by a guide molecule. First, it must have affinity either for the target protein itself or for a molecule in close proximity to the target. Second, it must be capable of being associated with HRP, either directly or indirectly. Examples of indirect coupling to HRP include binding of an HRP-coupled antibody to the guide molecule, or biotinylation of the guide molecule and subsequent binding of streptavidin-HRP. The guide molecule must also retain sufficient biological activity to preserve its association with the target (hence, antibodies that compete with the guide ligand for receptor-binding cannot be used to attach HRP to the ligand).

The ProxiMol^{TM} method in practice. Use of the ProxiMol^{TM} method for selections of target antigens *in situ* is straightforward (*Figure 4*). For a cell surface antigen, the guide molecule (usually either a biotinylated target ligand or an HRP-conjugated antibody) is added to the intact cells, together with phage from the $> 10^{11}$ human antibody library. Following incubation, unbound phage is washed off. HRP-conjugated secondary reagent is added if necessary and incubated further, before washing again. Biotin tyramine and hydrogen peroxide are added, incubated for 10 min and then washed. Bound phages are eluted, usually with triethylamine (TEA) and biotinylated phage isolated using streptavidin-magnetic beads, which are used to infect *E. coli*. Screening will reveal phage antibodies specific to the target of interest, plus phages that bind to other proximal proteins or to the guide molecule itself.

Scope of ProxiMol^{TM} technique applications. Selections using the ProxiMol^{TM} technique have been performed on a range of different targets and different protein formats. The guide molecules used, in addition to antibodies and natural protein

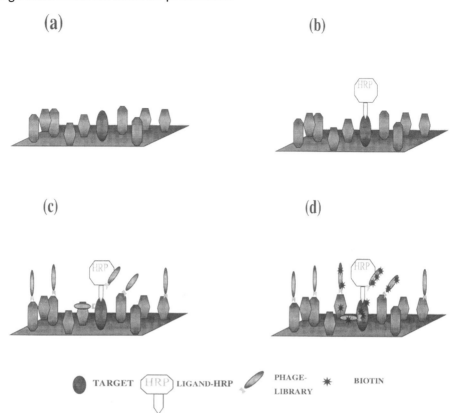

Figure 4. The 'Proximol'™ method of guiding antibody selection. (a) The antigen of interest in its normal environment on the cell surface. (b) A guide ligand is conjugated to HRP and bound specifically to the ligand. (c) The phage antibody library is added and diverse antibodies bind specifically to the ligand, to the target, and to all other cell surface markers. (d) Biotin-tyramine reagent is added and a reactive biotin species is generated; phages binding within about 25 nm of the guide ligand are covalently labeled with biotin. These phage antibodies, recognizing the target and near neighbors, are recovered using streptavidin-coated magnetic beads.

ligands such as cytokines, include carbohydrate ligands and conjugated fatty acids. The target proteins have been on intact cells, both adherent and in suspension, frozen tissue sections, and isolated membranes in addition to purified proteins.

Range of biotin deposition. The biotin molecules are deposited in close proximity to the HRP molecule due to the short half-life of the biotin tyramine free radicals. To establish the range of biotin deposition, anti-gene 3 antibodies were used as the guide molecule to biotinylate phage m13. Deposited biotin molecules were detected by electron microscopy using streptavidin-coated 5 nm gold beads. The range of biotin deposition was up to 25 nm, with an average distance of 10–15 nm. This gives us confidence that deposition of the biotin occurs within a few protein diameters of the guide molecule.

The ProxiMol^{TM} technique for epitope walking: enlarging a repertoire of clones that bind to TGFβ1. Previous selections against TGFβ1 identified a dominant epitope/clone called 31G9. The ProxiMol^{TM} method was used to expand the repertoire of anti-TGFβ1 phage antibodies and to target new epitopes. The ProxiMol^{TM} method selection was carried out on TGFβ1 immobilized to a BIAcore chip; the guide molecule was HRP-conjugated 31G9. Library phages were bound as above and unbound phages were washed off before addition of hydrogen peroxide and biotin tyramine. The resulting biotinylated phages were isolated and used to infect *E. coli.* Following two rounds of selection, 13.5% of clones isolated were positive for TGFβ1 and all were novel clones. In contrast, in the absence of a guide molecule, all positive clones were identical to 31G9 (Osbourn *et al.*, 1998). Thus, ProxiMol^{TM} method selections can allow antibody repertoires to be expanded to purified protein targets by allowing the selection of phage antibodies to nondominant epitopes.

The ProxiMol^{TM} method selection of antibodies to specific cell-surface adhesion molecules. The most powerful applications of the ProxiMol^{TM} technique are in the selection of antibodies to specific cell-surface molecules. In order to target selectins, adhesion molecules expressed on the surface of activated endothelial cells, both selectin ligand and antibody, were used.

Human umbilical vein endothelial cells (HUVEC) were activated with TNFα, to induce the cell surface expression of P- and E-selectins. The ProxiMol^{TM} method selection was carried out on intact cell monolayers using sialyl Lewis X, the ligand for E- and P-selectins, as the guide molecule. Following two rounds of selection, 13.7% of clones were positive for E-selectin, while 6.3% of clones were positive for P-selectin (Osbourn *et al.*, 1998). In contrast, when anti-E-selectin-HRP was used as the guide molecule, only phages specific for E-selectin were isolated. Thus, it is possible to target native membrane-bound antigens *in situ* on whole cells, by ligand-specific guided selection.

The ProxiMol^{TM} method to obtain antibodies to a cell surface 7-TM receptor. Transmembrane proteins that are deeply embedded in the plasma membrane, such as those with seven transmembrane domains (7-TM) are very difficult to purify in a native state and thus represent an ideal target for *in situ* selections. The chemokine receptor CCR5 is expressed on T lymphocytes and has been identified as a co-receptor for HIV entry. ProxiMol^{TM} method selections were performed on human T lymphocytes using biotinylated MIP-1α, a CCR5 ligand, as the guide molecule. MIP-1α and phages were incubated with cells, then streptavidin-HRP was added and the ProxiMol^{TM} method carried out as above. The output from the first round of selection was analyzed by ELISA on whole cells and on protein. Out of 95 clones chosen randomly, 30 recognized T cells. Of these, 13 bound to CCR5, and the remaining 17 bound to other T cell surface epitopes that may represent CCR5-associated or proximal molecules. Some of these antibodies have been used to screen cDNA expression libraries and have yielded several potentially interesting proteins, including a phosphotyrosine phosphatase. Thus, the ProxiMol^{TM} method selection has enabled us not only to target a specific cell surface receptor, but has also allowed us to identify other proteins potentially associated with this receptor.

Step-back selections. Use of a receptor ligand as the guide molecule for ProxiMol^{TM} method selections limits the isolation of antibodies to those that do not bind at the ligand binding site, and that do not neutralize the receptor–ligand interaction. To overcome this limitation, a two-stage selection method may be utilized. The ProxiMol^{TM} method is performed as usual with the receptor ligand, and biotinylated phage is recovered. Rather than using this phage to infect *E. coli*, it is used as the guide molecule in a second round of the ProxiMol^{TM} method with a fresh aliquot of library. This enables phage to be isolated that 'step-back' into the binding site of the original guide molecule/ligand. This necessarily yields a very diverse output as all phages binding near to the entire output of the first round are selected. Screening of the first round of ProxiMol^{TM} method output in order to identify receptor-binding phage, and subsequent use of this phage as the guide molecule for the step-back selection, can increase the specificity of the selection for the target.

CCR5 step-back selections were performed using MIP-1α, as above, as the guide molecule for the first ProxiMol^{TM} reaction. The phage output from this selection was then used as the guide molecules for the step-back selection. Clones that bound CCR5 were identified by ELISA. Clones that inhibited the binding of MIP-1α were identified by flow cytometry using biotinylated MIP-1α and scFv (*Figure 5a*). A clone that was identified as inhibiting MIP-1α binding was also shown to inhibit the MIP-1α-induced calcium flux in CCR5 expressing cells (*Figure 5b*). Thus, with a simple two-step selection strategy, antibodies were targeted specifically to the ligand-binding site of a cell surface 7-TM receptor.

4.4 ProxiMol^{TM} method applications in proteomics research

Use of antibody clones from ProAb^{TM} output as 'tags' to guide new selections. Phage antibodies identified as giving interesting profiles in ProAb^{TM} screening are unlikely themselves to have activity in a biological assay. These antibodies are, however, useful 'tags' for disease-specific proteins or epitopes, and as such can be used as guide molecules in ProxiMol^{TM} method selections. Because the ProAb^{TM} database has indicated tissue specificity for the phage antibodies, appropriate tissue sections can be used as the source of antigen for ProxiMol^{TM} method selections. As guide molecule, the simplest approach is to biotinylate the ProAb^{TM} phage antibody; alternatively, biotinylated scFv could be used.

Screening of the ProxiMol^{TM} technique output can be tailored to the desired biological activity of the antibody. The lowest-throughput screen would be on tissue sections; however, a staining pattern that matched that of the original ProAb^{TM} antibody could be identified. A higher-throughput screen would utilize ELISA or flow cytometry on a defined cell type. This screening strategy could also utilize cells at defined stages of activation, possibly giving an insight into the state of cells present in the original tissue section. Phage or scFv can then be utilized directly in cell-based

(a)

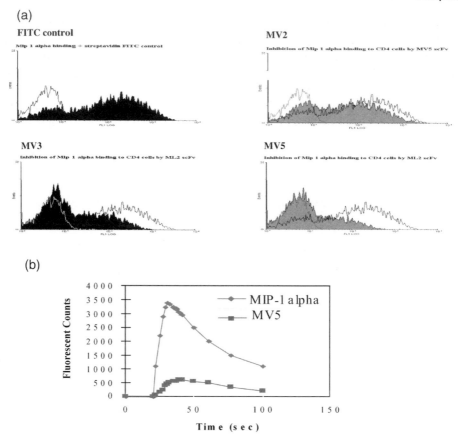

(b)

Figure 5. Biologically active antibodies from 'step-back' ProxiMol^{TM} method selections. (a) ScFv from step-back selections inhibit MIP-1α binding to CD4-positive cells in flow-cytometry. The shaded area of each peak represents MIP-1α binding to the cells (revealed by addition of biotinylated MIP-1α to the cells, followed by addition of streptavidin labeled with fluorescein isothiocyanate, FITC). The unshaded area of each peak represents the signal obtained when streptavidin-FITC is added in the absence of the ligand, biotinylated MIP-1α. Compared with the control (top left panel), binding of MIP-1α is inhibited by the three scFv clones – MV2, MV3 and MV5. (b) The scFv MV5 blocks calcium signaling induced by MIP-1α binding to its receptor, CCR5; 50 nM MIP-1α was added to a CCR5-transfected CHO cell line, and the rise in intracellular free calcium ion concentration was monitored by fura-2 fluorescence using standard methods.

assays *in vitro*. In this way, ProAb^{TM} antibodies can be used to isolate biologically active antibodies, and possibly to define the activation state of the target cells.

The ProxiMol^{TM} technique to identify new members of protein or protein/DNA complexes. The biotin-labeling step in the Proximol^{TM} technique has a range equivalent to three or four protein diameters. Thus, the Proximol^{TM} technique usually generates reagents to other proteins near the target. The work on CCR5 (above) is a good example; when MIP-1α was used as guide molecule, phage antibodies were obtained that did not recognize CCR5 itself. Some of these antibodies were used to screen an expression library, and were found to recognize proteins that might be expected to exist in the membrane in proximity to CCR5. This led to the identification of several potentially interesting proteins, including a phosphotyrosine phosphatase. Thus, ProxiMol^{TM} method selection has enabled us not only to target a specific cell surface receptor, but has also allowed us to identify other proteins putatively associated with this receptor. This raises the interesting possibility that maps of multiprotein complexes could be drawn, based on successive rounds of ProxiMol^{TM} method selections and screening.

Acknowledgments

The work described above depended on the team efforts of more than a score of scientists at Cambridge Antibody Technology. We wish to thank the following individuals for helpfully supplying specific expertise towards writing this article: Simon Brocklehurst and all the BioInformatics team; George Masih for his achievements in peptide synthesis; John McCafferty, Tony Pope and all members of the MAb isolation team; and Tony Warford for expert direction on immuno-cytochemistry.

References

Adams, M.D., Kelley, J.M., Gocayne, J.D., Dubnick, M., Polymeropoulos, M.H., Xiao, H., Merril, C.R., Wu, A., Olde, B., Moreno, R.F. *et al.* (1991) Complementary DNA sequencing, expressed sequence tags and human genome project. *Science* **252**: 1651–1656.

Barbas, C.F., Kang, A.S., Lerner, R.A. and Benkovic, S.J. (1991) Assembly of combinatorial antibody libraries on phage surfaces: the gene III site. *Proc. Natl Acad. Sci. USA* **88**: 7978–7982.

Bobrow, M.N., Harris, T.D., Shaughnessy, K.J. and Litt, G.J. (1989) Catalyzed reporter deposition, a novel method of signal transduction. Application to immunoassays. *J. Immunol. Methods* **125**: 279–285.

Bobrow, M.N., Litt, G.J., Shaughnessy, K.J., Mayer, P.C. and Conlon, J. (1992) The use of catalysed reporter deposition as a means of signal amplification in a variety of formats. *J. Immunol. Methods* **150**: 145–149.

Boguski, M.S., Lowe, T.M. and Tolstoshev, C.M. (1993) dbEST – database for "expressed sequence tags". *Nature Genet.* **4**: 332–333.

Foote, J. and Eisen, H.N. (1995) Kinetic and affinity limits on antibodies produced during immune responses. *Proc. Natl Acad. Sci. USA* **92**: 1254–1256.

Griffiths, A.D., Malmqvist, M., Marks, J.D., Bye, J.M., Embleton, M.J., McCafferty, J., Baier, M., Holliger, K.P., Gorick, B.D., Hughes-Jones, N.C., Hoogenboom, H.R. and Winter, G. (1993) Human anti-self antibodies with high specificity from phage display libraries. *EMBO J.* **12**: 725–734.

Kay, B., Winter, J. and McCafferty, J. (1996) *Phage Display of Peptides and Proteins: a Laboratory Manual.* Academic Press, London.

Marks, J.D., Hoogenboom, H.R., Bonnert, T.P., McCafferty, J., Griffiths, A.D. and Winter, G. (1991) By-passing immunization: human antibodies from V-gene libraries displayed on phage. *J. Mol. Biol.* **222**: 581–597.

McCafferty, J., Griffiths, A.D., Winter, G. and Chiswell, D.J. (1990) Phage antibodies: filamentous phage displaying antibody variable domains. *Nature* **348**: 552–554.

McCafferty, J., FitzGerald, K.J., Earnshaw, J., Chiswell, D.J., Link, J., Smith, R. and Kenten, J.

(1994) Selection and rapid purification of murine antibody fragments that bind a transition-state analog by phage display. *Appl. Biochem. Biotechnol.* **47:** 157–173.

McCafferty, J., Hoogenboom, H.R. and Chiswell, D.J. (1996) *Antibody Engineering – a Practical Approach.* IRL Press, Oxford.

Osbourn, J.K., Derbyshire, E.J., Vaughan, T.J., Field, A.W. and Johnson, K.S. (1998) Pathfinder (ProxiMol) selection: *in situ* isolation of novel antibodies. *Immunotechnology* **3:** 293–302.

Perelson, A.S. and Oster, G.F. (1979) Theoretical studies of clonal selection: minimal antibody repertoire size and reliability of self non-self discrimination. *J. Theor. Biol.* **81:** 645–670.

Vaughan, T.J., Williams, A.W., Pritchard, K., Osbourn, J.K., Pope, A.R., Earnshaw, J.C., McCafferty, J., Hodits, R., Wilton, J. and Johnson, K.S. (1996) Human antibodies with sub-nanomolar affinities isolated from a large non-immunized phage-display library. *Nature Biotechnol.* **14:** 309–314.

Winter, G. and Milstein, C. (1991) Man-made antibodies. *Nature* **349:** 293–299.

Yanisch, P.C., Vieira, J. and Messing, J. (1985) Improved M13 phage cloning vectors and host strains: nucleotide sequences of the M13mp18 and pUC19 vectors. *Gene* **33:** 103–119.

Chapter 13

Glycobiology and proteomics

Nicolle H. Packer and Lucy Keatinge

1. Introduction

The development of proteomics has had and will continue to have a dramatic impact on the study of diverse biological systems. Two-dimensional gel electrophoresis has emerged as the method of choice to explore the parallel effects of changes on the protein complement of cells. It is now possible to identify and characterize femtomoles of protein from gels and blots quickly, accurately and in a high-throughput manner. What is striking is the slow rate at which this approach has been translated into the concomitant analysis of the subset of the cell's proteins which are glycosylated. One of the distinguishing features between the genome and proteome in eukaryotic cells is that most of the gene products are post-translation-ally modified and that glycosylation is a predominant modification. It is becoming clear that just identifying a protein from the amino acid sequence that is derived from the known DNA sequence may often be insufficient to associate a protein with a particular function. The modifications that occur to a protein both during and after its expression are increasingly implicated in protein function, and hence analysis of these modifications is critical to the understanding of the different protein expression patterns which are now readily observable by proteomic technologies.

The changes in glycosylation that are observed are a reflection of the activity of the glycosyl transferases that synthesize the oligosaccharides and attach them to the protein, and the glycosidases which trim and degrade the glycans. If it is to be useful, the aim of glycoproteomics must be to identify the glycoproteins that are markers of a particular phenotype, to characterize the specific changes in glycosylation that correlate with a change in phenotype and to target the metabolic step(s) that are responsible for this change.

A comparison of the two-dimensional electrophoresis (2-DE) separation of human serum proteins using Coomassie blue to visualize total protein and a new fluorescent glycoprotein visualization technique that we have developed (see legend to *Figure 1*) shows that most of the abundant proteins are glycosylated and appear as trains of glycoforms differing in pI and/or molecular mass. There is a general acceptance that we need to know why there is so much heterogeneity in the glycosylation of proteins and that we need to be able to analyze these differences. Despite this, comparatively little attention has been paid to the analysis of the glycan component of proteins in the context of the parallel processing used in proteomics.

Proteomics, edited by S.R. Pennington and M.J. Dunn
© 2001 BIOS Scientific Publishers, Oxford.

(a)   (b)

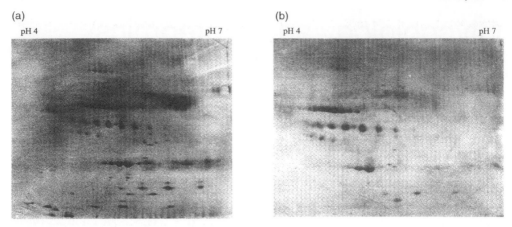

Figure 1. Fluorescent staining of glycoproteins on PVDF membrane: (a) Coomassie blue stain; (b) fluorescein carbohydrate stain. One micoliter of serum was loaded by rehydration, separated by 2-DE and electroblotted to PVDF membrane. The fluorescent stain used periodate oxidation of the sugars as contained in the Bio-Rad glycoprotein detection kit, but used a solution of 0.05 mg/ml fluorescein semicarbazide hydrazide (Molecular Probes) in 50% (v/v) methanol in 100 mM sodium acetate pH 5.5, instead of sandwich antibody color development. Fluorescence was visualized with a Fluor S imager (Bio-Rad).

There are two reasons, we believe, for this delay in focusing on the glycoprotein complement of the cell – firstly, although there is substantial literature on the changes in glycosylation that occur during such processes as development and disease, we have yet to gain an understanding as to how these glycosylation changes are reflected in the biology of an organism. Secondly, the analytical methods for studying glycans have traditionally been based in the laboratories of carbohydrate chemists and as such have given the impression of requiring specialized knowledge and skills, which protein chemists have been reluctant to exploit. This chapter will try to address these two issues in the context of the science of proteomics.

To establish the role of glycosylation in the metabolism of an organism it is not sufficient just to characterize the glycan structures on the proteins. Hence, although analysis of the changes in glycosylation of a protein can be useful for the diagnosis or monitoring of metabolic alterations, such as those that occur in disease, these changes do not give an insight into the function that the glycan confers on the protein. Furthermore, the removal of the glycan from a glycoprotein does not always affect the observed activity of the protein, and some recombinant glycoproteins have activity without the native glycosylation. Alternatively, there are many situations in which specific glycosylation is crucial to the correct folding and action of the protein, and variations in the glycosylation are indicative of abnormalities in function. The review of Varki (1993) gives a good overview of these apparently paradoxical observations.

The structural diversity inherent in the branched sugar moieties characteristic of the glycan element of glycoproteins enables subtle changes in protein shape, charge and volume that have the potential to modify its function both temporally and spatially. This diversity has required relatively sophisticated physical and chemical techniques to assign composition, linkage, branching, anomeric configuration and attachment sites. The analysis of glycoproteins, however, is simplified because of the limited number of constituent monosaccharide residues and linkages present in the attached glycans. The general carbohydrate analytical chemistries that have been developed for the determination of the structure of complex polysaccharides can be selectively exploited for the analysis of the glycans on proteins. This has enabled some commercially available analytical tools for the analysis of structurally conserved glycans to be developed (Oxford GlycoSciences, UK; Glyko Inc., USA; Bio-Rad, USA; Beckman, USA). Together with the availability of specific exoglycosidases and the emergence of biological capillary electrophoresis and mass spectrometry, the glycan heterogeneity on proteins can now be characterized. The challenge becomes one of adapting these techniques to support analysis of the amounts of glycoprotein that are obtained by proteomic approaches.

In this chapter we give some examples of the areas in which glycosylation has been found to have an important role, and which will hopefully convince the reader that the analysis of protein glycoforms is fundamental to the understanding of the proteome of the cell. We then describe the glycoanalytical technology that is available for adaptation for the high-throughput requirements of glycoproteomics research.

2. What is a glycoprotein?

Glycoproteins are formed by the enzymic addition of one or more saccharide units to the nascent polypeptide chain. Modifications range from single saccharides on proteins, such as those found on collagens and nuclear glycoproteins, through to densely packed, complex, branched structures found on mucins, and the complex glycosaminoglycan side chains found on proteoglycans. The three main types of oligosaccharide attachment to the proteins are N-linked, O-linked and glycosyl phosphatidylinositol (GPI) anchored structures. All three classes of modifications can be found on the same protein. Protein isoforms that have the same amino acid sequence but different glycosylation profiles are often obvious on 2-DE separations as they appear as trains of spots which can differ in pI and/or molecular mass (*Figs. 1–3*).

2.1. N-linked structures

N-glycan assembly takes place in the endoplasmic reticulum and Golgi where a common structure of $GlcNAc_2Man_9Glc_3$ is trimmed and modified variously by glycosidases and glycosyltransferases. Almost always the resultant N-linked oligosaccharide has the core structure of $GlcNAc_2Man_3$ attached to an asparagine residue present in a defined motif Asn-Xaa-Ser/Thr (where Xaa = any amino acid except Pro). The N-linked structures have been classified as high mannose, hybrid or complex oligosaccharides depending on the composition of the monosaccharides. Different glycan chains can be attached to the same site (microheterogeneity) and site occupancy can vary (macroheterogeneity).

While the N-linked pentasaccharide core structure is the same on the proteins of all species characterized so far, the trimming and reconstruction of the oligosaccharides varies across and within species in terms of composition, branching, size and anomeric configuration [for example $\alpha(1-3)$ or $\beta(1-2)$].

2.2. O-linked structures

O-linked oligosaccharides are commonly attached via the hydroxyl group of serine and threonine, but are also found on hydroxylysine and hydroxyproline amino acids. There is no single defined amino acid motif that predicts the attachment to a hydroxyl residue, although a neural network is being developed by Hansen *et al.* (1998) to predict mucin-type glycosylation. Mucin-type O-glycosylation is initiated by tissue-specific addition of a GalNAc-residue to a serine or a threonine of the fully folded protein. This event appears to be dependent on the primary, secondary and tertiary structure of the (glyco)protein. Further elongation by the addition of single monosaccharide residues by specific transferases is highly regulated (Van den Steen *et al.*, 1998a). In contrast to the more common mucin-type O-glycosylation, some specific types of O-glycosylation, such as the O-linked attachment of fucose and glucose, are dependent on the amino acid sequence of the protein and eight different reducing terminal attachments to serine or threonine have been described (Hounsell *et al.*, 1996). Most mammalian O-glycosylation is found on secreted proteins, but single GlcNAc moieties are found on cytoplasmic and nuclear-pore proteins (Hart *et al.*, 1996) and phosphodiester linkages between the sugar moiety and protein have been described on slime molds and several parasites, and probably are more widespread (Haynes, 1998).

2.3. GPI-anchored structures

Glycosyl phosphatidylinositol structures anchor functionally diverse proteins into the hydrophobic lipid bilayer of cell membranes (Kinoshita *et al.*, 1997). They have an identical saccaharide core consisting of Man_3GlcNH_2, which is a bridge between the phospholipid in the membrane and the protein. The anchor is attached in a post-translational event that takes place after *C*-terminal cleavage of the signal peptide in the endoplasmic reticulum.

3. Why are glycoproteins of interest in proteomics?

Proteomics is a protein-based approach to holistic biology in which the expressed gene-products are analyzed rather than the DNA or mRNA from which they are made. More often than not, the biologically active form of a mature protein has been modified post-translationally and these modifications are highly dynamic. The roles that have been assigned to the glycan component of proteins in mammals include modification of immunological responses, protein recognition and targeting, protein folding, regulation of serum clearance rates and signaling. Comprehensive reviews discussing the function of carbohydrates *in vivo* are available in the literature (Dwek, 1995; Kukuruzinska and Lennon, 1998; Lis and Sharon, 1993; Varki, 1993).

Experimental systems that are of great interest in human biology include the involvement of glycoproteins in fertilization and in the emergence of disease. These two examples are discussed here to highlight the need to understand the function of post-translational modifications to expressed gene-products.

3.1. Glycoproteins and fertilization

Glycoproteins play a key role in developmental processes which, in animals, begin with recognition between the oocyte and sperm cell and culminate in the development of a mature organism. Critical to mammalian fertilization is the extracellular glycoprotein coat surrounding the egg, the zona pellucida (ZP). After penetrating the follicle cells, sperm binds to the ZP, which is composed of three sulfated glycoproteins ZP1, ZP2 and ZP3, the glycan portion of which is crucial for recognition and binding (Zara and Naz, 1998). β-1,4-Galactosyltransferase on the sperm surface is the best studied of the candidate receptors for egg-coat ZP3. Overexpression of this enzyme in transgenic mice affects many aspects of the sperm–egg interactions (Youakim *et al.*, 1994). Furthermore, sperm from the few mice that survived deletion of β-1,4-galactosyltransferase gene could only penetrate the zona pellucida poorly and were unable to fertilize the oocyte *in vitro* (Lu and Shur, 1997). The glycoproteomic approach allows changes in the glycoproteins involved in fertilization to be monitored and targeted in contraception and fertility studies. An interesting candidate glycoprotein for a contraceptive are the progesterone-regulated uterine glycodelins, that exhibit sequence homology to the β-lactoglobulins, and differ from each other only by their unique oligosaccharide structures. The endometrium-derived glycodelin A inhibits sperm–egg binding, whereas the differently glycosylated seminal plasma-derived glycodelin S does not (Seppala *et al.*, 1998). The gene for glycodelin A and S is the same (placental protein 14) but the post-translational glycosylation reflects the function and tissue localization of the protein.

3.2. Glycoproteins and disease

In the initial stages of a disease, glycoproteins often exhibit changes in structure and function that relate to the structure of the attached glycans. These changes can chart the progress of the disease. There is substantial literature in this area and two books (Brockhausen and Kuhns, 1997; Montreuil *et al.*, 1996) have been published describing the role of glycosylation in disease processes in humans. A recent review (Brockhausen *et al.*, 1998) discusses the mechanisms of the glycoslyltransferase defects underlying these glycoprotein alterations in human disease. What is lacking is our understanding of the difference between those glycosylation changes that are associated with and result from the disease process, and those changes that may actually cause the disease. Clearly, the latter are potentially of more interest as they serve as sites of therapeutic intervention.

The proteomic approach to determining the factors involved in a disease process initially involves identification of proteins that are differentially expressed between the diseased and normal tissue. These proteins can then become the focus of efforts to diagnose and/or control the disease. Many glycoproteins, and specifically changes in their glycosylation, have been shown to reflect a range of diseases in humans (*Table 1*). Where possible in this table, review articles have been cited and while the list is not exhaustive, it does give an idea of the diversity of glycosylation changes that have been associated with abnormal metabolism. Most of the glycoproteins reported as differently glycosylated in disease are contained in serum, for the obvious reason that this tissue is important in clinical diagnosis procedures and is easily obtainable. We suspect that glycosylation changes displayed in samples of diseased solid tissues will reflect a similar diversity.

Glycosylation differences are also being used to seek candidates for glycan-based cancer vaccines. For example, mucins on normal epithelial cells have densely packed

Table 1. Examples of the types of glycosylation changes associated with human disease

Disease	Glycoprotein	Tissue	Alteration	Reference
Carbohydrate-deficient syndrome	Transferrin, antithrombin, orosomucoid	Serum	Different N-linked glycoforms	Winchester *et al.* (1995)
Hepatic cancer	Alpha-fetoprotein	Serum	Different N-linked glycoforms	Seregni *et al.* (1995); Taketa (1998)
Inflammation, cancer, AIDS	Orosomucoid (α1-acid glycoprotein)	Serum	Different N-linked glycoforms	Van Dijk *et al.* (1995); Mackiewicz and Mackiewicz (1995)
Immune disorders	CD43	T cells	Different O-linked glycans	Ellies *et al.* (1994)
Rheumatoid arthritis	IgG	Serum	N-linked lowered terminal galactose	Delves (1998)
Creutzfeldt–Jakob disease	Prion protein	CSF	Different degree of glycosylation	Furakawa *et al.* (1998)
Schistosomiasis	Circulating cathodic/anodic antigens	Serum, urine	Different O-linked glycans	Van Dam *et al.* (1996)
Diabetic pregnancy, choriocarcinoma	hCG	Urine	Hyperbranching of N- and O-linked glycans	Elliott *et al.* (1997)
Alcohol abuse	Transferrin	Serum	Desialylation	Tagliaro *et al.* (1998)
Cancer	Mucin	Tumors	Overexpression or exposure of specific short chain O-linked glycans	Taylor-Papadimitriou and Epenetos (1994); Kim *et al.* (1996)
Lysosomal storage diseases	Lysosomal pathways of degradation of oligosaccharides	Urine	Secretion of abnormal carbohydrates	Starr *et al.* (1994)
Menopause	FSH, LH	Serum	Increase in acidity, decrease in complexity	Anobile *et al.* (1998)

complex glycan structures on their surface. On carcinoma cells, cryptic carbohydrate epitopes (e.g. Tn and Sialyl Tn antigens) are exposed on the mucins and represent targets for anticancer therapeutic vaccines. Different strategies for presenting these specific carbohydrates attached to synthetic peptides for immunotherapy are being clinically trialled (Kudryashov et al., 1998; Lo-Man et al., 1999) and are an exciting area of research.

Two-dimensional electrophoresis is widely regarded as the method of choice for differentially displaying changes in protein expression. Many proteins are displayed simultaneously and differences observed can be correlated with phenotypic differences. Proteomics does not necessarily dictate that a single protein will be the only target, but in all likelihood a network of pathways will be affected in any altered biological system. So proteomics requires that many members of a population be screened to understand the changes that occur in protein expression at a population level as well as at an individual level. If a glycoprotein is targeted as being significant to the change in phenotype, it can then be used as a diagnostic marker for the disease, either directly by using gel electrophoresis or by other assays derived from a knowledge of the specific attributes of the identified change in glycosylation of the protein. 2-DE has been successfully used to identify glycoprotein differences, in extracellular collagenase secreted by osteoblast primary cell lines (Hankey et al., 1992), in multiple serum proteins in carbohydrate-deficient syndrome (Yuasa et al., 1995) and in a mucin in serum from pancreatic cancer patients (Ching and Rhodes, 1988). Numerous serum glycoproteins have been reported as markers for renal cell carcinoma (Delahunt et al., 1996).

As seen from *Table 1*, many of the changes that occur in glycosylation of target proteins are either changes in molecular mass of the protein due to different degrees of glycosylation and/or changes in pI of the protein due to changes in relative sialylation. Phosphorylation and sulfation of the glycans will also have an effect on the pI of the glycoprotein. These inherent properties of the glycoprotein make the array of proteins by 2-DE ideally suited to detect diagnostic changes in glycosylation. At the present time no other technology has the potential for this type of differential display of glycoproteins.

4. Glycoanalysis in proteomics

Carbohydrate analysis was initially focused on determining the chemical and physical structure of the complex, very high molecular mass polysaccharides. The techniques for these types of analyses have been adapted to the analysis of the smaller, less complex glycans on glycoproteins and have become quicker, easier and more readily available to the protein biochemist. The advent of mass spectrometry that is able to achieve mass analysis of small amounts of carbohydrates and glycopeptides, together with the cloning of many specific exoglycosidases, has enabled many oligosaccharides on glycoproteins to be analyzed for their composition, linkage and sequence (Packer and Harrison, 1998). Most of these techniques have been optimized on purified glycoproteins expressed in recombinant systems, which provide relatively large amounts of protein (tens of nanomoles or hundreds of micrograms of protein) to analyze. In contrast, relatively little analysis to date has been carried out on glycoproteins that have been separated by 2-DE (typically tens of picomoles or hundreds of nanograms of protein).

4.1. Recombinant glycoprotein analysis

Proteins produced by recombinant DNA methods are an expanding area in the biotechnology industry. Most therapeutic and diagnostic proteins are glycosylated, therefore the need for rapid separation and identification of glycoforms is necessary. Increasingly, regulatory authorities require the precise determination of the oligo-saccharide moiety, as changes can induce an undesired immunological response.

Expression systems vary in their ability to glycosylate and do so differentially, depending on the cell line and culture conditions (Lifely *et al.*, 1995; Wright and Morrison, 1997). The conditions of the fermentation process determine how and when various carbohydrate structures are produced. Liquid chromatography has been the validated method for monitoring the batch-to-batch variations and down-stream processing effects, but electrophoresis is becoming more popular because of its capabilities of high-throughput, comparative analysis and resolution (Taverna *et al.*, 1998). Rapid routine analysis and production monitoring is made simpler using electrophoretic techniques such as isoelectric focusing (IEF), capillary electro-phoresis and 2-DE.

The presence or absence of glycosylation also modulates the effectiveness of some recombinant proteins. For example, human granulocyte colony-stimulating factors (rHu G-CSF; Hoglund, 1998), human lymphotoxin (Funahashi *et al.*, 1993), interleukin-2 (Voshol *et al.*, 1996), human thyroperoxidase (Fayadat *et al.*, 1998) and α1-acid glycoprotein (Shiyan and Bovin, 1997) are all recombinant proteins that require glycosylation for *in vivo* activity. Other recombinant proteins that are glycosylated in their native form, such as interleukin-3 (Ziltener *et al.*, 1994), thrombopoietin (Jagerschmidt *et al.*, 1998) and human growth hormone (Becker and Hsiung, 1986), have activity even when nonglycosylated. Proteomic approaches, in which the *in vivo* changes of these glycoproteins are monitored, can provide an understanding of the native glycoform heterogeneity and function. A knowledge of this is important before the *in vitro* expression of recombinant forms of a biologically significant glycoprotein is attempted.

4.2. Gel-separated glycoproteins

Determination of the glycosylation structural changes on proteins which have been separated by either 1-D or 2-DE has been challenging, as the amounts of purified proteins available are limited. Mass spectrometry [both matrix-assisted laser desorption ionization (MALDI-MS) and electrospray ionization (ESI-MS)] and capillary zone electrophoresis (CZE) are the two most promising technologies being used to solve the question of the types of heterogeneity of oligosaccharide structures present on glycoprotein isoforms.

It is becoming clear that the physical and chemical properties of glycoproteins make their analysis by current proteomic technology difficult. The attachment of glycans, especially the large N-linked oligosaccharides, confers different attributes to the protein by making the molecule bulky and hydrophilic. To analyze glycoproteins that have been separated by 2-DE, the proteomic processes of spatial resolution (array), characterization and bioinformatics need to be optimized specifically for glycoproteins.

Spatial resolution (array). Two-dimensional electrophoresis of plasma glycoproteins is well established as the serum protein map is published on the World Wide Web (Sanchez *et al.*, 1995). In our work, we have found that loading of first-dimension immobilized pH gradient (IPG) strips with a purified recombinant glycoprotein resulted in a very low amount of protein entering the gel by passive rehydration techniques. Cup loading of glycoproteins onto the IPG strips increased the amount of glycoprotein entering the strip, as did rehydration of the recombinant glycoprotein mixed with a small amount of protein extracted from *Escherichia coli*. These effects are shown in *Figure 2* using the standard bovine fetuin glycoprotein. There is not the same difficulty in loading recombinant nonglycosylated purified proteins. The reason for this effect of glycosylation can only be speculated upon, but it may reflect the effect that the oligosaccharides have on the apparent size and charge of the protein and it does serve to highlight the differences in the physical properties of glycoproteins.

Once the glycoproteins have been separated, the next step is to visualize them. Many who have worked on staining glycoproteins with the standard Coomassie blue and silver-staining procedures have noted that the glycoproteins either do not stain or stain poorly. The best general method for locating the glycoproteins in a mixture of proteins arrayed on a 2-DE gel is to stain specifically for the glycan moiety. Sensitive lectin (Golaz *et al.*, 1995) and hapten recognition protocols are available and there are several products on the market (Bio-Rad glycan detection kit, Boehringer-Mannheim DIG detection kit) based on periodate oxidation of the sugars and a sandwich antibody or biotin/streptavidin color development. We have modified this approach by using a fluorescent rather than a color tag on the oxidized sugars (see *Figure 1*) that allows faster, sensitive visualization of glycoproteins on PVDF blots.

Characterization. The reason for the discrete separation of single glycoform spots on 2-DE gels is not obvious when the heterogeneity of oligosaccharide structures is considered. A recombinant form of the truncated immunoglobulin superfamily glycoprotein CD4, which contains one site of N-glycosylation, separates into three distinct glycoforms on 2-DE gels (*Figure 3*). Differential sialylation is the easiest explanation of this pI difference. However in separate experiments, the mass spectrometric analysis of the N-linked oligosaccharides released from another glycoprotein, human α1-antitrypsin, showed that it is not this straightforward, as there was heterogeneity of glycosylation in each of the separated spots (Packer *et al.*, 1998). Monosaccharide analysis of each spot of the α1-antitrypsin showed the same

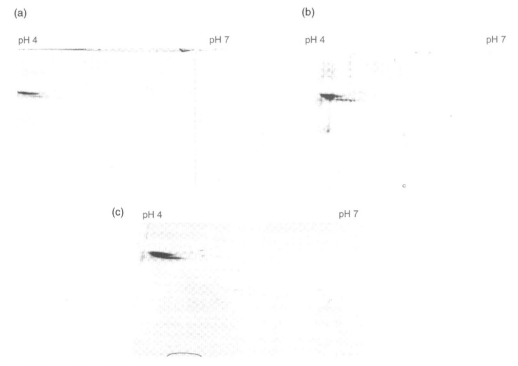

Figure 2. Two-dimensional electrophoresis of purified glycoproteins. Glycoprotein was solubilized in 5 M urea, 2 M thiourea, 2 mM TBP, 2% CHAPS, 2% sulfobetaine 3–10, 0.5% ampholytes 3–10, 40 mM Tris, loaded onto pH 4–7 IPG strips, separated by 2-DE and stained with Coomassie blue. (a) 10 μg bovine fetuin by first-dimension rehydration technique (Rabilloud *et al.*, 1994); (b) 10 μg bovine fetuin using first-dimension anode cup load +1% (pH 2.5–5) Pharmalytes (Pharmacia); (c) 10 μg bovine fetuin + 50 μg *E. coli* total proteins by first-dimension rehydration technique.

ratio of sialylation to neutral sugars on each glycoform, with apparently increased glycosylation correlating with a decrease in pI (Packer *et al.*, 1996).

The rapid development of mass spectrometry, both MALDI and ESI, has revolutionized the analysis of proteins and has the potential to do the same for glycans (Costello, 1997; Packer and Harrison, 1998). The techniques of deglycosylation, normal-phase HPLC of derivatized released oligosaccharides, enzymatic sequencing and ESI and MALDI-MS have been developed on recombinant glycoproteins and are starting to be applied to fully characterize the glycan structures on glycoproteins separated *in-gel* (Kuster *et al.*, 1997, 1998).

Glycopeptides substituted with N-linked oligosaccharides are difficult to see by MALDI-MS because their mass, together with that of the peptide, is usually greater than 3500 Da and, together with their tendency to be heterogeneous, they result in broad signals only visible in linear mode. O-linked oligosaccharides on the whole tend to be smaller than N-linked oligosaccharides and they can sometimes be seen by mass spectrometry as tryptic glycopeptide fragments of in-gel digested glycoproteins

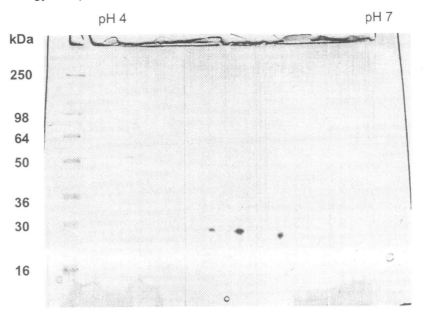

Figure 3. Two-dimensional electrophoresis of the isoforms of CD4.c truncated glycoprotein. CD4.c truncated glycoprotein (1 nmol) was separated by 2-DE using first-dimension pH 4–7 IPG strips. The protein was anode cup loaded with 1% (pH 2.5–5) Pharmalytes (Pharmacia) and was separated in the second dimension using 10–20% SDS–PAGE. The unmodified protein has a calculated molecular mass of 20 429 Da and pI of 5.66.

(Packer and Harrison, 1998). It has also been possible to characterize O-linked glycans released by alkaline β-elimination from glycophorin A separated by gel electrophoresis and blotted to PVDF membrane (Packer *et al.*, 1998) and from radioactively labeled viral protein in-gel (Medina *et al.*, 1998). The analysis of glycoproteins in-gel has been extended to include hydrazine release and mass spectrometry of O-linked sugars from in-gel separated gelatinase B variants (Van den Steen *et al.*, 1998b) and to N-linked site mapping using ^{18}O-labeling and peptide mass fingerprinting (Kuster and Mann, 1999).

Capillary zone electrophoresis is also proving to be useful in the high-throughput monitoring of recombinant glycoprotein isoforms. This technology is capable of resolving very small differences in pI and structure and is starting to be coupled to MS for characterization of these differences. Two detailed reviews (Susuki and Honda, 1998; Taverna *et al.*, 1998) describe the techniques and their use in the mapping of recombinant oligosaccharides and glycopeptides. The difficulty with applying this technology to the analysis of the amounts of protein separated by 2-DE is the requirement for a relatively concentrated solution (μM) of analyte.

Our choice for the high-throughput analysis of proteins is to use, in the first instance, the sensitive, rapid and easy MALDI-MS automated approach for identification of a protein from a known genome. The identification of the protein by this method is achieved by using only a small fraction of the tryptic digest of a Coomassie blue-stained protein spot. Once the protein has been identified from the databases, most of the sample is still available for microcharacterization of the modifications to the protein (*Figure 4*). We use the slower technique of LC-ESI-MS as a second tier approach to characterize specific proteins of interest for their modifications. To fully characterize glycosylation both the microheterogeneity and macroheterogeneity of the oligosaccharides should ideally be determined.

In our hands, ESI has proved to be more successful than MALDI-TOF in ionizing glycoproteins/glycopeptides, in particular those with heterogeneous N-linked glycosylation. This may be a reflection of the matrices used in MALDI-MS, the heterogeneity of the glycans and the large molecular mass of the N-linked oligosaccharides or the difficulty in observing sialylated oligosaccharides without derivatization (Powell and Harvey, 1996). An example of the difference in spectra of a glycoprotein obtained by the two types of mass spectrometry is seen in *Figure 5*, in which the truncated, CHO-produced recombinant rat CD4 (unmodified M_r 20 429.31 Da), which contains one site of N-glycosylation, was analyzed. Eight branched N-linked oligosaccharides have been characterized after desialylation of the complete molecule (Ashford *et al.*, 1993), which contains two N-linked glycosylation sites. The MALDI-MS indicated extensive heterogeneity (*Figure 5a*) but no discrete masses could be resolved at this high mass range. ESI-MS was able to resolve nine possible glycoforms on the single site (*Figure 5b*). Using the unmodified protein mass the broad main peak can be probably be assigned as a mixture of a hybrid oligosaccharide structure [22 497 Da =

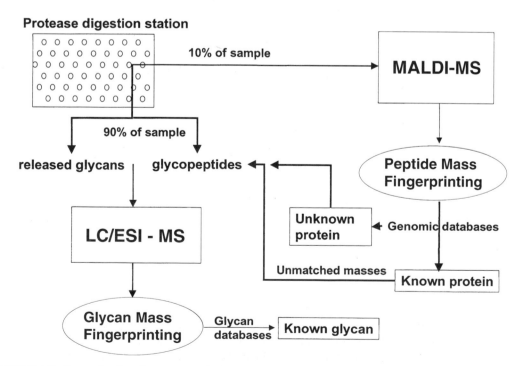

Figure 4. Strategies for glycoproteomics. Mass spectrometric approach to the automated high-throughput analysis of glycoproteins separated by 2-DE.

(a)

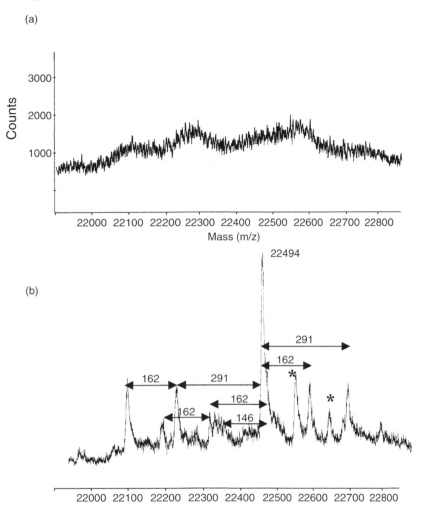

Figure 5. Mass spectra of CD4.c glycoprotein. (a) MALDI-TOF (Perseptive Voyager): 20 pmol CD4.c protein, cleaned up over a Poros 50 R1 (equivalent to C4 reversed-phase properties) minicolumn (Kussman *et al.*, 1997) eluted with 2 μl of 2% TFA , 80% MeCN containing 10 mg/ml sinapinic acid directly onto a target plate. (b) ESI-TOF-MS (MicroMass): 100 pmol of CD4.c (same batch) separated from salt using a HPLC C4 column (Aquapore BU-300 Å 100 × 2.1 mm, PE Brownlee) using a gradient of 0.05% TFA to 80% MeCN and 0.05% TFA over 30 min. The protein eluted after approximately 22 min as monitored at 214 nm. (*TFA adducts of the major peak.)

$\text{core}(\text{GlcNAc})_2(\text{Man})_3 + (\text{Gal})_1(\text{GlcNAc})_1(\text{Man})_5]$ and a complex oligosaccharide structure $[22\,489 \quad \text{Da} = \text{core}(\text{GlcNAc})_2(\text{Man})_3 + (\text{Gal})_2\text{GlcNAc})_2(\text{Fuc})_1(\text{NeuAc})_1]$. The mass differences shown result from other related heterogeneous structures.

Mass spectrometry is able to determine the possible monosaccharide composition of the different glycoforms. The technique is also able to provide information on the sequence and linkage of the residues by the characteristic way in which they fragment by collision-induced dissociation. Most data on fragmentation in the literature has been obtained by fast-atom-bombardment of permethylated oligo-saccharides. The development over the last 10 years of the two new ionization techniques of MALDI-TOF and ESI, which are able to measure at high sensitivity over the large mass range applicable to biomolecules, has resulted in both forms of ionization being explored for their ability to fragment glycoconjugates. The best data has been obtained on underivatized oligosaccharides after release from the protein/peptides. Fragmentation of the glycosidic bonds occurs with post-source decay MALDI-TOF (Naven *et al.*, 1997) and in-source collision-induced dissocia-tion CID-ESI-TOF (Packer and Harrison, 1998) or ESI-MS-MS (Oxley *et al.*, 1998). At the moment electrospray ionization has become the method of choice for the structural characterization of protein glycosylation (Burlingame, 1996), as the interface with microbore and capillary LC allows the collision-induced fragmenta-tion of separated glycoforms. The future will see many other developments in mass spectrometry as different ionization techniques are married with better detection methods – and it seems likely that some will be better than others for the analysis of protein glycoforms.

The routine application of mass spectrometry to glycoproteomics still requires a number of problems to be solved. For MALDI-MS analysis, further work on the investigation of different matrices specific for the larger more hydrophilic glycopep-tides is required and the use of a selective upstream enrichment of glycopeptides would prove advantageous. For ESI-MS, glycopeptides can be located in an LC separation by in-source CID, but the fragmentation of the oligosaccharide on the peptide is difficult to interpret. The sensitivity of detection of the uncharged released oligosaccharides needs to be further increased and may require derivatization to tag the sugars with a readily ionizable group. This type of chemical reaction needs to be complete and simple to allow adaptation to high-throughput analysis. Similarly, any column or electrophoretic separation steps that are introduced before the mass spectrometer need to be fast and able to be automated.

Complete characterization of the structure of the glycoforms on proteins, how-ever, requires more than just measurement of the mass of the sugars. Compared with amino acid masses used in peptide mass fingerprinting, there are many mono-saccharide residues that have the same molecular mass, and linkage and sequence cannot be determined easily by mass spectrometry. Nevertheless, glycoprotein analysis is facilitated by a degree of conservation in the range of structures and the constituent monosaccharides within species, so that a knowledge of the precedents found within a particular tissue and/or species will restrict the possible glycan structures that can account for a specific mass.

Informatics. The field of bioinformatics has received a huge impetus from the mass of data being generated in the genomic and proteomic fields. Databases and the tools to mine them are used extensively in the interpretation of DNA and protein sequence and identification. Concurrent with the improvements in analysis, there is a need for the construction of databases and informatic tools for the interpretation of glycosylation data.

Carbank (http://128.192.9.29/carbbank/CarbBank.html), which has not received funding for data entry since 1995, and O-Glycbase (http://www.cbs.dtu.dk/databases/OGLYCBASE/), which uses neural networks to predict O-glycosylation sites, are the only publicly available databases containing glycoprotein oligosaccharide information. We have recently collated a glycoprotein O-linked glycosylation database (Cooper *et al.*, 1999) and are preparing more extensive bioinformatic tools and databases for use in glycoproteomic research.

There is a need for an integrated information package to be developed for glycoanalysis much the same way as SWISS-PROT and similar databases have provided the protein chemists with repositories of information and the tools to mine them. Mass measurements, of either glycopeptides or released sugars, can easily be converted to possible monosaccharide compositions. These compositions can be matched to previously reported structures, and experiments using specific exoglycosidases can be designed to test the structural possibilities. Sequence can also be deduced from a knowledge of the structure-specific mass spectrometric fragmentation patterns (glycan mass fingerprinting). There is a great deal of available knowledge about glycoproteins; this now needs to be assembled so that it can be used to help interpret the data which will be acquired from glycoproteomic analysis.

5. Conclusion

The post-translational modifications to proteins, including glycosylation, will prove to be the 'fine control' of metabolic changes that occur in biological systems and which only become apparent when the gene product has been modified in the tissue in which the protein is active. The extensive diversity available in the glycan structure makes it a perfect moiety to modulate subtle control of metabolism. Such a function of specific glycosylation has been postulated by Clark *et al.* (1997) to underlie the protection of the human embryo from the maternal immune response. Furthermore, the acquisition of these specific glycans may be the means by which pathogens can subvert the host's immune system. The array of glycoproteins displayed by 2-DE provides a map by which the changes that occur in development and disease can be explored. The tools for the analysis of recombinant glycoproteins, where the amount of protein is not a limiting factor, have been developed. Extension of these and other methods to provide picomolar sensitivity is now required to enable high-throughput identification and characterization of glycoproteins to parallel existing methods for protein identification and characterization. The extent to which the information obtained from this type of data can be exploited will depend on both the continued development of improved, more sensitive methodologies and the provision of interfaces to bioinformatic tools constructed specifically for glycobiology.

Acknowledgments

The authors wish to thank Neil Barclay for kindly supplying the purified recombinant truncated CD4 glycoprotein, Ben Herbert for help with 2-DE, and Keith Williams and Andrew Gooley for their helpful discussion of the manuscript.

References

Anobile, C.J., Talbot, J.A., McCann, S.J., Padmanabhan, V. and Robertson, W.R. (1998) Glyco-form composition of serum gonadotrophins through the normal menstrual cycle and in the post-menopausal state. *Mol. Hum. Reprod.* **4**: 631–639.

Ashford, D.A., Alafi, C.D., Gamble, V.M., MacKay, D.J.G., Rademacher, T.W., Williams, P.J., Dwek, R.A., Barclay, A.N., Davis, S.J., Somaoza, C., Ward, H.A. and Williams, A.F. (1993) Site-specific glycosylation of recombinant rat and human soluble CD4 variants expressed in Chinese hamster ovary cells. *J. Biol. Chem.* **268**: 3260–3267.

Becker, G.W. and Hsiung, H.M. (1986) Expression secretion and folding of human growth hormone in *Escherichia coli*. Purification and characterization. *FEBS.* **204**: 145–150.

Brockhausen, I. and Kuhns, W. (1997) *Glycoproteins and Human Disease.* R.G. Landes, New York.

Brockhausen, I., Schutzbach, J. and Kuhns, W. (1998) Glycoproteins and their relationship to human disease. *Acta Anat.* **161**: 36–78.

Burlingame, A.L. (1996) Characterization of protein glycosylation by mass spectrometry. *Curr. Opin. Biotechnol.* **7**: 4–10.

Ching, C.K. and Rhodes, J.M. (1988) Identification and partial characterization of a new pancreatic cancer-related serum glycoprotein by sodium dodecyl sulfate–polyacrylamide gel electrophoresis and lectin blotting. *Gastroenterology* **95**: 137–142.

Clark, G.F., Dell, A., Morris, H.R., Patankar, M., Oehninger, S. and Seppala, M. (1997) Viewing AIDS from a glycobiological perspective: potential linkages to the human fetoembryonic defence system hypothesis. *Mol. Hum. Reprod.* **3**: 5–13.

Cooper, C.A., Wilkins, M.R., Williams, K.L and Packer, N.H. (1999) *Electophoresis* **20** (in press).

Costello, C.E. (1997) Time, life ... and mass spectrometry. New techniques to address biological questions. *Biophys. Chem.* **68**: 173–188.

Delahunt, B., Maxwell, J., Moroni-Rawson, P., Renner, R. M., Nacey, J.N. and Jordan, T.W. (1996) Serum protein variation in patients with renal cell carcinoma. *J. Urol. Pathol.* **5**: 97–107.

Delves, P.J. (1998) The role of glycosylation in autoimmune disease. *Autoimmunology* **27**: 239–253.

Dwek, R.A. (1995) Glycobiology: towards understanding the function of sugars. *Biochem. Soc. Trans.* **23**: 1–25.

Ellies, L.G., Jones, A.T., Williams, M.J. and Ziltener, H. J. (1994) Differential regulation of CD43 glycoforms on CD4+ and CD8+ T lymphocytes in graft-versus-host disease. *Glycobiology* **6**: 885–893.

Elliott, M., Kardana, A., Lustbader, J. and Cole, L. (1997) Carbohydrate and peptide structure of the alpha- and beta-subunits of human chorionic gonadotrophin from normal and aberrant pregnancy and choriocarcinoma. *Endocrine* 15–32.

Fayadat, L., Niccoli-Sire, P., Lanet, J. and Franc, J.L. (1998) Human thyroperoxidase is largely retained and rapidly degraded in the endoplasmic reticulum. Its N-glycans are required for folding and intracellular trafficking. *Endocrine* **139**: 4277–4285.

Funahashi, I., Wantanabe, H., Abo, T., Indo, K. and Miyaji, H. (1993) Tumour growth inhibition in mice by glycosylated recombinant human lymphotoxin: analysis of tumour-regional mono-nuclear cells involved with its action. *Br. J. Cancer* **67**: 447–455.

Furakawa, H., Doh-ura, K., Kikuchi, H., Tateishi, J. and Iwaki, T. (1998) A comparative study of abnormal prion protein isoforms between Gerstmann–Straussler–Scheinker syndrome and Creutzfeldt–Jakob disease. *J. Neurol. Sci.* **158**: 71–75.

Golaz, O., Gravel, P., Walzer, C., Turler, H., Balant, L. and Hochstrasser, D. F. (1995) Rapid detection of the main human plasma glycoproteins by two-dimensional polyacrylamide gel electrophoresis lectin affinoblotting. *Electrophoresis* **16**: 1187–1189.

Hankey, D.P., Nicholas, R.M. and Hughes, A.E. (1992) Two-dimensional polyacrylamide gel electrophoresis reveals differences between osteoblast and fibroblast extracellular proteins. *Electrophoresis* **13**: 329–332.

Hansen, J.E., Lund, O., Tolstrup, N., Gooley, A.A., Williams, K L. and Brunak, S. (1998) NetOglyc: prediction of mucin type O-glycosylation sites based on sequence context and surface accessibility. *Glycoconj. J.* **15**: 115–130.

Hart, G.W., Kreppel, L.K., Comer, F.I., Arnold, C.S., Snow, D.M., Ye, Z., Cheng, X., Della Manna, D., Caine, D.S., Earles, B.J., Akimoto, Y., Cole, R.N. and Hayes, B.K. (1996) O-GlcNAcylation of key nuclear and cytoskeletal proteins: reciprocity with O-phosphorylation and putative roles in protein multimerization. *Glycobiology* **7**: 711–716.

Haynes, P.A. (1998) Phosphoglycosylation: a new structural class of glycosylation? *Glycobiology* **8**: 1–5.

Hoglund, M. (1998) Glycosylated and non-glycosylated recombinant human granulocyte colony-stimulating factor (rhG-CSF) – What is the difference? *Med. Oncol.* **15**: 229–233.

Hounsell, E.F., Davies, M.J. and Renouf, D.V. (1996) O-linked protein glycosylation structure and function. *Glycoconj. J.* **13**: 19–26.

Jagerschmidt, A., Fleury, V., Anger-Leroy, M., Thomas, C., Agnel, M. and O'Brein, D.P. (1998) Human thrombopoietin structure-function relationships: identification of functionally important residues. *Biochem. J.* **333**: 729–734.

Kim, Y.S., Gum, J.R. and Brockhausen, I. (1996) Mucin glycoproteins in neoplasia. *Glycoconj. J.* **13**: 693–707.

Kinoshita, T., Ohishi, K. and Takeda, J. (1997) GPI-anchor synthesis in mammalian cells: genes, their products, and a deficiency. *J. Biochem. (Tokyo)* **122**: 251–257.

Kudryashov, V., Kim, H.M., Ragupathi, G., Danishefsky, S.J., Livingston, P.O. and Lloyd, K.O. (1998) Immunogenicity of synthetic conjugates of Lewis(y) oligosaccharide with proteins in mice: towards the design of anticancer vaccines. *Cancer. Immunol. Immunother.* **45**: 281–286.

Kukuruzinska, M.A. and Lennon, K. (1998) Protein N-glycosylation: molecular genetics and functional significance. *Crit. Rev. Oral Biol. Med.* **9**: 415–448.

Kussman, M., Nordhoff, E., Rahbek-Nielsen, H., Haebel, S., Rossel-Larsen, M., Jakobsen, L., Gobom, J., Mirgorodskaya, E., Kroll-Kristensen, A., Palm, L. and Roepstorff, P. (1997) Matrix-assisted laser desorption/ionisation mass spectrometry sample preparation techniques designed for various peptide and protein analytes. *J. Mass Spec.* **32**: 593–601.

Kuster, B. and Mann, M. (1999) 18O-labeling of N-glycosylation sites to improve the identification of gel-separated glycoproteins using peptide mass mapping and database searching. *Anal. Chem.* **71**: 1431–1440.

Kuster, B., Wheeler, S.F., Hunter, A.P., Dwek, R.A. and Harvey, D.J. (1997) Sequencing of N-linked oligosaccharides directly from protein gels: in-gel deglycosylation followed by matrix-assisted laser desorption/ionization mass spectrometry and normal-phase high-performance liquid chromatography. *Anal. Biochem.* **250**: 82–101.

Kuster, B., Hunter, A.P., Wheeler, S.F., Dwek, R.A. and Harvey, D.J. (1998) Structural determination of N-linked carbohydrates by matrix-assisted laser desorption/ionization-mass spectrometry following enzymatic release within sodium dodecyl sulphate–polyacrylamide electrophoresis gels: application to species-specific glycosylation of alpha1-acid glycoprotein. *Electrophoresis* **19**: 1950–1959.

Lifely, M.R., Hale, C., Boyce, S., Keen, M.J. and Phillips, J. (1995) Glycosylation and biological activity of CAMPATH-1H expressed in different cell lines and grown under different culture conditions. *Glycobiology* **5**: 813–822.

Lis, H. and Sharon, N. (1993) Protein glycosylation: structural and functional aspects. *Eur. J. Biochem.* **218**: 1–27.

Lo-Man, R., Bay, S., Vichier-Guerre, S., Deriaud, E., Cantacuzene, D. and Leclerc, C. (1999) A fully synthetic immunogen carrying a carcinoma-associated carbohydrate for active specific immunotherapy. *Cancer Res.* **59**: 1520–1524.

Lu, Q. and Shur, B.D. (1997) Sperm from beta 1,4-galactosyltransferase-null mice are refractory to ZP3-induced acrosome reactions and penetrate the zona pellucida poorly. *Development* **124**: 4121–4131.

Mackiewicz, A. and Mackiewicz, K. (1995) Glycoforms of serum alpha 1-acid glycoprotein as markers of inflammation and cancer. *Glycoconj. J.* **12**: 241–247.

Medina, L., Grove, K. and Haltiwanger, R. S. (1998) SV40 large T antigen is modified with O-linked N-acetylglucosamine but not with other forms of glycosylation. *Glycobiology* **8**: 383–391.

Montreuil, J., Vliegenthart, J.F.G. and Schachter, H. (1996) Glycoproteins and disease. In: *New Comprehensive Biochemistry*, Vol. 30. Elsevier, New York.

Naven, T.J., Harvey, D.J., Brown, J. and Critchley, G. (1997) Fragmentation of complex carbohydrates following ionisation by matrix-assisted laser desorption with an instrument fitted with time-lag focusing. *Rapid Commun. Mass Spectrom.* **11:** 1681–1686.

Oxley, D., Munro, S.L., Craik, D.J. and Bacic, A. (1998) Structure and distribution of N-glycans on the S7-allele stylar self-incompatibility ribonuclease of *Nicotiana alata. J. Biochem. (Tokyo).* **123:** 978–983.

Packer, N.H. and Harrison, M.J. (1998) Glycobiology and proteomics: is mass spectrometry the holy grail? *Electrophoresis* **19:** 1872–1882.

Packer, N.H., Wilkins, M.R., Golaz, O., Lawson, M.A., Gooley, A.A., Hochstrasser, D.F., Redmond, J.W. and Williams, K.L. (1996) Characterization of human plasma glycoproteins separated by two-dimensional gel electrophoresis. *Bio/Technology* **14:** 66–70.

Packer, N.H., Lawson, M.A., Jardine, D.R., Sanchez, J.-C. and Gooley, A.A. (1998) Analysing glycoproteins separated by two-dimensional gel electrophoresis. *Electrophoresis* **19:** 981–988.

Powell, A.K. and Harvey, D.J. (1996) Stabilization of sialic acids in N-linked oligosaccharides and gangliosides for analysis by positive ion matrix-assisted laser desorption/ionization mass spectrometry. *Rapid Commun. Mass Spectrom.* **10:** 1027–1032.

Rabilloud, T., Valette, C. and Lawrence, J.J. (1994) Sample application by in-gel rehydration improves the resolution of two-dimensional electrophoresis with immobilized pH gradients in the first dimension. *Electrophoresis* **15:** 1552–1558.

Sanchez, J.-C., Appel, R.D., Golaz, O., Pasquali, C., Ravier, F., Bairoch, A. and Hochstrasser, D.F. (1995) Inside SWISS-2DPAGE database. *Electrophoresis* **16:** 1131–1151.

Seppala, M., Koistinen, H., Koistinen, R., Mandelin, E., Oehninger, S., Clark, G.F., Dell, A. and Morris, H.R. (1998) Glycodelins: role in regulation of reproduction, potential for contraceptive development and diagnosis of male infertility. *Hum. Reprod.* **13(Suppl. 3):** 262–291.

Seregni, E., Botti, C. and Bombardieri, E. (1995) Biochemical characteristics and clinical applications of alpha-fetoprotein isoforms. *Anticancer Res.* **15:** 1491–1499.

Shiyan, S. D. and Bovin, N.V. (1997) Carbohydrate composition and immunomodulatory activity of different glycoforms of alpha1-acid glycoprotein. *Glycoconj. J.* **14:** 631–638.

Starr, C.M., Klock, J.C., Skop, E., Masada, I. and Giudici, T. (1994) Fluorophore-assisted electrophoresis of urinary carbohydrates for the identification of patients with oligosaccharidosis- and mucopolysaccharidosis-type lysosomal storage diseases. *Glycosyl. Dis.* **1:** 165–176.

Susuki, S. and Honda, S. (1998) A tabulated review of capillary electrophoresis of carbohydrates. *Electrophoresis* **19:** 2539–2560.

Tagliaro, F., Crivellente, F., Manetto, G., Puppi, I., Deyl, Z. and Marigo, M. (1998) Optimized determination of carbohydrate-deficient transferrin isoforms in serum by capillary zone electrophoresis. *Electrophoresis* **19:** 3033–3039.

Taketa, K. (1998) Characterization of sugar chain structures of human alpha-fetoprotein by lectin affinity electrophoresis. *Electrophoresis* **19:** 2595–2602.

Taverna, M., Tran, N.T., Merry, T., Hovarth, E. and Ferrier, D. (1998) Electrophoretic methods for process monitoring and the quality assessment of recombinant glycoproteins. *Electrophoresis* **19:** 2572–2594.

Taylor-Papadimitriou, J.J. and Epenetos, A.A. (1994) Exploiting altered glycosylation patterns in cancer: progress and challenges in diagnosis and therapy. *TiBTech.* **12:** 227–233.

van Dam, G., Bogitsh, B., van Zeyl, R., Rotmans, J. and Deelder, A. (1996) *Schistosoma mansoni*: in vitro and in vivo excretion of CAA and CCA by developing schistosomula and adult worms. *J. Parasitol.* **82:** 557–564.

Van den Steen, P., Rudd, P.M., Proost, P., Martens, E., Paemen, L., Kuster, B., van Damme, J., Dwek, R.A. and Opdenakker, G. (1998a) Oligosaccharides of recombinant mouse gelatinase B variants. *Biochim. Biophys. Acta* **1425:** 587–598.

Van den Steen, P., Rudd, P.M., Dwek, R.A. and Opdenakker, G. (1998b) Concepts and principles of O-linked glycosylation. *Crit. Rev. Biochem. Mol. Biol.* **33:** 151–208.

van Dijk, W., Havenaar, E.C. and Brinkman-van der Linden, E.C. (1995) Alpha 1-acid glycoprotein (orosomucoid): pathophysiological changes in glycosylation in relation to its function. *Glycoconj. J.* **12:** 227–233.

Varki, A. (1993) Biological roles of oligosaccharides: all of the theories are correct. *Glycobiology* **3:** 97–130.

Voshol, H., Dullens, H.F., Otter, W.D. and Vliegenthart, J.F. (1996) The glycosylation profile of interleukin-2 activated human lymphocytes correlates to their anti-tumour activity. *Anticancer Res.* **16:** 155–159.

Winchester, B., Clayton, P., Mian, N., di-Tomaso, E., Dell, A., Reason, A. and Keir, G. (1995) The carbohydrate-deficient glycoprotein syndrome: an experiment of nature in glycosylation. *Biochem. Soc. Trans.* **23:** 185–188.

Wright, A. and Morrison, S.L. (1997) Effect of glycosylation on antibody function: implications for genetic engineering. *TiBTech.* **15:** 26–32.

Youakim, A., Hathaway, H., Miller, D., Gong, X. and Shur, B. (1994) Overexpressing sperm surface beta 1,4-galactosyltransferase in transgenic mice affects multiple aspects of sperm–egg interactions. *J. Cell Biol.* **126:** 1573–1583.

Yuasa, I., Ohno, K., Hashimoto, K., Iijima, K., Yamashita, K. and Takeshita, K. (1995) Carbohydrate-deficient glycoprotein syndrome: electrophoretic study of multiple serum glycoproteins. *Brain Devl.* **17:** 13–19.

Zara, J. and Naz, R.K. (1998) The role of carbohydrates in mammalian sperm–egg interactions: how important are the carbohydrate epitopes? *Front. Biosci.* **3:** d1028–1038.

Ziltener, H.J., Clark-Lewis, I., Jones, A.T. and Dy, M. (1994) Carbohydrate does not modulate the *in vivo* effects of injected interleukin-3. *Exp. Hematol.* **22:** 1070–1075.

Chapter 14

Proteomics in an academic environment

Michael A. Cahill and Alfred Nordheim

1. Introduction

The affiliations of authors of this book are distributed roughly equally between academia and industry. An examination of the research budgets available to these two groups is likely to indicate that academic expenditure is only a fraction of that available to industry. Given that the introduction of new technologies, as required to support proteomics, is expensive, it has been suggested by some that academic proteomics is unlikely to be competitive. In spite of this we and many others are attempting to establish a proteome infrastructure, in our case 'from scratch', in a newly established Interfaculty Institute for Cell Biology at the University of Tuebingen. Our laboratory has traditionally been recognized in the fields of mammalian transcriptional regulation and signal transduction (Arsenian *et al.*, 1998; Cahill *et al.*, 1996; Heidenreich *et al.*, 1999; Hipskind *et al.*, 1991). The potential of applying proteomics to this area prompted us to embrace the technology in a dedicated manner. Establishment of a proteomics facility has now been in progress for 30 months in our institute, and at the time of writing the proteome team has two postdocs, two students and two technical assistants. We are coordinating technological improvements, as well as projects from the clinical and biological disciplines appropriate for our interfaculty status. We have successfully secured funding for the purchase of isoelectric focusing equipment, a phosphorimager, an automated protein digestion station, a matrix-assisted laser desorption time of flight (MALDI-TOF) mass spectrometer (MS), a robot for transfer of protein digests onto the MALDI sample stage, and a quadrupole time of flight hybrid mass spectrometer. We have also established strong contacts with our local clinicians and with some businesses. Several proteomics-related patents have been, and we have submitted one paper for publication which is currently under review.

2. Why not do proteome research in an academic environment?

Proteome analysis is a technology with the potential to change the face of medical science. Some schools of thought argue that proteomic research should not be conducted in academic environments because of the high costs involved. While we

Proteomics, edited by S.R. Pennington and M.J. Dunn
© 2001 BIOS Scientific Publishers, Oxford.

recognize that the cost of equipment and consumables involved in protein analysis, and especially in high-throughput automation, means that much of the pharmaceutical-related proteome work will probably be performed by commercial organizations, there is still significant potential and value to be extracted in establishing state-of-the-art proteomics facilities in academic institutions.

Another argument put forward suggests that high-throughput activities are repetitive, automated and intellectually undemanding, whereas academic work should be original and novel. The fallacy of this argument will be immediately apparent to anybody in an academic environment who has prepared and screened 100 plasmid DNA mini-preparations to identify the correct DNA vector construct that subsequently enabled critical experiments to be performed.

A more serious problem for the viability of proteomics facilities in academic institutions is that trained staff have skills that are highly desirable to industry, and that industry is able to lure these staff away from academic laboratories by offering higher salaries and seemingly greater security. This is particularly a problem in countries, including Germany, where academic salary levels are predetermined and not negotiable, so that academic groups often have difficulties in retaining well-qualified highly trained staff.

3. Why do proteome research in an academic environment?

There are a number of factors which, when considered together, argue strongly for a valuable role for the involvment of universities and nonprofit organizations in the field of proteomics. Consider firstly the argument that academic and nonprofit institutes cannot afford proteomics.

3.1. Can universities afford proteomics?

The ongoing costs of proteome analysis are not inherently more expensive than other types of research, for example mouse molecular genetics, once the costs of establishing a suitable facility have been met. Even these costs are not overly high relative to other types of research. We have estimated that the costs to establish a proteome laboratory as opposed to a similarly staffed laboratory focusing on conventional molecular biology as within a factor of three. It is clear that the facilities to identify proteins cost money; however, when the quality and volume of data generated by proteome research are considered, as well as its potential impact, there is a strong argument that proteomics is more cost effective than other areas of life sciences research!

The cost of running 2-DE gels. The expenditure of our proteome laboratory from March to December in 1998 is shown *Figure 1* (in Euro). During this period approximately 1300 gels of 13% polyacrylamide were run, with standard dimensions of approximately 20 × 20 cm, containing piperazine diacrylamide (PDA) as cross-linker. The vast majority of gels were stained with the diamine silver protocol (Hochstrasser *et al.*, 1988; Rabilloud 1992). Much of the PDA (a large cost factor)

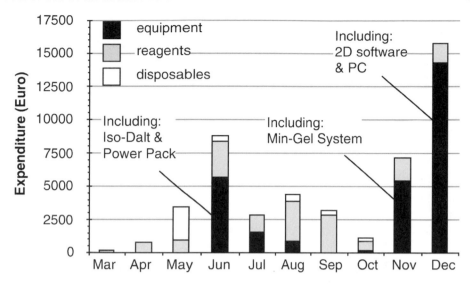

Figure 1. Laboratory expenditure (equipment, reagents and disposables) for proteomics activities at the Department of Molecular Biology, University of Tübingen, from March to December 1998.

was purchased before March, which would escalate the actual costs during the indicated period by perhaps 2000 Euro. Of the approximately 40 000 Euro spent, the majority was spent on equipment, which any newly established laboratory must purchase. During the period of peak activity from June to December, running costs were relatively constant at about 2500 Euro/month. This was for a staff of two postdocs, two students and one technician, which comes to about 500 Euro per month per staff member, and falls well within the expenditure ranges normally allocated to consumables by funding bodies. Thus, we conclude that the cost of running 2-DE gels in the laboratory is comparable to the cost of other scientific endeavors.

The cost of analyzing 2-D data. With conventional published techniques for proteome analysis, the most expensive commodity in our laboratory is the time of staff employed. Proteomics is currently labor-intensive. Due to gel-to-gel variability, we find it is necessary to run three to six replicates of any particular sample, preferably from duplicate samples. These gels are then documented digitally and aligned using commercial 2-DE image analysis software. We use both the ImageMaster 2-D Elite software (purchased from Amersham Pharmacia BioTech), and the MELANIE II software from the Swiss 2D-PAGE group (purchased from Bio-Rad). The software runs on a dual processor 350 MHz Pentium II personal computer with a 10 GB hard disk operating under Windows NT 4.0 with 500 MB RAM, set to 3000 MB virtual memory. The graphic processing time is acceptable with this configuration, yet neither software is satisfactory in our experience. It has been argued by some that this is unfortunately also the case for the technical support offered by some companies. Other users have also expressed similar sentiments to us, so let the buyer beware! No matter which of these software products are purchased, the investigator will have to spend many hours per gel to achieve a satisfactory data set where all proteins are aligned and matched on all gels within the experiment. For instance, with a cerebrospinal fluid (CSF) project being pursued in our group we are using immobilized pH gradients (IPGs) with a pH range of 4–7, and observe between 1000 and 1400 spots per gel. Automatic spot detection recognizes many pairs of proteins between gels as one spot, but unfortunately not consistently between different gels. Therefore the researcher must manually confirm or invalidate each spot to ensure its correctness. In our hands this system with the use of the automatic spot detection feature followed by editing, which takes over 4 h of staff time per gel, is only marginally quicker than adding all spots manually, which takes 4.5 h per gel. In the CSF study there were 15 control patients, and five with a well-diagnosed illness. Five gels were run from each patient to generate a total experiment of 100 gels. This required 400 h for computer analysis, which is by far the most expensive component of the analysis. Similar methods are employed by others (Andersen et al., 1997).

Indeed, from our discussions with industry representatives, this labor cost, which is structured differently in business than in academia, is a major detractant from the appeal of industrial proteomics. There are software programs that are not publicly available but that perform better than the commercially available ones and several organizations are pursuing strategies to circumvent the image analysis bottleneck. However, for many applications a university is, with current publicly accessible technology, in a better position to perform proteomics gel analyses than many industry concerns. Furthermore, with their inherent diversity of expertise, such institutions may be better placed to improve existing software and/or develop new software.

The cost of protein identification. The reagents required for protein identification in particular enzymes are expensive, but no more so than those required to undertake routine molecular biology. By far the most demanding cost is the purchase of state-of-the-art instrumentation for protein identification. Mass spectrometers are widely recognized as the method of choice for this purpose (Burlingame et al., 1998; Gygi et al., 1999; James, 1997; Yates, 1998). In particular, MALDI-TOF mass spectrometers offer the potential for high-throughput identifications. Scientists from the company Hoffman–La Roche reported at the 3rd Siena 2D Electrophoresis Meeting in September 1998 that they can process 500 proteins for identification per day by MALDI-MS. A major priority of mass spectrometer manufacturers is to offer this level of throughput as routine in the future. Such an automated system, from gel spot excision to MALDI mass spectrum, would cost approximately 500 000 Euro. Additionally, a climatized and preferably very clean laboratory equipped with at least two expert staff would be desirable. If a cDNA sequence for a particular protein exists in a database this facility should be able to identify the protein. Soon all human proteins should fall into this category.

For proteins whose sequence is not known, such as those from less well-characterized organisms or proteins translated from many low-abundance human mRNAs, *de novo* determination of at least part of the amino acid sequence of the protein is necessary for identification. This is possible with the postsource decay (PSD) function of modern MALDI mass spectrometers (Chaurand *et al.*, 1999), or alternatively electrospray mass spectrometers can be employed such as triple quadrupole, ion-trap, or Q-ToF machines (Burlingame *et al.*, 1998; Kuster and Mann, 1998; Yates, 1998; see also Chapter 5). Another often underrated but eminently useful option is the application of micro-HPLC peptide fractionations, followed by Edman degradation. Modern capillary HPLC Edman sequencers have a sensitivity limit of 10–20 fmol; however the real problem area is loss of sample during HPLC separation of the enzymatic digestion mixture into individual peptides that can be sequenced. Hence, protein identification requires potentially an arsenal of machines that range in price from 50 000 to 400 000 Euro each, and are in a high-risk category for quickly becoming outdated technology. At face value there certainly seems to be a compelling case that such a 'Machine Park' should be a centralized facility to maximize return on the capital investment. For an investment of a half-million Euro in equipment a highly functional and competitive facility could be established, and for one million or more well spent it would be an excellent one. It is the affordability of this level of expenditure that critics of university-based proteomics call into question. Nevertheless, effective if not state-of-the-art proteomics facilities can be established using nonautomated protein digestion, and less expensive MALDI-MS instrumentation, and this may be sufficient for some laboratories (Zubritsky, 1998). In our opinion it is not unreasonable to propose that a country the size of Germany should be able to support multiple such centers in academic circles. The benefits to the scientific community and to medical and biotechnological research will more than justify the costs.

The cost of information management. The generation of large and complex datasets requires concomitantly sophisticated database archiving and analysis software, not to mention powerful computers. In principle, proteomics data must be generated, documented and archived in a database. Several large companies are developing their own software and databases. These bioinformatics engines are designed to cogently document and interconnect as many relevant parameters as possible in database format. Powerful algorithms then attempt to extract patterns from the primary data, which are then used for data interpretation. This in turn should allow modeling and simulation, which eventually would produce experimentally verifiable predictions and reveal the true complexity of the system.

   Although the specialized commercial 2-DE analysis software packages mentioned above exist, their performance falls far short of these industrial products. Nor to our knowledge is any similar database management and analysis system under development. A coordinated program involving the expenditure of millions of Euro would be required to produce such a product, and presently it seems that only the larger industries are willing to embark on such a dedicated course. This is probably the area in which it can most justifiably be argued that universities cannot perform excellent proteome analysis without industrial assistance.

3.2. Can industry afford proteomics without academia?

Although the research and development budgets of the large pharmaceutical companies are considerable, they are not unlimited and are increasingly under pressure. For instance, painting with a very broad brush, pharmaceutical companies currently have to generate a return of 15–20% p.a. in the USA to keep pace with the Dow Jones index. Effectively, investors expect this profit return to be realized annually, otherwise they will remove their investment capital and place it elsewhere. Board members must decide how much to invest in the future, through research and development, and how much will satisfy the shareholders. In this fiscal environment the limited research funds of a pharmaceutical company also have to be spread over many other high-cost technologies that compete with proteomics. What is relevant is whether the allocation to proteomics within commercial organizations is likely to be sufficient to support development of the technology, which presently requires significant optimization. The cost-effectiveness of industrial investments will be improved if they use more efficient technology. Historically, much new technology has been initiated and/or developed by academia. New technologies are transferred to industrial settings where they are optimized for specialized applications. Particularly, it is likely that academic groups will develop innovative new strategies and processes that will improve industrial efficiency, but only if they have access to suitable facilities. For this to evolve, it is crucial that a critical mass of academic centers of expertise be established and maintained. These centers will provide the additional advantage of generating valuable public domain information.

3.3. Is something rotten in the state of proteomics? Some perceived limitations

There are a number of shortcomings we perceive with current proteomics technology, and in the following section we detail some of the main technological challenges that currently impede proteomic studies. These aspects have been considered by others (Herbert et al., 1997; Jungblut et al., 1996; Wilkins et al., 1996), yet to our knowledge the significance of the current limits of proteomics has not been fully evaluated and sufficiently documented.

Protein solubility. Very hydrophobic and very large proteins are poorly represented in 2-DE gels, as discussed previously (Wilkins et al., 1996, 1998). Recent advances in chaotropes and surfactants have significantly improved this situation; however the problem is far from solved (Blisnick et al., 1998; Chevallet et al., 1998; Rabilloud, 1998, 1999; Rabilloud et al., 1997). The following points are relevant for the 85% of proteins that are soluble.

Throughput. Due to the inherent variability of 2-DE gels, multiple gels have to be performed on the same sample. Additionally, as mentioned above, commercially available 2-DE analysis software is only mildly effective at spot identification and cross matching between gels in automatic mode. Since the process of performing 2-DE is time-consuming, this is a major bottleneck in achievable throughput. We note that this bottleneck could be largely alleviated by the use of multicolor visualization of proteins from different samples electrophoresed on the same gel. The possibility of differential isotopic labeling of proteins (Langen et al., 1998) is very interesting in this light.

Sensitivity. With the exception of cases where limited material is available, such as where clinical samples are analyzed, most proteomics studies are not currently limited by sensitivity, but instead by issues connected to 2-DE gel separation and problems of detection over the broad range of protein abundances (the dynamic range of abundance issue, see below). Consider the following: similar 2-DE patterns are observed when visualizing 2 mg of whole cell protein with Coomassie blue, 100 μg protein with diamine silver, or 50 ng protein with a phosphorimager after metabolic labeling to high specific activity using ^{35}S-methionine. However, there is no advantage in visualizing 2 mg of the ^{35}S-methionine-labeled protein with a phosphorimager since practically the entire gel provides a signal and the discrimination between individual proteins is lost. Thus, sensitivity is not limiting in most cases. The amount of protein loaded is in fact optimized for the resolution capacity of the gel at a given sensitivity, which is limited by the linear dynamics of abundance of proteins (see below), and by the number of resolution windows within the gel system. To assess the validity this concept, consider the related problems of dynamic range of protein abundance and gel resolution in the following.

Dynamic range of abundance. One central problem is that the 2-DE gels currently used for proteomic research typically visualize between 2000 and 10 000 proteins per gel, depending on laboratory and sample type (Fey et al., 1997; Görg et al., 1988; Jungblut et al., 1996; Klose and Kobalz, 1995). It is estimated that the human genome encodes between 65 000 to 100 000 functional genes (Borrebaeck, 1998). Of these, between 5000 and 15 000 are thought to be expressed in single cell-types and, with multiply spliced mRNAs and post-translational modifications, the number of different spots to be expected from the proteome of one cell may be around 50 000. This is evidenced by the observation that even when up to 10 000 species are present on a 2-DE gel, these are mainly housekeeping and structural proteins (Jungblut et al., 1996). It is conceivable that the dynamic range of protein abundances within cells could be as high as 10^7. Perhaps just a few molecules of some proteins (e.g. telomerase?) are required per cell to provide biological function, whereas upwards of 20 million molecules exist of proteins which constitute only 1% of total protein. Therefore, an ideal analysis system must be capable of handling a dynamic range of seven orders of magnitude or more, and over 50 000 separation windows. We estimate that currently only some 5–20% (2000–10 000/50 000) of the most soluble protein species are detectable by 2-DE separation and protein visualization.

Resolution/protein separation. If 50 000 spots were uniformly distributed on a 20 × 20 cm gel and the spots occupied an average of 2 mm^2, they would be present at five spots per mm^2. Since the distribution of proteins over molecular weight and isoelectric points actually resembles skewed curves (Cavalcoli et al., 1997), it is obvious that certain regions of the gel are hopelessly overcrowded with poorly resolved low-abundance protein spots. This problem may be partially solved by the use of narrower range pH gradients in isoelectric focusing; however this strategy does not address the linear dynamics issue, and exacerbates the throughput problem.

The protein identification issue. Using conventional techniques, most laboratories typically require picomole (10^{-12} mol) to hundreds of femtomole amounts of protein to achieve identification of the protein. This is obviously only possible for the most abundant 5% of protein spots, assuming there are 50 000 spots per cell. At recent conferences some laboratories have claimed routine success with 40 fmol, or even 50 amol starting amounts. However, these figures are atypical.

Comparison with 'genomic' approaches. Having expounded the weaknesses of proteomics, a brief comparison with genomics techniques is appropriate. Nucleic acid-based analyses are superior in sensitivity to protein studies, largely due to the ability to amplify with PCR technology. However, it should be noted that high-throughput screening discovery studies are also limited by the sheer volumes necessary to detect low abundance mRNAs in EST sampling or SAGE studies. The sensitivity of detection of fluorescent molecules commonly used in microarray studies is also limiting when compared with radioactive methods. It is presently unclear what percentage of genes can be visualized by microchip technologies (Bowtell, 1999; Carulli *et al.*, 1998; DeRisi and Iyer, 1999; Drmanac *et al.*, 1998; Iyer *et al.*, 1999; Lipshutz *et al.*, 1999; Watson *et al.*, 1998). Chip technology can quickly and economically assay the abundance of mRNAs; however it provides no information on protein abundance. It is anticipated that the combination of both genomic and proteomic strategies will be necessary to obtain a global insight into cell and whole organism biology. It is evident that the performance of proteomics technologies will have to be significantly improved.

3.4. Academic proteomics is good for industry

From the above description of the limitations of proteomics, it may be justifiable to argue that considerable technological advances must be made before proteomics reaches a stage where industries can rapidly and cost-effectively screen a large percentage of cellular proteins, as they can at present with cellular mRNAs. The realization of this goal would provide much more instructive biological information than is provided by powerful modern genomics techniques alone. However, it is unlikely that individual commercial organizations will invest the resources necessary for such technological improvements. The task will most likely be undertaken by public-funded research centers and multidisciplinary research projects within such institutions.

4. University proteomics offers many advantages

4.1. Universities are flexible and rapid-response industrial resources. The price is right!

In the modern international climate of financial rationalization large firms continuously seek to subcontract development projects to universities. It is recognized by industry that the university system can generate more results per dollar for small to medium amounts of investment capital. Without establishing an entire new laboratory and associated infrastructure, a company can fund a short-term, high-risk project in a university, and exit if necessary at an appropriate time point without suffering excessive financial losses. Universities that establish suitable facilities and foster appropriate expertise stand to profit considerably from the resulting flow of both technology and cash from industry.

4.2. Universities offer access to clinical samples

Many universities are associated with world-class clinical research departments, offering industry collaborators unprecedented access to significant biological and medical expertise. Excellent patient case histories and practiced sample handling ensure the highest quality of results, enabling direct comparisons between relevant samples. Physical proximity to the clinics and a reliable communication base with the clinical specialists mean that sample material can be transferred to academic laboratories within minutes of leaving the operation theater, or can be proficiently and systematically frozen for later use. Personal contact with the medical specialists enables the exact requirements to be specified, and continuously ensure that the necessary quality controls function satisfactorily. It is this type of infrastructure, maintained at minimal cost to potential investors or collaborative partners, that industry seeks. It is indeed likely that industry will out-compete universities in the labor-intensive target molecule discovery field. However, for applications such as rapid on-the-spot assessment of the effects of trial drugs on test patients in phase I to phase III clinical trials, or of the onset of tissue transplant rejection at time points long before the histologist can detect it, proteome facilities attached to specialized hospital clinics offer an ideal entry position for industry partners. The importance of this consideration is difficult to overstate.

4.3. Universities don't just study medical applications

It has been cynically noted that the main aim of proteomics might be to bankrupt the health insurance system. Aside from this function, universities offer the only vehicle whereby some fundamental biological systems could be examined by proteomics. For instance much understanding into embryology and development could be obtained by the directed application of advanced proteomic and genomic studies to model organisms. The model organisms used to study these systems, such as the fruit fly, the zebra fish or the frog, are currently considered to be financially too exotic by the mainstream pharmaceutical industry. It is this type of project which could be a *forte* of academic proteomics centers, alongside and fully complementary to human biology projects.

4.4. University basic research will be publicly accessible

One consequence of the biological data management system as it exists today is that data is worth money. Biological knowledge can be transformed into marketable pharmaceutical products. This is not compatible with the deposition of information on public databases. Typically, an industrial concern will download the entire contents of public resources and use this as a foundation upon which to expand a private company database. Since a significant fraction of the mass-screening, rapid-throughput, proteome analysis will probably eventually be performed by the industrial sector, most of this valuable biological information will be unavailable to the public.

4.5. Universities train students

Basic economic tenets dictate that for any highly technical industrial process to be viable, a suitably trained skilled workforce is necessary. The cost of this training is considerable, so it is highly advantageous for a company to be able to recruit staff with suitable qualifications. The teaching of qualifications is the main role of universities, and for them to proficiently impart facultative skills in the field of proteomics, it is clear that universities must be actively involved in proteome research. Indeed much of the necessary infrastructure is pre-existing, and requires the directed coordination of the protein chemistry, cell biology and bioinformatics curricula towards proteomics.

5. Academic 'spin-off' companies

Historically, German universities have been inexpert in patenting their results to reap the commercial benefits of their research. It is widely recognized that this situation must change. Another aspect of proteomics activity in academic institutions is one we have experienced at first-hand during the period of preparation of this manuscript. That is, because of the potential economic importance of proteomics there exists intense economic interest in technological developments, such that many academic groups produce 'spin-off' companies to commercialize the expertise they have acquired. The potential economic significance of this sort of activity becomes apparent when considering some genomics companies, where for instance the technologies underpinning the firms Hyseq, Affymetrix and Incyte all arose out of academic groups. We see no such reason why similar phenomenon should not occur with proteomics, and several authors of this book (including ourselves, in the form of a new company called ProteoMed GmbH) are doing their best to realize such goals.

6. Conclusion

In summary, contrary to some opinions, we believe that proteome research in academic institutions is both desirable and necessary, offering a diversity of benefits to both industry and the academic system. From our perspective, it would be unfortunate if European universities abandoned the appropriate continued investment in this powerful technology, having provided in large part the research that underpinned its evolution. Industry has recognized that proteome research provides a huge unrealized potential. At least part of the necessary technology development will probably arise from academic institutions. Arguably, this is the most compelling reason why proteome projects should be supported in academic institutions.

References

Andersen, H.U., Fey, S.J., Larsen, P.M., Nawrocki, A., Hejnaes, K.R., Mandrup-Poulsen, T. and Nerup, J. (1997) Interleukin-1beta induced changes in the protein expression of rat islets: a computerized database. *Electrophoresis* **18:** 2091–2103.

Arsenian, S., Weinhold, B., Oelgeschlager, M., Ruther, U. and Nordheim, A. (1998) Serum response factor is essential for mesoderm formation during mouse embryogenesis. *Embo J.* **17:** 6289–6299.

Blisnick, T., Morales-Betoulle, M.E., Vuillard, L., Rabilloud, T. and Braun Breton, C. (1998) Non-detergent sulphobetaines enhance the recovery of membrane and/or cytoskeleton-associated proteins and active proteases from erythrocytes infected by *Plasmodium falciparum. Eur. J. Biochem.* **252:** 537–541.

Borrebaeck, C.A. (1998) Tapping the potential of molecular libraries in functional genomics. *Immunol. Today* **19:** 524–527.

Bowtell, D.D. (1999) Options available – from start to finish – for obtaining expression data by microarray. *Nature Genet.* **21:** 25–32.

Burlingame, A.L., Boyd, R.K. and Gaskell, S.J. (1998) Mass spectrometry. *Anal. Chem.* **70:** 647R–716R.

Cahill, M.A., Janknecht, R. and Nordheim, A. (1996) Signalling pathways: jack of all cascades. *Curr. Biol.* **6:** 16–9.

Carulli, J.P., Artinger, M., Swain, P.M., Root, C.D., Chee, L., Tulig, C., Guerin, J., Osborne, M., Stein, G., Lian, J. and Lomedico, P.T. (1998) High throughput analysis of differential gene expression. *J. Cell Biochem.* **30–31** (Suppl.): 286–296.

Cavalcoli, J.D., VanBogelen, R.A., Andrews, P.C. and Moldover, B. (1997) Unique identification of proteins from small genome organisms: theoretical feasibility of high throughput proteome analysis. *Electrophoresis* **18:** 2703–2708.

Chaurand, P., Luetzenkirchen, F. and Spengler, B. (1999) Peptide and protein identification by matrix-assisted laser desorption ionization (MALDI) and MALDI-post-source decay time-of-flight mass spectrometry. *J. Am. Soc. Mass Spectrom.* **10:** 91–103.

Chevallet, M., Santoni, V., Poinas, A., Rouquie, D., Fuchs, A., Kieffer, S., Rossignol, M., Lunardi, J., Garin, J. and Rabilloud, T. (1998) New zwitterionic detergents improve the analysis of membrane proteins by two-dimensional electrophoresis. *Electrophoresis* **19:** 1901–1909.

DeRisi, J.L. and Iyer, V.R. (1999) Genomics and array technology. *Curr. Opin. Oncol.* **11:** 76–79.

Drmanac, S., Kita, D., Labat, I., Hauser, B., Schmidt, C., Burczak, J.D. and Drmanac, R. (1998) Accurate sequencing by hybridization for DNA diagnostics and individual genomics. *Nature Biotechnol.* **16:** 54–58.

Fey, S.J., Nawrocki, A., Larsen, M.R., Gorg, A., Roepstorff, P., Skews, G.N., Williams, R. and Larsen, P.M. (1997) Proteome analysis of *Saccharomyces cerevisiae*: a methodological outline. *Electrophoresis* **18:** 1361–1372.

Görg, A., Postel, W. and Günther, S. (1988) The current state of two dimensional electrophoresis with immobilised pH gradients. *Electrophoresis* **9:** 531–546.

Gygi, S.P., Han, D.K.M., Gingras, A.-C., Sonenberg, N. and Aebersold, R. (1999) Protein analysis by mass spectrometry and sequence database searching: tools for cancer research in the post-genomic era. *Electrophoresis* **20:** 310–319.

Heidenreich, O., Neininger, A., Schratt, G., Zinck, R., Cahill, M.A., Engel, K., Kotlyarov, A., Kraft, R., Kostka, S., Gaestel, M. and Nordheim, A. (1999) MAPKAP kinase 2 phosphorylates serum response factor in vitro and in vivo. *J. Biol. Chem.* **274:** 14434–14443.

Herbert, B.R., Sanchez, J.-C. and Bini, L. (1997). Two-dimensional electrophoresis: the state of the art and future directions. In: *Proteome Research: New Frontiers in Functional Genomics* (ed. M.R. Wilkins, K.L. Williams, R.D. Appel, and D.F. Hochstrasser). Springer, Berlin, pp. 13–33.

Hipskind, R.A., Rao, V.N., Mueller, C.G., Reddy, E.S. and Nordheim, A. (1991) Ets-related protein Elk-1 is homologous to the c-fos regulatory factor p62TCF. *Nature* **354:** 531–544.

Hochstrasser, D.F., Patchornik, A. and Merril, C.R. (1988) Development of polyacrylamide gels that improve the separation of proteins and their detection by silver staining. *Anal. Biochem.* **173:** 412–423.

Iyer, V.R., Eisen, M.B., Ross, D.T., Schuler, G., Moore, T., Lee, J.C.F., Trent, J.M., Staudt, L.M., Hudson, J., Jr., Boguski, M.S., Lashkari, D., Shalon, D., Botstein, D. and Brown, P.O. (1999) The transcriptional program in the response of human fibroblasts to serum. *Science* **283:** 83–87.

James, P. (1997) Of genomes and proteomes. *Biochem. Biophys. Res. Commun.* **231:** 1–6.

Jungblut, P., Thiede, B., Zimny-Arndt, U., Muller, E.C., Scheler, C., Wittmann-Liebold, B. and Otto, A. (1996) Resolution power of two-dimensional electrophoresis and identification of proteins from gels. *Electrophoresis* **17:** 839–847.

Klose, J. and Kobalz, U. (1995) Two-dimensional electrophoresis of proteins: an updated protocol and implications for a functional analysis of the genome. *Electrophoresis* **16:** 1034–1059.

Kuster, B. and Mann, M. (1998) Identifying proteins and post-translational modifications by mass spectrometry. *Curr. Opin. Struct. Biol.* **8:** 393–400.

Langen, H., Fountalakis, M., Evers, S., Wipf, B. and Berndt, P. (1998). ^{15}N- and ^{13}C- labeling of cells for identification and quantification of proteins on 2D gels. In: *From Genome to Proteome: 3rd Siena 2D Electrophoresis Meeting*, Siena, Italy.

Lipshutz, R.J., Fodor, S.P., Gingeras, T.R. and Lockhart, D.J. (1999) High density synthetic oligonucleotide arrays. *Nature Genet.* **21:** 20–24.

Rabilloud, T. (1992) A comparison between low background silver diammine and silver nitrate protein stains. *Electrophoresis* **13:** 429–439.

Rabilloud, T. (1998) Use of thiourea to increase the solubility of membrane proteins in two-dimensional electrophoresis. *Electrophoresis* **19:** 758–760.

Rabilloud, T. (1999) Solubilization of proteins in 2-D electrophoresis. An outline. *Methods Mol. Biol.* **112:** 9–19.

Rabilloud, T., Adessi, C., Giraudel, A. and Lunardi, J. (1997) Improvement of the solubilization of proteins in two-dimensional electrophoresis with immobilized pH gradients. *Electrophoresis* **18:** 307–316.

Watson, A., Mazumder, A., Stewart, M. and Balasubramanian, S. (1998) Technology for micro-array analysis of gene expression. *Curr. Opin. Biotechnol.* **9:** 609–614.

Wilkins, M.R., Sanchez, J.C., Williams, K.L. and Hochstrasser, D.F. (1996) Current challenges and future applications for protein maps and post-translational vector maps in proteome projects. *Electrophoresis* **17:** 830–838.

Wilkins, M.R., Gasteiger, E., Sanchez, J.C., Bairoch, A. and Hochstrasser, D.F. (1998) Two-dimensional gel electrophoresis for proteome projects: the effects of protein hydrophobicity and copy number. *Electrophoresis* **19:** 1501–1505.

Yates, J.R. III (1998) Mass spectrometry and the age of the proteome. *J. Mass Spectrom.* **33:** 1–19.

Zubritsky, E. (1998) MALDI-TOFMS: biomolecules and beyond. *Anal. Chem.* **70:** 733A–737A.

Chapter 15

Proteomics as a tool for plant genetics and breeding

Hervé Thiellement, Christophe Plomion and Michel Zivy

1. Introduction

Since the original description of the technique of two-dimensional gel electrophoresis (2-DE) of proteins, by O'Farrell (1975) and Klose (1975), plant biologists have used this technique for various purposes, including for genetic or physiological studies. The opportunity to reveal several hundreds of gene products on one single gel (see *Figure 1*) has permitted examination of the differences in protein patterns between genotypes, and in this way genetic distances have been estimated and phylogenetic relationships established. Variations in gene expression have also been studied during development and in response to various treatments, and this has led to the identification of the proteins whose expression is regulated during these conditions. Mutants have been characterized at the protein level and pleiotropy estimated. The genes encoding the identified proteins have been mapped and their products show a variation in isoelectric point or molecular mass (i.e. Position Shift or PS variants, *Figure 2*). The expressed genes can be added to the genetic maps currently established with nucleic acid markers. In addition, thanks to dedicated software (reviewed in Appel *et al.*, 1997), a protein amount can be treated as a quantitative trait amenable to quantitative trait loci (QTL) analysis procedures, which localize on the genetic map the factors controlling the variability of this amount.

Technological advances in characterizing the protein spots from 2-DE gels are currently being developed in plant biology. Furthermore, several plant protein databases are in development. This initiative in plant proteomics will be significant, since the first plant genome (*Arabidopsis thaliana*), will be entirely sequenced at the end of 2000 and the integration of proteome and genome data will be important. The recently developed 'transcriptome' tools, such as high-density cDNA filters, cDNA microarrays or DNA chips (Castellino, 1997; Lemieux *et al.*, 1998; Marshall and Hodgson, 1998), will be usefully complemented by proteomics, although it is noteworthy that the amounts of a protein and of its mRNA are not always correlated (Anderson and Anderson, 1998; Anderson and Seilhammer, 1997; Haynes *et al.*, 1998). Further, since protein turnover and post-translational modifications cannot be studied at the cDNA or mRNA level, the

Proteomics, edited by S.R. Pennington and M.J. Dunn
© 2001 BIOS Scientific Publishers, Oxford.

Figure 1. 2-DE gel from leaf proteins of *Arabidopsis thaliana.* The first dimension is in IPG strips with nonlinear pH gradient from 3.5 to 10; the second SDS dimension is in homogeneous acrylamide/PDA 12%T; the protein spots are silver stained.

importance of on-going proteomics is and will continue to be of critical importance.

Several reviews on plant proteomics have recently been published (de Vienne *et al.*, 1996, 1999; Thiellement *et al.*, 1999) and here examples of the various uses of proteomics in plant biology will be described, with a special emphasis on the contribution that plant proteomics has made to the progress of plant genetics.

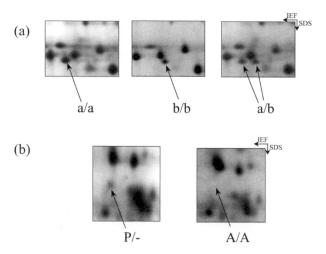

Figure 2. (a) Cuttings from 2-DE gels showing a position shift (PS) variation in the needles of an F2 maritime pine progeny. Panel a1, homozygote genotype 'a/a'; panel a2, homozygote genotype 'b/b'; panel a3, heterozygote genotype 'a/b'. (b) Cuttings from 2-DE gels showing a presence/ absence (P/A) variation in the needles of an F2 maritime pine progeny. Panel b1, phenotype 'present', genotype P/-; panel b2, phenotype 'absent', homozygote null genotype 'A/A'. IEF, first dimension; SDS, second dimension.

2. Genetic diversity analysis

2.1. Inter- and intra-specific genetic differentiation

The extent of genetic variability in natural populations has always been a subject of great interest to population geneticists, evolutionists and plant breeders. While isozymes and DNA-based markers have been and are still intensively used, proteins revealed by 2-DE were seldom utilized for population genetics studies. However, 2-DE allows far more markers to be revealed than isozyme electrophoresis, for which few assays are available, and de Vienne *et al.* (1996) reported that 2-DE revealed as many alleles per polymorphic loci as isozymes. Compared with DNA markers, 2-DE reveals the genetic variability only of expressed genes. For the last 20 years, experiments have been performed at various taxonomic levels to assess genetic differences using 2-DE protein patterns, and some examples are detailed below.

Interspecific variability. Cultivated wheats are allopolyploids species possessing either two (for hard wheat, *Triticum turgidum* AABB) or three (for soft wheat, *Triticum aestivum* AABBDD) genomes. These genomes originate from a common ancestor that has since diverged, and today there are numerous diploid or polyploid wild species with different so-called 'homoeologous' genomes that still can be intercrossed, more or less easily. The number of bivalent chromosome figures at meiosis led wheat geneticists to hypothesize which of the present species may have given rise to the different genomes of the cultivated wheats. The A genome originated from *Triticum urartu* and the D genome from *T. taushii* (reviewed by Kerby and Kuspira, 1987). However, there was still a conflicting debate about which species of the *Sitopsis* section contributed the B genome and the cytoplasm of the cultivated wheats. To answer this question, 2-DE patterns of the different species of this section (*T. longissimum, T. sharonense, T. bicorne, T. searsii* and *T. speltoides*) were cross-compared and compared with Chinese spring (CS), a bread wheat cultivar. By analyzing the number of protein spots found in common, similarity indices were computed between each pair of genotypes. From the resulting similarity matrix, dendograms were drawn, reflecting the phylogenetic relationships in the *Sitopsis* section. In addition to the perfect accordance between the 2-DE-based dendrogram and the classical taxonomy, it was found that *T. speltoides* is the wild species that is most closely related to the B genome of cultivated wheats (Bahrman *et al.*, 1988a). This finding was confirmed by analysis of the cytoplasmic proteins: the large subunit of Rubisco and two forms of the beta-ATPase (Bahrman *et al.*, 1988b) have the same allelic forms in CS and *T. speltoides*, while the other *Sitopsis* species show other alleles of these chloroplastic genes.

The study of phylogenetic relationships by comparing 2-DE patterns has been extended to species of the *Triticum* genus possessing different genomes (Thiellement *et al.*, 1989). In another experiment between different *Triticeae* genus, that is the A and D genomes of *Triticum*, the H genome of *Hordeum* (barley) and the S genome of *Secale* (rye), the dendrograms obtained reflected well the known phylogenetic relationships (Zivy *et al.*, 1995), but the number of protein spots found in common dropped to 30% between *Triticum* and *Hordeum*, indicating that the limits of the method may have been reached.

In European oaks, the relationships between the two closely related species *Quercus petraea* and *Quercus robur* were examined by Barreneche *et al.* (1996). They used 2-DE to study the genetic differentiation between the two species by comparing 23 oaks from six European countries partly covering the natural geographic range of white oaks in Europe. Total proteins from seedlings were analyzed and 530 protein spots scored, among which 101 were polymorphic. The dissimilarity between the two species was 0.36, whereas the within-species dissimilarities were 0.35 for *Q. petraea* and 0.33 for *Q. robur*. Such very close interspecific and intraspecific distances confirmed the low level of genetic differentiation between both species, already reported with isozymes, random amplified polymorphic DNAs (RAPDs) and chloroplastic DNA.

Intraspecific variability. In maritime pine (*Pinus pinaster*), Bahrman *et al.* (1994) used 2-DE to study the relationship between seven populations (provenances) of the natural range, and to evaluate the genetic variability existing within and between geographical origins. Taking advantage of the possibility of distinguishing between allelic forms of protein loci in the megagametophyte (a nutritive haploid tissue surrounding the embryo of conifer seeds), a total of 968 protein spots were scored, from which 84% were variable. Based on this information, three main groups of pine (namely Atlantic, Mediterranean and North African) could be distinguished, a genetic structure that was in agreement with terpene data. In another study, Petit *et al.* (1995) showed that proteins revealed by 2-DE displayed a similar level of genetic differentiation among populations to terpenes and isozymes, indicating the absence (or similar level) of selection acting on the three types of loci.

Burstin *et al.* (1994) studied the genetic diversity among 21 maize inbred lines by using 2-DE, isozyme and RFLP markers. Distances computed from 21 enzyme loci, 59 RFLP probes, 70 position shift (PS) loci, and the quantitative variation of 103 proteins were compared and related to coefficients of coancestry. No difference was found between PS and isozyme loci for the average number of alleles per polymorphic locus (respectively 2.4 and 2.2), which was higher for RFLP loci (4.6). Among related lines, all distances were significantly correlated to the coefficient of coancestry except the one computed according to isoenzyme loci. This is most likely due to the low number of isozyme loci tested: the correlation to the coefficient of coancestry was higher (0.85) when all PS, isoenzyme and RFLP loci were taken into account simultaneously than when they were taken into account separately. Among unrelated lines, the correlation between quantitative variation of polypeptides with all other markers dropped to 0.14. Several explanations can account for this result: (i) evolutionary forces act differently upon genes controlling protein amounts than on other loci; (ii) epistatic and pleiotropic interactions can result in a nonlinear relationship between the number of varying proteins and the number of polymorphic genes controlling their amount; (iii) the significant correlation to monogenic markers, when only related lines are considered, can be due to linkage disequilibrium.

David *et al.* (1997) studied the genetic differentiation of 11 wheat populations originating from a single population that evolved independently during 8 years of cultivation in different locations. Thirty-nine out of 162 polypeptides taken into account in this analysis proved to be polymorphic between the populations. Multivariate analysis showed that all populations differentially evolved from the original one, and that natural selection rather than random drift was responsible for these differentiations.

2.2. *Distinction of varieties, lines and cultivars*

In the 1980s, Zivy *et al.* (1983, 1984) distinguished three different wheat varieties by qualitative and quantitative differences of proteins revealed by 2-DE. The use of alloplasmic lines (the same nuclear genome introduced by repeated backcrosses in alien cytoplasms) revealed genetic differences for cytoplasmically encoded proteins (e.g. the large subunit of Rubisco). Improvements in 2-DE techniques, and particularly in the modification of procedures used to extract protein from plant tissues (Bahrman *et al.*, 1985; Colas des Francs *et al.*, 1985), and the use of larger-dimension gels have increased the number of genetic variations detected (Damerval *et al.*, 1986). Although only a fraction of the mutations in the coding sequence are detectable on 2-DE gels, the great number of gene products has allowed many authors to characterize and distinguish genotypes unambiguously, even when they belong to related populations. This was published in wheat (Anderson *et al.*, 1985; Dunbar *et al.*, 1985; Picard *et al.*, 1997), barley (Görg *et al.*, 1988, 1992), rice (Abe *et al.*, 1996; Saruyama and Shimbashi, 1992), sugarcane (Ramagopal, 1990) and in pepper (Poch et *al.*, 1992, 1994). In woody plants, Bon (1989) has applied 2-DE to meristem culture to separate clones of *Sequoiadendron giganteum*, *Sequoia semper-virens*, *Pseudotsuga menziesii*, *Picea abies*, *Pinus pinaster*, *Eucalyptus gunii* and *Populus nigra*.

When genetic distinction was examined using 2-DE, the authors found enough spot differences to realize their objectives. Data obtained allowed in many cases to establish genetic distances that were in accordance with other estimations based either on molecular or on phenotypic characteristics.

3. Genome expression

3.1. *Mutant characterization*

Examination of the 2-DE pattern of a mutant compared with the wild type may support evaluation of the effects of a mutation. It may, for instance, enable examination of the pleiotropy of a mutant which has been described already. Alternatively, it may allow identification of the protein(s) encoded or influenced by the mutated gene. In the model plant *Arabidopsis thaliana*, 2-DE patterns of a series of mutants affected in the first steps of development have been examined (Santoni *et al.*, 1994). The amount of one protein, characterized as an isoform of actin, was correlated to the length of the hypocotile. In tomato, the comparison of 2-DE patterns between the wild type and an Fe-deficient mutant led to the identification of several enzymes whose amounts were different and which were involved in anaerobic metabolism and stress defense (Herbik *et al.*, 1996). In experiments with moss, Kasten *et al.* (1997) compared the proteins of a chloroplastic mutant with the wild type, whether supplemented with cytokinin or not. It was concluded that the hormone affects both nuclear and cytoplasmically encoded proteins. The analysis of the protein patterns of a series of *Arabidopsis* mutants affected in early development, and of hormone-treated wild types, led to a biochemical classification that was consistent enough to predict that an uncharacterized mutant was probably a cytokinin overproducer (Santoni *et al.*, 1997). This hypothesis was later confirmed by cytokinin dosage (Faure *et al.*, 1998). The comparison of several late-flowering mutants of *Arabidopsis* revealed protein spots that appeared or disappeared compared to the wild type (Tacchini *et al.*, 1995). However, in the F2 offsprings of the crosses between mutants and wild type, none of the variable protein spots cosegregated with the flowering phenotype. In addition to the necessity of genetic confirmations, these experiments demonstrated that 2-DE analysis can reveal the genetic heterogeneity of the ecotypes or lines used in mutagenesis experiments. However, it must be noted that this proteome analysis examined revealed only a fraction of the proteins.

The power of proteome analysis in the characterization of mutants is further evidenced by studying pleiotropic mutations. Gottlieb and de Vienne (1988) compared two near-isogenic lines of pea differing by the *r* gene that determines round (*RR*) or wrinkled (*rr*) seeds. In the proteomes of the mature seeds, nearly 10% of the protein spots differed in amounts, confirming the numerous known physiological differences between the two types of seeds. In maize, Damerval and de Vienne (1993) studied the pleiotropic effects of the *Opaque2* (*O2*) gene, which codes for a transcription factor. The comparison of 2-DE patterns of *O2* and wild-type maize lines in several unrelated backgrounds has permitted the identification of specific targets of the *O2* gene. Several enzymes belonging to various metabolic pathways were identified, confirming that *O2* is a regulatory gene connecting different grain metabolism pathways (Damerval and Le Guilloux, 1998).

It is therefore apparent that proteomics is indeed useful and powerful for distinguishing genotypes, even in closely related backgrounds. However, since this analysis takes place relatively far from the DNA sequence, the differences observed may be numerous even when genetic differences are few. At the same time, this allows one to decipher the multiple effects of a single mutation. In addition, only the protein level is pertinent for looking at post-translational modifications that are also the subject of genetic variations.

3.2. *Variability between organs and developmental stages*

Several studies have demonstrated that organ-specific proteins are more variable than nonorgan-specific proteins and that the level of genetic variability depends on the organ or tissue considered. In maritime pine, Bahrman and Petit (1995) examined the genetic variation of proteins in three organs (needle, bud and pollen) of 18 unrelated trees. Of 902 polypeptides scored for their presence/absence variations, staining intensities and position shift variations, 27.2% were polymorphic. Among spots common to all three organs, only 18.1% were variable, while organ-specific polypeptides were characterized by a very high level of polymorphism (56% on average, ranging from 44.4% for spots specific to needles, 58.1% for pollen polypeptides and 70.4% for those of the buds).

A higher level of variability for organ-specific polypeptides was also found in maize. By comparing the proteins expressed in three organs (mesocotyl and second leaf's sheath and blade of 3-week-old seedlings) between two inbred lines, Leonardi et al. (1988) found a genetic qualitative or quantitative variation for 3.6% of 357 proteins showing no variation between organs, while 22.6% were found variable among 629 proteins showing organ-specific variations. Later, this comparison was extended to two additional genotypes (de Vienne et al., 1988): 59% of the proteins showing a variation between organs were genetically variable, compared with 18% of the stable proteins. The authors suggested that the higher level of genetic variation of organ-specific proteins might be related to a larger number of genes controlling their expression: the higher this number is, the higher the number of possible targets for mutations affecting protein amount. This hypothesis was supported by results in maritime pine where position shift variants were twice as frequent and quantitative variants five times as frequent among organ-specific spots than among spots found in every analyzed organ (Bahrman and Petit, 1995).

Reduced 'functional constraints' (Kimura, 1983) might also explain the increased level of allelic variability of organ-specific polypeptides that are expressed in a single cellular environment, compared to 'house keeping' ubiquitous proteins, as suggested by Klose (1982).

3.3. Identification and characterization of abiotic stresses responsive proteins

A great deal of interest has been focused on plant responses to abiotic stresses, mainly because of possible applications to breeding programs of cultivated species.

Heat shock proteins. In plants, as in other organisms, elevated temperatures induce the synthesis of heat shock proteins (HSPs). Their synthesis is correlated to the acquisition of thermal tolerance, that is the ability to withstand higher temperatures. Plants differ from other organisms in that they synthesize a great number of low molecular weight HSPs (LMW-HSPs), for example, Nover and Scharf (1984) showed the induction of 48 HSPs in tomato cell cultures. The correlation between thermal tolerance and HSP synthesis has been studied by comparing the effects of different treatments, or by studying their induction during development: Albernathy et al. (1989) correlated HSP synthesis with the evolution of thermal tolerance during the first stages of wheat seed imbibition. The synthesis of HSPs in different organelles has also been studied: Lund et al. (1998) showed that a particular LMW-HSP (HSP22) was strongly induced in maize mitochondria.

As HSPs were: (i) easy to identify as induced by HS treatment (no sequencing was necessary for identifying them); (ii) revealed on 2-DE patterns, not only by radioactive labeling but also by silver staining; and (iii) *a priori* known to be correlated to thermal tolerance, genetic studies were undertaken early to investigate the correlation between HSP polymorphism and the genetic variation of tolerance to high temperatures. Zivy (1987) investigated the genetic variability of HSPs in wheat. A high level of polymorphism was detected: more than one-third of the 35 detected HSPs were found to be qualitatively or quantitatively variable between three inbred lines, while only 13% of the other proteins were variable among the same genotypes. Later, El Madidi and Zivy (1993) found that thermal tolerance was not correlated with qualitative variation of HSPs but with the quantitative variation of two LMW-HSPs common to the seven tested genotypes.

Several other genetic studies were conducted with the objective of looking for causal relationships between the variation of expression of particular HSPs and the genetic variability of thermal tolerance (Krishnan *et al.*, 1989; Vierling and Nguyen, 1990). Vierling and Nguyen (1992) detected only quantitative variations between a heat-tolerant and a heat-sensitive line of *Triticum monococcum*. Later, Jorgensen and Nguyen (1995) studied the inheritance of LMW HSPs in maize seedlings. A high level of polymorphism was detected: between two lines, 15 of the 29 detected LMW HSPs were qualitatively variable, and three showed quantitative variations.

Cold acclimation proteins. Exposure to cold can induce freezing tolerance, but this mechanism may not have the same biochemical mechanism as tolerance to high temperatures, since no specific set of proteins is *a priori* known to be induced by cold. Thus, 2-DE was used for the description of protein response to cold and to identify polypeptides whose regulation could be correlated to cold acclimation: Meza-Basso *et al.* (1986) showed the induction of proteins by low temperature in rape seedlings. Later, Guy and Haskell (1988) detected polypeptides associated with cold acclimation in spinach seedlings and showed that the synthesis of these proteins was increased during the period of freezing tolerance acquisition, and reduced during reacclimation. Cabane *et al.* (1993) identified chilling-acclimation-related proteins in soybean, one of them being a heat shock protein (HSP70). Van Berkel *et al.* (1994) detected quantitative changes for 26 proteins in response to cold treatment of potato tubers. Sabehat *et al.* (1996) showed that long-term chilling tolerance of tomato fruit acquired by heat shock treatment is correlated to the persistence of HSPs. A genetic approach was initiated by Danyluk *et al.* (1991), who studied the response to cold treatment of three varieties of *Triticum aestivum*: one set of 18 proteins was transiently induced, and a second set of 53 induced proteins stayed at a high level of expression during the 4 weeks required to induce freezing tolerance. Thirty-four of the latter were expressed at a higher level in the most freezing-tolerant line.

Proteins related to drought. Water deficit induces numerous morphologic, physiologic and biochemical responses in plants, for example, reduced growth of aerial parts, stomatal closure, leaf rolling, leaf senescence and osmotic adjustment. These responses are at least partially controlled by abscissic acid (ABA), a phytohormone whose concentration increases in plants subjected to drought. Many cellular functions are affected and different types of proteins have been found to be induced by this stress. Some of them have functions directly related to water stress (e.g. aquaporines), others are related to cellular damage (HSPs, proteins related to oxidative stress, proteases and antiproteases), and others are related to different metabolic functions (reviewed in Ingram and Bartels, 1996). The first water stress-induced proteins were identified in dessicating embryos (Aspart *et al.*, 1989; Dure *et al.*, 1981; Sanchez-Martinez *et al.*, 1986). These proteins were called late embryogenesis-abundant (LEA) proteins. Essentially of unknown function, they are highly hydrophilic, and were classified in different groups according to arrangements of repeated domains (Dure *et al.*, 1989). Several of them were later shown to be induced by drought, ABA, or salinity stress in different organs. One of them, maize RAB17 (a group 2 LEA protein responsive to ABA) was shown to be differently phosphorylated in the nucleus and in the cytoplasm (Jensen *et al.*, 1998; Goday *et al.*, 1994).

In some instances, 2-DE has been used without protein identification, as a first approach for visualizing protein changes in diffcrent conditions (Chen and Tabaeizadeh, 1992; Hurkman and Tanaka, 1988; Ramani and Apte, 1997). Ramagopal (1987) showed that the response to salinity stress is different in roots and shoots of barley seedlings. Leone *et al.* (1994) showed that different sets of proteins are induced as a function of the severity of osmotic stress in potato cell suspensions. Leymarie *et al.* (1996) studied the effect of auxin on the protein response of *Arabidopsis thaliana* to drought, using auxin-insensitive mutants.

In several other studies, the induced proteins have been identified. For example, Claes *et al.* (1990) isolated a cDNA for a glycin-rich protein by using DNA primers synthesized according to microsequences of a polypeptide induced by salt stress in rice. Similarly, Reviron *et al.* (1992) identified a protease inhibitor induced by drought in rape leaves, and Moons *et al.* (1995, 1997) identified in rice a novel class of glycin-rich proteins, a group 3 LEA protein, and a series of group 2 LEA proteins (also called dehydrins, such as RAB17) by microsequencing and Western blotting. Among three rape varieties, the most tolerant to salinity synthesized higher amounts of these proteins. Rey *et al.* (1998) showed the induction by water stress of a thioredoxin-like protein in potato chloroplasts, and Moons *et al.* (1998) the induction of pyruvate orthophosphate dikinase in rice roots. Costa *et al.* (1998) found 38 proteins induced by drought in needles of maritime pine. Among them were enzymes involved in the response to oxidative stress, in the heat shock response and in lignin biosynthesis. Riccardi *et al.* (1998) found 78 proteins affected by drought in the growing part of maize leaves, 50 of which were up-regulated (see *Figure 3*). Partial sequence information was obtained for 19 of the proteins, allowing the identification of proteins involved in several metabolic pathways.

4. Genetic mapping and candidate proteins

The last few years have seen a dramatic increase in the application of molecular genetics methods to genetic mapping and quantitative trait dissection analysis (QTL mapping; Kearsey and Farquhar, 1998). While DNA-based techniques are considered the technique of choice for quickly saturating a genome, 2-DE not only

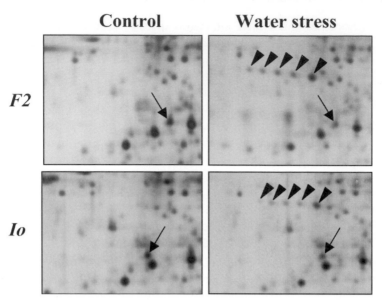

Figure 3. Proteins induced by water stress in roots of two maize inbred lines, F2 and Io. Small arrows show a train of induced proteins. Apparent molecular weight of all members of the train is slightly higher in Io than in F2, indicating that most likely all these proteins are encoded by a single polymorphic gene and differ by post-translational modifications. The large arrows show the position shift of a noninduced protein.

provides useful molecular markers for mapping the expressed genome, but also may provide 'candidate proteins' to understand the biological function of QTLs.

4.1. Genetic mapping of protein markers

Spots presenting qualitative variations, either position shifts (PS) or presence/absence (P/A) (see *Figure 2*), have been shown to be under monogenic control (reviewed by de Vienne *et al.*, 1996). PS variants are those cases where two spots with similar aspects and staining intensity are characterized by the following criteria: (i) they are located relatively close to each other on 2-D gels, having close pI and/or apparent molecular masses; (ii) they present a monogenic and codominant Mendelian mode of inheritance in a segregating pedigree [1:1 segregation ratio in haploid and double haploid lines (Plomion *et al.*, 1995; Zivy *et al.*, 1992); 1:2:1 segregation ratio in F2 selfed families (Costa *et al.*, 1999; Damerval *et al.*, 1994; Plomion *et al.*, 1997), and (iii) they correspond to the same protein. This latter criterion has been confirmed by microsequencing such allelic pairs in maize and pine (Plomion *et al.*, 1997; Touzet *et al.*, 1995a). If it is clear that a PS variation corresponds to an allelic difference in the primary structure of a protein, this parallelism is less straightforward for spots showing P/A variations. In 2-DE gels obtained from F2 offspring, such proteins revealed a clear monogenic dominant mode of inheritance (3:1 segregation ratio for presence/absence of a spot), showing that a single gene is responsible for this polymorphism. If a P/A variant corresponds to a PS variant, where a member of the pair is not detected, then the P/A locus corresponds to the structural gene of the protein. Conversely, if a P/A variant corresponds to quantitative differences with a spot below the level of detection, then the P/A locus map a major regulatory gene controlling the expression level of the protein.

In maritime pine, Bahrman and Petit (1995) showed that the level of variation in PS loci was positively associated with molecular weight, that is more such variants were found for high than for low molecular masses. A higher level of polymorphism in larger proteins was expected, as the 'variation in the substitution rate per polypeptide is due to difference either in the number of amino acids or in the substitution rate per site' (Nei, 1987). However, no correlation was found between the level of variation of P/A variants and molecular masses. These two observations (monogenic mode of inheritance and absence of relationship with the molecular mass) suggest that a P/A variant may result from a polymorphism in a major regulatory element.

Genetic localization of protein markers has mainly been reported for wheat, maize and maritime pine, although a few PS loci have also been mapped in other crops, including barley (Zivy et al., 1992) and pea (de Vienne et al., 1996). In wheat, Colas des Francs and Thiellement (1985) reported the chromosomal localization of 35 proteins comparing euploid and ditelosomic lines. In maritime pine, Bahrman and Damerval (1989) and Gerber et al. (1993) reported linkage analysis for 119 and 65 loci, respectively. A third-generation inbred pedigree of this species was recently used to map 68 proteins from haploid (Plomion et al., 1995) and diploid (Costa et al., 1999; Plomion et al., 1997) tissues. In maize, a composite linkage map showing the distribution of 65 PS loci was presented by de Vienne et al. (1996). In pine and maize, protein loci were found on each chromosome (*Figure 4*), interspersed with other markers (RFLPs in maize, RAPDs and AFLPs in pine).

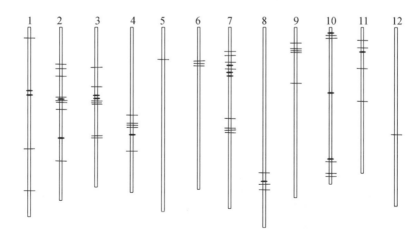

Figure 4. Schematic representation of the map location of 15 position shift variants (—) and 53 presence/absence variants (–), obtained from pine tree megagametophyte and needle 2-DE gels. The map has been constructed with RAPD and AFLP markers and is accessible via the Internet (http://www.pierroton.inra.fr/genetics/pinus/).

RFLP markers developed from cDNA libraries generally reveal several unlinked loci (Devey *et al.*, 1994; Helentjaris *et al.*, 1988; Song *et al.*, 1991) rendering the identification of the gene that is expressed, compared to silent or pseudogenes very difficult. Separation of proteins by 2-DE is thus an interesting technique for constructing maps of expressed genes and establishing which cDNA locus is expressed in different organs or during different developmental stages. For instance, a protein showing a PS variation in leaf, kernel and coleoptile of maize was identified using microsequencing and amino acid composition analysis as a phosphoglycerate mutase. The cDNA probe, used in RFLP analysis, was mapped at four different loci, on three chromosomes. The PS locus in each organ colocalized with only one (probably the expressed one) of these loci (de Vienne *et al.*, 1999).

Until now, no more than 100 protein loci have been mapped in a single plant species. The number of such markers could be substantially increased by analyzing some physiologically contrasted organs or tissues (root, leaves, bud, pollen, xylem, etc.) of the same individuals, until full coverage of the genome is obtained. Such maps of expressed genes will be of invaluable help for the candidate gene/protein strategy of QTL characterization.

4.2. Protein quantity loci (PQL) and candidate proteins

Candidate genes. During the last 10 years, the identification of the genes responsible for genetic variation in agronomically important traits has mainly been investigated using linkage mapping with anonymous markers in segregating progeny, where linkage disequilibrium can be maximized (reviewed by Tanksley, 1993). While successful, it is important to point out the limitation of such a QTL detection analysis. QTLs can only be localized approximately, that is the confidence interval of a QTL on the genetic map is large (Mangin *et al.*, 1994). This means that QTL mapping does not allow identification of the underlying genes. Thus, trait dissection analysis, although already useful in marker-assisted breeding, can be viewed as a first step towards the identification of the genes controlling part of the quantitative trait variation. In addition, based on the total map length of the genome and on DNA content, the average physical equivalent of 1 cM distance varies, for example 230, 460, 2500, 3500 and 12 000 kbases in *Arabidopsis*, Eucalyptus, maize, wheat and pine, respectively. This means that characterization of QTL by positional cloning (Rommens *et al.*, 1989) would be very difficult in most annual plants and forest trees. The accessibility of the biological meaning of a QTL can proceed from the candidate gene (CG) and the candidate protein (CP) approaches.

'Functional CG' that is, sequenced genes that putatively affect trait expression, can be proposed based on *a priori* understanding of biochemical or developmental pathways affecting the trait of interest. In addition, any locus at which mutations with large effects occur can be considered as a CG, and alleles with small effects at this CG might be responsible for naturally occurring quantitative variation. For example, Touzet *et al.* (1995b) proposed a dwarf gene involved in the giberellic acid biosynthesis pathway as a candidate for height growth QTL. Colocation between QTL and such qualitative mutations have already been reported (Beavis *et al.*, 1991;

Doebley and Stec, 1993). The main limitation of such a 'functional CG' approach is that only a small proportion of the genes affecting the trait are presently known.

'Positional CG' can be proposed for any genes that map to the same locations as QTLs for traits of interest. However, this strategy relies on a well-characterized gene map, which is only available for a few species. Colocalizations between QTLs and known-function genes has been reported previously (Byrne *et al.*, 1996; Causse *et al.*, 1995; Goldman *et al.*, 1994; Prioul *et al.*, 1997, 1999). The main limitation of the 'positional CG' approach is that a colocalization between a CG and a QTL can be purely fortuitous.

Protein quantity loci as a possible means to validate candidate genes. Validation of CG is always necessary before such information can be implemented in breeding programs. Validation or verification rely on the application of functional genomics tools (Bouchez and Höfte, 1998), among which the CP approach represents an essential step. Indeed, a CG can rarely be validated if its product does not display variability of quantity and/or activity, that is if the colocalization between a CG and a QTL is not supported by an additional colocalization with a QTL controlling its activity and/or quantity.

Damerval *et al.* (1994) were the first to investigate the genetic determinism of quantitative variation of proteins separated by 2-DE using a QTL detection strategy. They used a linkage map constructed with RFLPs and PS loci segregating in a F2 progeny of maize to locate by interval mapping (Lander and Botstein, 1989) the regulatory factors or PQL that explain part of the spot intensity variation. For the 72 proteins analyzed, 70 PQLs were detected for 42 proteins, 20 of them having more than one PQL, which can be found distributed all over the genome. Recently, PQLs controlling the accumulation of maritime pine needle proteins were detected in a F2 progeny (Costa *et al.*, 1999) and the same conclusions were drawn. An example of this approach was recently given in maize (de Vienne *et al.*, 1999), where a co-localization between a CG for drought response (the ASR1 gene, an ABA/water stress/ripening related protein) colocalized on chromosome 10, with QTLs for xylem sap, ABA content, leaf senescence and anther-silking interval (a symptomatic trait of the drought effect), and a major PQL controlling the presence/absence variation of the ASR1 protein under drought conditions.

The PQL strategy can also result in the identification of CP, that is proteins whose genetic factors controlling quantity and/or activity are colocalized with QTLs defined for the agronomic trait, while the structural gene does not. In this connection, de Vienne *et al.* (1999) showed that three PQLs controlling the quantity of a single leaf protein and three QTLs of height growth in maize were colocalized.

5. Technological advances and plant protein databases

Recent improvements in realization of 2-DE and protein characterization has allowed considerable progress in plant proteome analysis. Technical improvements, such as IPG precast strips, allow a better reproducibility of 2-DE and make it easier to run a large number of samples, and this represents a major advance in making the PQL approach more accessible. It is now relatively easy to characterize and identify polypeptide spots by microsequencing, by electrospray ionization MS/MS or MALDI-TOF mass spectrometry, followed by sequence homology searching in protein or gene databases (reviewed by Li *et al.*, 1997). Also, in the last few years, progress in software for image analysis permit the quantification and the comparison of a large number of spots in a large number of samples (see Appel *et al.*, 1997). These advances showed 2-DE to be a powerful technique with an overview of protein patterns and able to address questions of physiological or genetic interest.

While proteomics is intensively studied in an increasing number of microorganisms, as well as in humans (Humphery-Smith *et al.*, 1997; Wilkins *et al.* 1997), proteome analysis has not received as much attention in plants. In *Table 1* the main features of plant protein databases presently accessible through the Internet are listed. Databases are currently available for dicot *Arabidopsis thaliana*, and monocot rice, important crop species such as maize, and the gymnosperm *Pinus pinaster*. These databases offer scanned gels and clickable images, with characterized spots highlighted by hyperlinks. Individual protein entries are linked to other protein databases such as SWISS-PROT or TrEMBL by active cross-references. In terms of analysis of genome expression, all these databases include comparisons of polypeptides patterns in various plant organs (root, leaf, stem, bud, seed) or tissues (callus, xylem, phloem), or between genotypes. In some cases, other information, such as subcellular localizations, physiological data (effects of seasonal variations or water deficit on maritime pine needle proteins), as well as genetic data (localization of the corresponding genes on a linkage map in maize and maritime pine), are also provided.

Table 1. Plant proteomes databases

Species	Number of proteins characterized (% identified)	References	World Wide Web addresses
Rice, *Oriza sativa*	50 (58%) 59 (56%) 51 (53%)	Komatzu *et al.* (1993) Tsugita *et al.* (1994) Tsugita *et al.* (1996)	http://www.rs.noda.sut.ac.jp/~kamom/2-DE/2d.html*
Thale cress, *Arabidopsis thaliana*	57 (63%) 62 (56.5%) 150 (50%)	Kamo *et al.* (1995) Tsugita *et al.* (1996) Santoni *et al.* (1998)	http://www.rs.noda.sut.ac.jp/~kamom/2-DE/2d.html* http://sphinx.rug.ac.be:8080/ppmdb/index.html
Maize, *Zea mays*	74 (48.5%) 80 (67.5%)	Touzet *et al.* (1996) Zivy (personal communication)	http://moulon.inra.fr/imgd
Pine tree, *Pinus pinaster*	65 (90%)	Costa *et al.* (1999)	http://www.pierroton.inra.fr/genetics/2D/

*This web site is no longer available.

6. Concluding remarks, and a glance at the future

Regarding population genetic studies, it is generally accepted that the choice of a molecular screening technology for analyzing the extent and distribution of genetic diversity in natural populations will depend on many factors (Karp *et al.*, 1995). Because each type of marker presents advantages and limitations, a variety of techniques needs to be applied, among which is 2-DE. The strength of this technique is that protein loci sample the genome differently than most PCR-based techniques, and it therefore provides a different level of information with respect to the diversity questions being addressed. At present, 2-DE has seldom been utilized for studying the phylogenetic relationships. However, demonstrated in *Triticeae*, it appears to be of considerable value in the comparison of species that may not be clearly distinguished by more classical cytological or molecular approaches. Moreover, when studying related species belonging to the same family, common proteins are found. If a protein appears to be of interest in a cultivated species, such as rape, and if the gene coding an electrophoretically identical protein is already sequenced in *Arabidopsis* (another *Brassicacea*), this may dramatically shorten the time taken to establish it as a gene of agronomic interest. One can envisage the development in the coming years of such comparative proteomics. The use of protein analysis in the candidate gene approach is still in its infancy and it will be developed predominantly for the benefit of the plant breeders.

The relative limitations of 2-DE in revealing only one to two thousand more abundant proteins will be resolved by improving the extraction procedures and application of multistep techniques, and by increasing the number and the size of 2-DE gels per sample (narrowing and expending the pH range, etc.), as described in other chapters of this book. As in medicine and microbiology, proteomics will rapidly become one of the major approaches in plant biology for deciphering gene function and contributing to plant breeding programs, as exemplified above for crop species and forest trees.

Acknowledgments

Thanks are due to our colleagues at the 'Réseau I.N.R.A. Protéomes Végétaux' for their help in collecting the bibliography and for fruitful discussions. The work on forest trees conducted in Pierroton (I.N.R.A., France) was partly funded by EU projects FAIR-CT95-0781 and FAIR-CT98-3953

References

Abe, T., Gusti, R.S., Ono, M. and Sasahara, T. (1996) Variations in glutelin and high molecular weight endosperm proteins among subspecies of rice (*Oryza sativa* L.) detected by two-dimensional gel electrophoresis. *Genes Genet. Syst.* **71**: 63–68.

Albernathy, R.H., Thiel, D.S., Petersen, N.S. and Helm, K. (1989) Thermotolerance is developmentally dependent in germinating wheat seed. *Plant Physiol.* **89**: 576–596.

Anderson, N.L. and Anderson, N.G. (1998) Proteome and proteomics: new technologies, new concepts, and new world. *Electrophoresis* **19**: 1853–1861.

Anderson, L. and Seilhammer, J. (1997) A comparison of selected mRNA and protein abundances in human liver. *Electrophoresis* **18**: 533–537.

Anderson, N.G., Tollaksen, S.L., Pascoe, F.H. and Anderson, L. (1985) Two-dimensional electro-phoresis of wheat seed proteins. *Crop Sci.* **25**: 667–674.

Appel, R.D., Palagi, P.M., Walther, D., Vargas, J.R., Sanchez, J-C., Ravier, F., Pasquali, C. and Hochstrasser, D.F. (1997) Melanie II – a third-generation software package for analysis of two-dimensional electrophoresis images: 1. Features and user interface. *Electrophoresis* **18**: 2724–2734.

Aspart, L., Meyer, Y., Laroche, M. and Penon, P. (1984) Developmental regulation of the synthesis of proteins encoded by stored mRNA in radish embryos. *Plant Physiol.* **76**: 664–673.

Bahrman, N. and Damerval, C. (1989) Linkage relationships of loci controlling protein amounts in maritime pine (*Pinus pinaster* Ait). *Heredity* **63**: 267–274.

Bahrman, N. and Petit, R. (1995) Genetic polymorphisms in maritime pine (*Pinus pinaster* Ait) assessed by two-dimensional gel electrophoresis of needle, bud and pollen proteins. *J. Mol. Evol.* **41**: 231–237.

Bahrman, N., Zivy, M., Damerval, C. and Baradat, P. (1994) Organisation of the variability of abundant proteins in seven geographical origins of maritime pine (*Pinus pinaster* Ait). *Theor. Appl. Genet.* **88**: 407–411.

Bahrman, N., de Vienne, D., Thiellement, H. and Hofmann, J.P. (1985) Two-dimensional gel electrophoresis of proteins for genetic studies in Douglas fir (*Pseudotsuga mensiesii*). *Biochem. Genet.* **23**: 247–255.

Bahrman, N., Zivy, M. and Thiellement, H. (1988a) Genetic relationships in the *Sitopsis* section of *Triticum* and the origin of the B genome of the polyploid wheats. *Heredity* **61**: 473–480.

Bahrman, N., Cardin, M-L., Seguin, M., Zivy, M. and Thiellement, H. (1988b) Variability of 3 cytoplasmically encoded proteins in the *Triticum* genus. *Heredity* **60**: 87–90.

Barreneche, T., Bahrman, N. and Kremer, A. (1996) Two dimensional gel electrophoresis confirms the low level of genetic differentiation between *Quercus robur* L and *Quercus petraea* (Matt) Liebl. *For. Genet.* **3**: 89–92.

Beavis, W.D., Grant, D., Albertsen, M. and Fincher, R. (1991) Quantitative trait loci for plant height in four maize populations and their associations with qualitative genetic loci. *Theor. Appl. Genet.* **83**: 141–145.

Bon, M.C. (1989) A two-dimensional electrophoresis procedure for single meristems of different forest species. *Electrophoresis* **10**: 530–532.

Bouchez, D. and Höfte, H. (1998) Functional genomics in plants. *Plant Physiol.* **118**: 725–732.

Burstin, J., de Vienne, D., Dubreuil, P. and Damerval, C. (1994) Molecular markers and protein quantities as genetic descriptors in maize. I. Genetic diversity among 21 inbred lines. *Theor. Appl. Genet.* **89**: 943–950.

Byrne, P.F., McMullen, M.D., Snooks, M.E., Musket, T.A., Theuri, J.M., Widstrom, N.W., Wiseman, B.R. and Coe, E.H. (1996) Quantitative trait loci and metabolic pathways: genetic control of a concentration of maysin, a corn earworm resistance factor, in maize silks. *Proc. Natl Acad. Sci.* **93**: 8820–8825.

Cabane, M., Calvet, P., Vincens, P. and Boudet, A. M. (1993) Characterization of chilling-acclimation-related proteins in soybean and identification of one as a member of the heat shock protein (HSP 70) family. *Planta* **190**: 346–353.

Castellino, A. (1997) When the chips are down. *Genome Res.* **7**: 943–946.

Causse, M., Rocher, J.P., Henry, A.M., Charcosset, A., Prioul, J.L. and de Vienne, D. (1995) Genetic dissection of the relationship between carbon metabolism and early growth in maize, with emphasis on key-enzyme loci. *Mol. Breed.* **3**: 259–272.

Chen, R. D. and Tabaeizadeh, Z. (1992). Expression and molecular cloning of drought-induced genes in the wild tomato *Lycopersicon chilense*. *Biochem. Cell Biol.* **70**, 199–206.

Claes B., Dekeyser R., Villaroel R, van den Bulke M., Bauw G., van Montagu M. and Caplan A. (1990) Characterization of a rice gene showing organ-specific expression in response to salt stress and drought. *Plant Cell* **2**: 19–27.

Colas des Francs, C. and Thiellement, H. (1985). Chromosomal localization of structural genes and regulators in wheat by 2D electrophoresis of ditelosomic lines. *Theor. Appl. Genet.* **71**: 31–38.

Colas des Francs, C., de Vienne, D. and Thiellement, H. (1985) Analysis of leaf proteins by two-dimensional gel electrophoresis: protease action as examplified by ribulose bisphosphate carboxylase/oxygenase degradation and procedure to avoid proteolyse during extraction. *Plant Physiol.* **78**: 178–182.

Costa, P., Bahrman, N., Frigerio, J. M., Kremer, A. and Plomion, C. (1998) Water-deficit-responsive proteins in maritime pine. *Plant Mol. Biol.* **38**: 587–596.

Costa, P., Pot, D., Dubos, C., Frigerio, J-M., Pionneau, C., Bodénès, C., Bertocchi, E., Cervera, M-T., Remington, D.L. and Plomion, C. (2000) A genetic map of Maritime pine based on AFLP, RAPD and protein markers. *Theor. Appl. Genet.* **100**: 39–48.

Damerval, C. and de Vienne, D. (1993) Quantification of dominance for proteins pleiotropically affected by *Opaque-2* in maize. *Heredity* **70**: 38–51.

Damerval, C. and Le Guilloux, M. (1998) Characterization of novel proteins affected by the *o2* mutation and expressed during maize endosperm development *Mol. Genes Genet.* **257**: 354–361.

Damerval C., de Vienne D., Zivy M. and Thiellement, H. (1986) Technical improvements in two-dimensional electrophoresis increase the level of genetic variation detected in wheat-seedling proteins. *Electrophoresis* **7**: 52–54.

Damerval, C., Maurice, A., Josse, J.M. and de Vienne, D. (1994) Quantitative trait loci underlying gene product variation: a novel perspective for analyzing regulation of genome expression. *Genetics* **137**: 289–301.

Danyluk, J., Rassart, E. and Sarhan, F. (1991) Gene expression during cold and heat shock in wheat. *Biochem. Cell Biol.* **69**: 383–391.

David J.L., Zivy M., Cardin M.L. and Brabant, P. (1997) Protein evolution in dynamically managed populations of wheat: adaptive responses to macro-environmental conditions. *Theor. Appl. Genet.*, **95**: 932–941.

de Vienne, D., Leonardi, A. and Damerval, C. (1988) Genetic aspects of variation of protein amounts in maize and pea. *Electrophoresis* **9**: 742–750.

de Vienne, D., Burstin, J., Gerber, S., Leonardi, A., Le Guilloux, M., Murigneux, A., Beckert, M., Bahrman, N., Damerval, C. and Zivy, M. (1996) Two-dimensional electrophoresis of proteins as a source of monogenic and codominant markers for population genetics and mapping the expressed genome. *Heredity* **76**: 166–177.

de Vienne, D., Leonardi, A., Damerval, C. and Zivy, M. (1999) Genetics of proteome variation as a tool for QTL characterization. Application to drought stress responses in maize. *J. Exp. Bot.* **50**: 303–309.

Devey, M.E., Fiddler, T.A., Liu, B.-H., Knapp, S.J. and Neale, B.D. (1994) An RFLP linkage map for loblolly pine based on a three-generation outbred pedigree. *Theor. Appl. Genet.* **88**: 273–278.

Doebley, J. and Stec, A. (1993) Inheritance of the morphological differences between maize and teosinte: comparison of the results for two F2 populations. *Genetics* **134**: 559–570.

Dunbar, B.D., Bundman, D.S. and Dunbar, B.S. (1985) Identification of cultivar-specific proteins of winter wheat (*Triticum aestivum* L.) by high resolution two-dimensional polyacrylamide gel electrophoresis and color-based silver stain. *Electrophoresis* **6**: 39–43.

Dure, L. III, Greenway, S.C. and Galau, G.A. (1981) Developmental biochemistry of cottonseed embryogenesis and germination: changing messenger ribonucleic acid populations as shown by *in vitro* and *in vivo* protein synthesis. *Biochemistry* **20**: 4162–4168.

Dure, L. III, Crouch, M., Harada, J., Ho, T.H.D., Mundy, J., Quatrano, R.S., Thomas, T. and Sung, Z.R. (1989) Common amino acid sequence domains among the LEA proteins of higher plants. *Plant Mol. Biol.* **12**: 475–486.

El Madidi, S. and Zivy, M. (1993) Variabilité génétique des protéines de choc thermique et thermo-tolérance chez le blé. In: *Le progrés génétique passe-t-il par le repérage et l'inventaire des génes?* (eds. H. Chlyah and Y. Demarly). John Wiley Eurotext, London, pp. 173–184.

Faure, J.-D., Vittorioso, P., Santoni, V., Fraisier, V., Prinsen, E., Barlier, I., Van Onckelen H., Caboche, M. and Bellini, C. (1998) The PASTICCINO genes of *Arabidopsis thaliana* are involved in the control of cell division and differentiation. *Development* **125**: 909–918.

Gerber, S., Rodolphe, F., Bahrman, N. and Baradat, P. (1993) Seed-protein variation in maritime pine (*Pinus pinaster* Ait) revealed by two-dimensional electrophoresis: genetic determinism and construction of a linkage map. *Theor. Appl. Genet.* **85**: 521–528.

Goday, A., Jensen, A.B., Culianez-Macia, F.A., Mar Alba, M., Figueras, M., Serratosa, J., Torrent, M. and Pagès, M. (1994) The maize abscisic acid-responsive protein Rab17 is located in the nucleus and interacts with nuclear localization signals. *Plant Cell* **6**: 351–360.

Goldman I.L., Rocheford, T.R. and Dudley, J.W. (1994) Quantitative trait loci influencing protein and starch concentration in the Illinois long term selection maize strains. *Theor. Appl. Genet.* **87**: 217–224.

Görg, A., Postel, W., Domscheit, A. and Gunther, S. (1988) Two-dimensional electrophoresis with immobilized pH gradients of leaf proteins from barley (*Hordeum vulgare*): method, reproducibility and genetic aspects. *Electrophoresis* **9**: 681–692.

Görg, A., Postel, W., Baumer, M. and Weiss, W. (1992) Two-dimensional polyacrylamide gel electrophoresis, with immobilized pH gradients in the first dimension, of barley seed proteins: discrimination of cultivars with different malting grades. *Electrophoresis* **13**: 192–203.

Gottleib, L.D. and de Vienne, D. (1988) Assessment of pleiotropic effects of a gene substitution in Pea by two-dimensional polyacrylamide gel electrophoresis. *Genetics* **119**: 705–710.

Guy, C.L. and Haskell, D. (1988) Detection of polypeptides associated with the cold acclimation process in spinach. *Electrophoresis* **9**: 787–796.

Haynes, P.A., Gygi, S.P., Figeys, D. and Aebersold, R. (1998) Proteome analysis: biological assay or data archive? *Electrophoresis* **19**: 1862–1871.

Helentjaris, T., Weber, D. and Wright, S. (1988) Identification of the genomic locations of duplicate nucleotide sequences in maize by analysis of restriction fragment length polymorphisms. *Genetics* **118**: 353–363.

Herbik, A., Giritch, A., Horstmann, C., Becker, R., Balzer, H.-J., Bäulmein, H. and Stephan, U.W. (1996) Iron and copper nutrition-dependent changes in protein expression in a tomato wild type and the nicotianamine-free mutant *chloronerva*. *Plant Physiol.* **111**: 533–540.

Humphery-Smith, I., Cordwell, S.J. and Blackstock, W.P. (1997) Proteome research: complementarity and limitations with respect to the RNA and RNA worlds. *Electrophoresis* **18**: 1217–1242.

Hurkman, W.J. and Tanaka, C.K. (1988). Polypeptide changes induced by salt stress, water deficit, and osmotic stress in barley roots: a comparison using two-dimensional gel electrophoresis. *Electrophoresis* **9**: 781–787.

Ingram, J. and Bartels, D. (1996) The molecular basis of dehydratation tolerance in plants. *Annu. Rev. Plant Physiol. Plant Mol. Biol.* **47**: 377–403.

Jensen, A.B., Goday, A., Figueras, M., Jessop, A.C. and Pagès, M. (1998) Phosphorylation mediates the nuclear targeting of the maize Rab17 protein. *Plant J.* **13**: 691–697.

Jorgensen, J.A. and Nguyen, H.T. (1995) Genetic analysis of heat shock proteins in maize. *Theor. Appl. Genet.* **91**: 38–46.

Kamo, M., Kawakami, T., Miyatake, N. and Tsugita, A. (1995) Separation and characterization of *Arabidopsis thaliana* proteins by two-dimensional gel electrophoresis. *Electrophoresis* **16**: 423–430.

Karp, A., Kresovich, S., Bhat, K.V., Ayad, W.G. and Hodgkin, T. (1995) Molecular tools in plant genetic resources conservation: a guide to the technologies. IPGRI Technical bulletin no. 82, Rome, Italy, 47pp.

Kasten, B., Buck, F., Nuske, J. and Reski, R. (1997) Cytokinin affects nuclear- and plastome-encoded energy-converting plastid enzymes. *Planta* **201**: 261–272.

Kearsey, M.J. and Farquhar, A.G.L. (1998) QTL analysis in plants; where are we now? *Heredity* **80**: 137–142.

Kerby, K. and Kuspira, J. (1987) The phylogeny of the polyploid wheats *Triticum aestivum* (bread wheat) and *Triticum turgidum* (macaroni wheat). *Genome* **29**: 722–737.

Kimura, M. (1983) *The Neutral Theory of Molecular Evolution*. Cambridge University Press, Cambridge.

Klose, J. (1975) Protein mapping by combined isoelectric focusing and electrophoresis of mouse tissues. A novel approach to testing for induced point mutations in mammals. *Humangenetik* **26**: 231–243.

Klose, J. (1982) Genetic variability of soluble proteins studied by two-dimensional electrophoresis on different inbred mouse strains and on different mouse organs. *J. Mol. Evol.* **18**: 315–328.

Komatsu, S., Kajiwara, H. and Hirano, H. (1993) A rice protein library: a data-file of rice proteins separated by two-dimensional electrophoresis. *Theor. Appl. Genet.* **86:** 935–942.

Krishnan, M., Nguyen, H.T. and Burke, J.J. (1989) Heat shock protein synthesis and thermal tolerance in wheat. *Plant Physiol.* **90:** 140–145.

Lander, E.S. and Botstein, D. (1989) Mapping mendelian factors underlying quantitative traits using RFLP linkage maps. *Genetics* **121:** 185–199.

Lemieux, B., Aharoni, A. and Schena, M. (1998) Overview of DNA chip technology. *Mol. Breed.* **4:** 277–289.

Leonardi, A., Damerval, D and de Vienne, D. (1988) Organ-specific variability and inheritance of maize proteins revealed by two-dimensional electrophoresis. *Genet. Res. Camb.* **52:** 97–103.

Leone, A., Costa, A., Tucci, M. and Grillo, S. (1994) Comparative analysis of short- and long-term changes in gene expression caused by low water potential in potato (*Solanum tuberosum*) cell-suspension cultures. *Plant Physiol.* **106:** 703–712.

Leymarie, J., Damerval, C., Marcotte, L., Combes, V. and Vartanian, N. (1996) Two-dimensional protein patterns of *Arabidopsis* wild-type and auxin insensitive mutants, axr1, axr2, reveal interactions between drought and hormonal responses. *Plant Cell Physiol.* **37:** 966–975.

Li, G., Waltham, M., Anderson, N.L., Unsworth, E., Treston, A. and Weinstein, J.N. (1997) Rapid mass spectrometric identification of proteins from two-dimensional polyacrylamide gels after in gel proteolytic digestion. *Electrophoresis* **18:** 382–390.

Lund, A.A., Blum, P.H., Bhattramakki, D. and Elthon, T.E. (1998) Heat-stress response of maize mitochondria. *Plant Physiol.* **116:** 1097–1110.

Mangin, B., Goffinet, B. and Rebaï, A. (1994) Constructing confidence intervals for QTL location. *Genetics* **138:** 1301–1308.

Marshall, A. and Hodgson, J. (1998) DNA chips: an array of possibilities. *Nature Biotechnol.* **16:** 27–31.

Meza-Basso, L., Alberdi, M., Raynal, M., Ferrero-Cadinanos, M.L. and Delseny, M. (1986) Changes in protein synthesis in rapeseed (*Brassica napus*) seedlings during a low temperature treatment. *Plant Physiol.* **82:** 733–738.

Moons, A., Bauw, G., Prinsen, E., Van Montagu, M. and Van der Straeten, D. (1995) Molecular and physiological responses to abscisic acid and salts in roots of salt-sensitive and salt-tolerant *Indica* rice varieties. *Plant Physiol.* **107:** 177–186.

Moons, A., Bauw, G., Prinsen, E., Van Montagu, M. and Van der Straeten, D. (1995) Molecular and physiological responses to abscisic acid and salts in roots of salt-sensitive and salt-tolerant Indica rice varieties. *Plant Physiol.* **107:** 177–186.

Moons, A., Gielen, J., Vandekerckhove, J., Van der Straeten, D., Gheysen, G. and Van Montagu, M. (1997) An abscisic-acid- and salt-stress-responsive rice cDNA from a novel plant gene family. *Planta* **202:** 443–454.

Nei, M. (1987) *Molecular Evolutionary Genetics.* Columbia University Press.

Nover, L. and Scharf, K.D. (1984) Synthesis, modification and structural binding of heat-shock proteins in tomato cell cultures. *Eur. J. Biochem.* **139:** 303–313.

O'Farrell, P.H. (1975) High resolution two-dimensional electrophoresis of proteins. *J. Biol. Chem.* **250:** 4007–4021.

Petit, R.J., Bahrman, N. and Baradat, P. (1995) Comparison of genetic differentiation in maritime pine (*Pinus pinaster* Ait) estimated using isozyme, total protein and terpenic loci. *Heredity* **75:** 382–389.

Picard, P., Bourgoin-Grenéche, M. and Zivy, M. (1997) Potential of two-dimensional electrophoresis in routine identification of closely related durum wheat lines. *Electrophoresis* **18:** 174–181.

Plomion, C., Bahrman, N., Durel, C.-E. and O'Malley, D.M. (1995) Genomic mapping in *Pinus pinaster* (maritime pine) using RAPD and protein markers. *Heredity* **74:** 661–668.

Plomion, C., Costa, P. and Bahrman, N. (1997) Genetic analysis of needle protein in Maritime pine. 1. Mapping dominant and codominant protein markers assayed on diploid tissue, in a haploid-based genetic map. *Silvae Genet.* **46:** 161–165.

Posch, A., van den Berg, B.M., Postel, W. and Görg, A. (1992) Genetic variability of pepper (*Capsicum annuum* L.) seed proteins studied by 2-D electrophoresis with immobilized pH gradients. *Electrophoresis* **13:** 774–777.

Posch, A., van den Berg, B.M., Duranton, C. and Görg, A. (1994) Polymorphism of pepper

(*Capsicum annuum* L.) seed proteins studied by two-dimensional electrophoresis with immobilized pH gradients: methodical and genetic aspects. *Electrophoresis* **15**: 297–304.

Prioul, J.L., Quarrie, S., Causse, M. and de Vienne, D. *(*1997) Dissecting complex physiological functions into elementary components through the use of molecular quantitative genetics. *J. Exp. Bot.* **48**: 1151–1163.

Prioul, J.L., Pelleschi, S., Sene, M., Thévenot, C., Causse, M., de Vienne, D. and Leonardi, A. (1999) From QTLs for enzyme activity to candidate genes in maize. *J. Exp. Bot.* **50**: 1281–1288.

Ramagopal, S. (1987) Salinity stress induces tissue-specific proteins in barley seedlings. *Plant Physiol.* **84**: 324–331.

Ramagopal, S. (1990) Protein polymorphism in sugarcane revealed by two-dimensional gel analysis. *Theor. Appl. Genet.* **79**: 297–304.

Ramani, S. and Apte, S. K. (1997) Transient expression of multiple genes in salinity-stressed young seedlings of rice (*Oryza sativa* L.) cv. bura rata. *Biochem. Biophys. Res. Commun.* **233**: 663–667.

Reviron, M.P., Vartanian, N., Sallantin, M., Huet, J.C., Pernollet, J.C. and de Vienne, D. (1992) Characterization of a novel protein induced by progressive or rapid drought and salinity in *Brassica napus* leaves. *Plant Physiol.* **100**: 1486–1493.

Rey, P., Pruvot, G., Becuwe, N., Eymery, F., Rumeau, D. and Peltier, G. (1998) A novel thioredoxin-like protein located in the chloroplast is induced by water deficit in *Solanum tuberosum* L. plants. *Plant J.* **13**: 97–101.

Riccardi, F., Gazeau, P., de Vienne, D. and Zivy, M. (1998) Protein changes in response to progressive water deficit in maize. Quantitative variation and polypeptide identification. *Plant Physiol.* **117**: 1253–1263.

Rommens, J.M., Iannuzzi, M.C., Kerem, B.S., Drumm, M.L., Melmer, G., Dean, M., Rozmahel, R., Cole, J.L., Kennedy, D., Hidaka, N., Zsiga, M., Buchwald, M., Riordan, J.R., Tsui, L.C. and Collins, F.S. (1989) Identification of the cystic fibrosis gene: chromosome walking and jumping. *Science* **245**: 1059–1065.

Sabehat, A., Weiss, D. and Lurie, S. (1996) The correlation between heat-shock protein accumulation and persistence and chilling tolerance in tomato fruit. *Plant Physiol.* **110**: 531–537.

Sanchez-Martinez, D., Puigdomènech and Pagès, M. (1986) Regulation of gene expression in developing *Zea mays* embryos. Protein synthesis during embryogenesis and early germination of maize. *Plant Physiol.* **82**: 543–549.

Santoni, V., Bellini, C. and Caboche, M. (1994) Use of two-dimensional protein-pattern analysis for the characterization of *Arabidopsis thaliana* mutants. *Planta* **192**: 557–566.

Santoni, V., Delarue, M., Caboche, M. and Bellini, C. (1997) A comparaison of two-dimensional electrophoresis data with phenotypical traits in *Arabidopsis* leads to the identification of a mutant (cri1) that accumulates cytokinins. *Planta* **202**: 62–69.

Santoni, V., Rouquié, D., Doumas, P., Mansion, M., Boutry, M., Déhais, P., Sahnoun, I. and Rossignol, M. (1998) Use of proteome strategy for tagging proteins present at the plasma membrane. *Plant J.* **16**: 633–641.

Saruyama, H. and Shinbashi, N. (1992) Identification of specific proteins from seed embryos by two-dimensional gel electrophoresis for the discrimination between *indica* and *japonica* rice. *Theor. Appl. Genet.*, **84**, 947–951.

Song, K.M., Suzuki, J.Y., Slocum, M.K., Williams, P.H. and Osborn, T.C. (1991) A linkage map of *Brassica rapa* (syn. *campestris*) based on restriction fragment length polymorphism loci. *Theor. Appl. Genet.* **82**: 296–304.

Tacchini, P., Fink, A., Thiellement, H. and Greppin, H. (1995) Analysis of leaf proteins in late flowering mutants of *Arabidopsis thaliana*. *Physiol. Plant.* **93**: 312–316.

Tanksley, S.D. (1993) Mapping polygenes. *Annu. Rev. Genet.* **27**: 205–233.

Thiellement H., Seguin M., Bahrman N. and Zivy M. (1989). Homoeology and phylogeny of the A, S and D genomes of the *Triticinae*. *J. Mol. Evol.* **29**: 89–94.

Thiellement, H., Bahrman, N., Damerval, C., Plomion, C., Rossignol, M., Santoni, V., de Vienne, D. and Zivy, M. (1999) Proteomics for genetical and physiological studies in plants. *Electrophoresis* **20**: 2013–2026.

Touzet, P., Morin, C., Damerval, C., Le Guilloux, M., Zivy, M. and de Vienne, D. (1995a) Characterizing allelic proteins for genome mapping in maize. *Electrophoresis* **16**: 1289–1294.

Touzet, P., Winkler, R.G. and Helentjaris, T. (1995b) Combined genetic and physiological analysis of a locus contributing to quantitative variation. *Theor. Appl. Genet.* **91**: 200–205.

Touzet, P., Riccardi, F., Morin, C., Damerval, C., Huet, J-C., Pernollet, J-C., Zivy, M. and de Vienne D. (1996) The maize two-dimensional gel protein database: toward an integrated genome analysis program. *Theor. Appl. Genet.* **93**: 997–1005.

Tsugita, A., Kawakami, T., Uchiyama, Y., Kamo, M., Miyatake, N. and Nozu, Y. (1994) Separation and characterization of rice proteins. *Electrophoresis* **15**: 708–720.

Tsugita, A., Kamo, M., Kawakami, T. and Ohki, Y. (1996) Two-dimensional electrophoresis of plant proteins and standardization of gel patterns. *Electrophoresis* **17**: 855–865.

van Berkel, J., Salamini, F. and Gebhardt, C. (1994) Transcripts accumulating during cold storage of potato (*Solanum tuberosum* L.) tubers are sequence related to stress-responsive genes. *Plant Physiol.* **104**: 445–452.

Vierling, R.A. and Nguyen, H.T. (1990) Heat-shock protein synthesis and accumulation in diploid wheat. *Crop Sci.* **30**: 1337–1342.

Vierling, R.A. and Nguyen, H.T. (1992) Heat-shock protein gene expression in diploid wheat genotypes differing in thermal tolerance. *Crop Sci.* **32**: 370–377.

Wilkins, M.R., Williams, K.L., Appel, R.D., and Hochstrasser, D.F. (1997) *Proteome Research: New Frontiers in Functional Genomics.* Springer, Berlin, p. 243.

Zivy, M. (1987) Genetic variability for heat shock proteins in common wheat. *Theor. Appl. Genet.* **74**: 209–213.

Zivy, M., Thiellement, H., de Vienne, D. and Hofmann, J.-P. (1983) Study on nuclear and cytoplasmic genome expression in wheat by two-dimensional gel electrophoresis. I. First results on 18 alloplasmic lines. *Theor. Appl. Genet.* **66**: 1–7.

Zivy, M., Thiellement, H., de Vienne, D. and Hofmann, J.-P. (1984) Study on nuclear and cytoplasmic genome expression in wheat by two-dimensional gel electrophoresis. II. Genetic differences between two lines and two groups of cytoplasms at five developmental stages or organs. *Theor. Appl. Genet.* **68**: 335–345.

Zivy, M., Devaux, P., Blaisonneau, J., Jean, R. and Thiellement, H. (1992) Segregation distortion and linkage studies in microspore derived doubled haploid lines of *Hordeum vulgare* L. *Theor. Appl. Genet.* **83**: 919–924.

Zivy, M., El Madidi, S. and Thiellement, H. (1995) Distance indices in a comparison between the A, D, and R genomes of the *Triticeae* tribe. *Electrophoresis* **16**: 1295–1300.

Index